TRANSGENIC MODELS IN ENDOCRINOLOGY

ENDOCRINE UPDATES

Shlomo Melmed, M.D., Series Editor

1. E.R. Levin and J.L. Nadler (eds.): Endocrinology of Cardiovascular Function. 1998. ISBN: 0-7923-8217-X
2. J.A. Fagin (ed.): Thyroid Cancer. 1998. ISBN: 0-7923-8326-5
3. J.S. Adams and B.P. Lukert (eds.): Osteoporosis: Genetics, Prevention and Treatment. 1998. ISBN: 0-7923-8366-4.
4. B.-Å. Bengtsson (ed.): Growth Hormone. 1999. ISBN: 0-7923-8478-4
5. C. Wang (ed.): Male Reproductive Function. 1999. ISBN 0-7923-8520-9
6. B. Rapoport and S.M. McLachlan (eds.): Graves' Disease: Pathogenesis and Treatment. 2000. ISBN: 0-7923-7790-7.
7. W. W. de Herder (ed.): Functional and Morphological Imaging of the Endocrine System. 2000. ISBN 0-7923-7923-9
8. H.G. Burger (ed.): Sex Hormone Replacement Therapy. 2001. ISBN 0-7923-7965-9
9. A. Giustina (ed.): Growth Hormone and the Heart. 2001. ISBN 0-7923-7212-3
10. W.L. Lowe, Jr. (ed.): Genetics of Diabetes Mellitus. 2001. ISBN 0-7923-7252-2
11. J.F. Habener and M.A. Hussain (eds.): Molecular Basis of Pancreas Development and Function. 2001. ISBN 0-7923-7271-9
12. N. Horseman (ed.): Prolactin. 2001 ISBN 0-7923-7290-5
13. M. Castro (ed.): Transgenic Models in Endocrinology. 2001 ISBN 0-7923-7344-8

TRANSGENIC MODELS IN ENDOCRINOLOGY

edited by

Maria G. Castro, Ph.D.

The University of Manchester
United Kingdom

KLUWER ACADEMIC PUBLISHERS
Boston / Dordrecht / London

Distributors for North, Central and South America:
Kluwer Academic Publishers
101 Philip Drive
Assinippi Park
Norwell, Massachusetts 02061 USA
Telephone (781) 871-6600
Fax (781) 681-9045
E-Mail <kluwer@wkap.com>

Distributors for all other countries:
Kluwer Academic Publishers Group
Distribution Centre
Post Office Box 322
3300 AH Dordrecht, THE NETHERLANDS
Telephone 31 78 6392 392
Fax 31 78 6392 254
E-Mail <services@wkap.nl>

 Electronic Services <http://www.wkap.nl>

Library of Congress Cataloging-in-Publication Data

Transgenic models in endocrinology / edited by Maria G. Castro.
    p. ; cm. -- (Endocrine updates ; 13)
  Includes bibliographical references and index.
  ISBN 0-7923-7344-8 (hardback : alk. paper)
  1. Endocrine genetics--Research--Methodology. 2. Transgenic animals. 3. Transgenic
mice. 4. Endocrinology--Research--Methodology. 5. Molecular
endocrinology--Research--Methodology. I. Castro, Maria G. II. Series.
  [DNLM: 1. Disease Models, Animal. 2. Endocrine Diseases--genetics. 3. Mice,
Transgenic. WK 20 T772 2001]
  QP187.5 .T73 2001
  616.4'042--dc21

                                                                          2001023393

Printed on acid-free paper.

Printed in the United States of America

*The Publisher offers discounts on this book for course use and bulk purchases. For
further information, send email to <barbara.murphy@wkap.com>.*

CONTENTS

CONTRIBUTORS

Professor A. Bartke
Department of Physiology
Southern IL University School of Medicine
Life Sciences 2 Rm. 245
Carbondale, IL 62901
U.S.A.

Dr. N. Binart
INSERM U 344
Endo & Molec
Fac Med Necker
156 Rue de Vaugirard
75730 Paris Cedex 15
France

Professor M. G. Castro
Molecular Medicine Unit
School of Medicine
University of Manchester
Room 1.302 Stopford Bldg.
Oxford Road
Manchester M13 9PT
U.K.

Professor P.A. Cattini
Department of Physiology
University of Manitoba
30 William Avenue
Winnipeg MB R3E 3J7
Canada

Dr. V. Chandrashekar
Department of Physiology
Southern IL University School of Medicine
Life Sciences 2 Rm. 245
Carbondale, IL 62901
U.S.A.

Dr. K.T. Coschigano
Edison Biotechnology Inst.
Ohio University
101 Konneker Research
Laboratories, The Ridges
Athens, OH 45701-2979

Dr. S. Coste
Department of Molecular Microbiology
and Immunology
Oregon Health Sciences University
3181 Sam Jackson Park Rd.
Portland, OR 97201-3098

Rachel Cowen
Molecular Medicine Unit
School of Medicine
University of Manchester
Room 1.302 Stopford Bldg.
Oxford Road
Manchester M13 9PT
U.K.

Dr. A. David
Molecular Medicine Unit
School of Medicine
University of Manchester
Room 1.302 Stopford Bldg.
Oxford Road
Manchester M13 9PT
U.K.

Dr. H. Gainer
National Institutes of Health
Building 36, Rm. 4D-20
Bethesda, MD 20892
U.S.A.

Andres Hurtado-Lorenzo
Molecular Medicine Unit
School of Medicine
University of Manchester
Room 1.302 Stopford Bldg.
Oxford Road
Manchester M13 9PT
U.K.

Dr. R.A. Keri
Case Western Reserve University
School of Medicine
Department of Pharmacology
2109 Adelbert Road
Cleveland, OH 44106-4965
U.S.A.

Professor I. Lindberg
LSUMC
1901 Perido Street
New Orleans, LA 70112
U.S.A.

Dr. Mary Lynn Duckworth
Department of Physiology
University of Manitoba
30 William Avenue
Winnipeg MB R3E 3J7
Canada

Dr. A.E. Herbison
Department of Neurobiology
Babraham Institute
Babraham
Cambridge CB2 4AT
U.K.

Professor P.A. Kelly
INSERM U 344
Endo & Molec
Fac Med Necker
156 Rue de Vaugirard
75730 Paris Cedex 15
France

Professor J.J. Kopchick
Edison Biotech Institute
Ohio University
Konneker Research Labs.
Athens, OH 45701-2979
U.S.A.

Dr. M.J. Low
Vollum Institute, L-474
Oregon Health Science Univ.
3181 S.W. Sam Jackson Pk. Rd.
Portland, OR 97201-3098
U.S.A.

Professor P.R. Lowenstein
Molecular Medicine Unit
School of Medicine
University of Manchester
Room 1.302 Stopford Bldg.
Oxford Road
Manchester M13 9PT
U.K.

Dr. S. Murray
Department of Molecular Microbiology
and Immunology
Oregon Health Sciences University
3181 Sam Jackson Park Rd.
Portland, OR 97201-3098
U.S.A.

Dr. G. Owens
Case Western Reserve University
School of Medicine
Department of Pharmacology
2109 Adelbert Road
Cleveland, OH 44106-4965
U.S.A.

Mr. J.L. Smart
Vollum Institute, L-474
Oregon Health Sciences University
3181 S.W. Sam Jackson Park Road
Portland, OR 97201-3098
U.S.A.

Tom Southgate
Molecular Medicine Unit
School of Medicine
University of Manchester
Room 1.302 Stopford Bldg.
Oxford Road
Manchester M13 9PT
U.K.

Professor D. Murphy
Department of Medicine
University of Bristol
Marlborough Street
Bristol
BS2 8HW
U.K.

Professor J.H. Nilson
Case Western Reserve Univ.
School of Medicine
Department of Pharmacology
2109 Adelbert Road
Cleveland, OH 44106-4497
U.S.A.

Dr. W. Scott Young III
Chief of Section Neural Gene
Expression
Lab of Cell Mol Regulation
NIMH
Building 36/2D10
Bethesda, MD 20892-4068
U.S.A.

Dr. J.R. Smith-Arica
Molecular Medicine Unit
School of Medicine
University of Manchester
Room 1.302 Stopford Bldg.
Oxford Road
Manchester M13 9PT
U.K.

Dr. M. Stenzel-Poore
Dept. Molecular Microbiology
and Immunology, L-220
Oregon Health Sciences Univ.
3181 S.W. Sam Jackson Pk. Rd.
Portland, OR 97201-3098
U.S.A.

Daniel Stone
Molecular Medicine Unit
School of Medicine
University of Manchester
Room 1.302 Stopford Bldg.
Oxford Road
Manchester M13 9PT
U.K.

Sara J. Wells
Department of Medicine
University of Bristol
Marlborough Street
Bristol
BS2 8HW
U.K.

Dr. P. Umana
Molecular Medicine Unit
School of Medicine
University of Manchester
Room 1.302 Stopford Bldg.
Oxford Road
Manchester M13 9PT
U.K.

Dr. J.C. Williams
Molecular Medicine Unit
School of Medicine
University of Manchester
Room 1.302 Stopford Bldg.
Oxford Road
Manchester M13 9PT
U.K.

PREFACE

The dramatic recent expansion in genomic information has motivated the development of new approaches to characterize gene expression and function. A critical issue for both basic and clinical endocrinologists is the physiological role of genes involved in regulating endocrine functions. Transgenic technologies allow the translation of genotypic information into specific phenotypes by using gene overexpression or loss of specific gene functions. Murine functional genomics is thus of central importance in modern biomedical endocrine research. Although mice are at present, the preferred mammalian species for genetic manipulations because of the availability of pluripotent embryonic stem cells and inbred strains and the relatively low breeding and maintenance costs, transgenic rats have also been generated and used to study endocrine physiology. The two basic techniques used in the creation of transgenic animal models are integration of foreign DNA into a fertilized oocyte by random chromosomal insertion and homologous recombination in embryonic stem cells that are then introduced into zygotes. Transgenic mice and rats serve as sophisticted tools to probe protein function, as models of human disease, and as hosts for the testing of gene replacement and other therapies. Embryonic stem cell libraries for mouse gene deletion are being developed, which will make it possible to generate knockout mice rapidly and without the need to analyze gene structure, construct targeting vectors, and screen embryonic stem cell clones. A novel approach to transgenesis for the expression of DNA within adult differentiated neuroendocrine cells *in vivo* is using viral vectors. Several viral vectors have been used to transfer genes into postmitotic neurons, pituitary cells and other endocrine cells and tissues *in vitro* and *in vivo*. This approach offers the advantage of bypassing either lethal mutations or compensating events which might occur during development. These exciting new opportunities to apply genetic approaches in the mouse and rat has enabled many researchers to address many outstanding questions of *in vivo* endocrine gene function.

This book has been designed to reflect a multidisciplinary approach to transgenics in endocrinology, encompassing contributions from internationally recognised experts in many aspects of endocrine physiology. These range from the basic technologies underpinning the generation of transgenic rats and mice, to knock-out mice and also transgenics using viral vectors for gene delivery. The authors have aimed to exemplify the use of transgenic technologies and models for the study of a wide range of endocrine topics, including, reproductive functions, the stress axis, control of lactation, water and electrolyte balance, appetite control and pituitary tumour therapy.

All the experimental techniques described in this book which involve the use of animals have been carried out following local and national regulations. I would like to convey my gratitude to all of the authors for their efforts to produce such informative and updated contributions and also their enlightened discussions in relation to the potentials and pitfalls of the experimental models described in each chapter. Special thanks are due to Daniel Stone, a Ph.D. student in my laboratory for his tireless efforts and dedication in editing this book. Many thanks to Ros Poulton for excellent secretarial assistance. I am also very grateful to Barbara Murphy of Kluwer Publications for facilitating the work which led to the publication of this volume. I dedicate this book to Pedro and Elijah, the two most inspiring forces in my life.

Maria G. Castro

Manchester, England
January, 2001.

1

TRANSGENIC RATS AND THE FUNCTIONAL GENOMICS OF ENDOCRINE SYSTEMS

David Murphy and Sara J. Wells

University Research Centre for Neuroendocrinology, University of Bristol, Bristol Royal Infirmary, Marlborough Street, Bristol BS2 8HW, UK

GENES AND ENDOCRINE FUNCTION

Based on the gene density of the recently sequenced human chromosome 22 (1), it can be calculated that mammals have about 50,000 genes. The identification of these genes is a necessary prelude to any attempt to construct models of endocrine function based on integrated gene networks. Such a global approach demands that we identify all of the genes that are expressed in endocrine cells, that we determine when and where these genes are expressed, and, finally, that we determine their functions. Although information in databases will not, by itself, be sufficient to determine biological function, it will provide a foundation for the design of appropriate experiments. This remarkable wealth of information that molecular genetics has provided us with needs to be integrated into an understanding of the functioning of whole tissues, organs and organisms. Without such integration, molecular information is nothing more than a confusing catalogue of sequences and structures. The experimental tools exist in model organisms such as the rat, but not in humans, for assembling genes into pathways and thus identifying gene function from sequence. In particular, transgenic technologies enable rapid movement between genotype and phenotype through the generation of specific loss-of-function, overexpression or misexpression phenotypes (2).

TRANSGENESIS

The term transgenesis is used very broadly to describe the introduction of cloned DNA into any living cell. Over the past 10-15 years, profoundly important techniques have been developed that enable new genes to be introduced into whole mammalian organisms, and that enable specific alterations to be made in endogenous genes. The resulting animals are called transgenic animals – animals with an altered genetic composition caused by the transfer of cloned DNA.

Prior to the current revolution in transgenic research, the only method available to study the regulation and function of mammalian genes within the context of the whole organism was to utilise mutants that arose spontaneously in nature. The study of such mutations has been termed "forward genetics" - from phenotype to gene. However, this process - the molecular identification and characterisation of the mutated gene - is technically difficult. Further, mutations arise opportunistically in nature and those only affecting particular systems are difficult to screen for and identify.

Gene function and regulation can also be studied in cultured mammalian cells, and evaluation of gene expression in such systems is relatively straightforward. However, as there are often no appropriate cell lines available that correspond to the differentiated cell-type of interest, researchers have resorted to heterologous cell types of dubious physiological value. Furthermore, the activity of a specific gene at the cellular level does not yield satisfactory information about the regulation of the gene among the complex physiological interactions of the whole animal. Even the best cell culture systems cannot possibly simulate tissues and organ systems and predict responses to sophisticated environmental stimuli. No cell line, nor culture system, can ever model the plasticity evoked by the panoply of developmental and physiological stimuli to which an endocrine cell is normally exposed.

Transgenic animal systems combine the virtues of cell culture and congenic breeding strategies whilst avoiding the negative aspects of each system - a defined genetic lesion can be studied within the physiological and developmental integrity of the whole animal. This has enabled the development of the concept of "reverse genetics" - from gene to phenotype. Using transgenic techniques, a characterised genetic sequence can be evaluated within the context of the whole animal without any prior knowledge of its regulation or function.

In order to apply a reverse genetic approach to mammals, we must be able to introduce a defined genetic change into the germ line such that 1) every cell of an organism will carry the change, and 2) the change will be transmitted to subsequent generations. An unlimited number of genetically similar offspring carrying the same genetic modification are thus available for experimental analysis.

Two methods of making germline transgenic animals are relevant to this review - microinjection of fertilised one cell eggs, and manipulation of embryonal stem cells (3).

Microinjection of Fertilised One-cell Rodent Eggs
One-cell fertilised eggs are harvested from donor females that have usually been hormonally stimulated to produce a large number of eggs prior to mating. Cloned DNA fragments are then directly introduced into one of the two pronuclei using a fine injection needle. A few hundred molecules of DNA are introduced into the nucleus. Surviving eggs are transferred to the natural environment provided by a pseudopregnant recipient female. A pseudopregnant recipient is an estrus female that has been mated with a sterile or vasectomised male. The animal carries only unfertilised eggs, but is physiologically prepared to carry implanted eggs through pregnancy. Pups are born 20-21 days later, 10-50% of which are transgenic as identified by genome analysis.

The process of microinjection results in the integration of the transgene into the host chromosomes. Integration is an additive process; the host genome gains new information. The host genome is unchanged, except at the locus of integration, where deletions may occur, possibly interrupting endogenous genes. Integration takes place through non-homologous recombination of the transgene into the host chromosomes, and there is no specificity with respect to either the host or the transgene DNA. There is usually only a single integration site per nucleus, but that integration site might contain between 1 and 1000 tandomly arranged copies of the transgene. The integration process is essentially random; the experimenter has no control over the site of integration, nor the copy number of the transgene.

A transgene is expressed with a spatial and temporal pattern that is a function of the cis-acting elements that it contains. However, the chromosomal position of the transgene can affect expression, resulting in transcriptional repression or ectopic activation. The structure of a microinjection transgene depends entirely on the aims of a particular experiment. A transgene should operate like any other gene in the targeted cells of an organism, and hence, structural elements must be appropriately recognized by the transcriptional, post-transciptional, and translational machinery of the host.

Manipulation of ES cells (Including Knockouts)
Embryonal stem (ES) cells are pluripotential cells derived from the primitive ectoderm of the mouse blastocyst. As with any mammalian cell in culture, DNA can be introduced into ES cells. ES cells can be genetically altered and these altered cells can be introduced into a blastocyst where they will contribute to the normal development of a chimeric mouse. If the germ line of the mouse has been colonised by the ES cell descendants, then, at the next generation, a pure line of heterozygous genetically altered mice can be derived. ES cells can be altered in two ways:

- Non-homologous recombination.

This is by far the most common mode of integration of exogenous DNA into the chromosomes of mammalian cells. Integration is at random, and there is no specificity with respect to either the host or input DNA. The host gains new information, with the rest of the genome remaining essentially unchanged, except for deletions or other rearrangements around the integration site.

- Homologous recombination.

Homologous recombination involves an exchange of information between the input DNA and homologous host sequences, mediated by Watson-Crick base-pairing and host recombination enzymes. Homologous events are very rare compared to non-homologous events. Screening systems have been developed that enable clones bearing these rare homologous event to be isolated. Using appropriately designed targeting vectors, specific mutations (for example, null mutants, known as knockouts) can be introduced into target genes in ES cells. The mutated ES clone can then be introduced into a blastocyst, where it will colonise the inner cell mass (ICM) and contribute to the development of a chimera. Breeding of a germline chimera with a wild-type mate results in the generation of a pure line that is heterozygous for the mutation, and brother-sister matings of these will give rise to homozygous mutant animals. The phenotype elicited by targeted mutation can then be studied.

ENDOCRINOLOGISTS AND TRANSGENESIS

Physiologists and endocrinologists have been slow to take advantage of the substantial benefits that transgenesis can afford. There are a number of reasons for this:

- the daunting prospect of establishing transgenic facilities. Most endocrinologists still lack the requisite skills and facilities. Funds are not readily available for this expensive type of endeavour, particularly if the applicant has no relevant background.
- the lack of appropriate collaborators.
- the cultural and scientific gulf between endocrinologists and molecular biologists.
- the language (jargon) barrier.
- the use, by molecular biologists, of an inappropriate model organism - the mouse - rather than that favoured by endocrinologists - the rat.
- the generation of transgenic models that, in addressing questions posed by molecular biologists, do not necessarily meet the needs of endocrinologists.
- serious doubts over the ability of the germline approach to deliver meaningful results because of developmental compensation (see below).

THE RAT – A KEY MODEL ORGANISM

Most transgenic studies in mammals have been performed on mice. But it is the rat that continues to be the species of choice for studies in neuroscience and physiology. The mouse is relatively inappropriate - in contrast, the anatomy of the rat brain is well mapped, and the structure, function and regulation of the rat CNS have been the subject of detailed study for many years and numerous effective behavioural paradigms have been developed. The large size of the rat makes it easily accessible for a whole range of physiological measurement and intervention, but its reproductive capacity and gestation time are equivalent to the mouse. For example, real-time monitoring of the canulated freely moving, conscious animal (4), whilst routine in the rat, is considerably more difficult in the mouse. Rats can be transformed by microinjection of fertilised one-cell eggs with cloned DNA fragments (3,5).

The Rat Genome Project
Although not as advanced as its mouse and human equivalents, the Rat Genome Project, a collaboration of a number of different Institutes, is

developing genetic and physical maps of the rat genome (6). The aims of the Rat Genome Project are to:

- create a database of 8000 simple sequence length polymorphisms (SSLPs) in 48 genetically and physiologically important inbred rat strains. This data provides investigators with a means of quickly selecting informative markers in any rat cross, resulting in substantial savings of both time and resources.
- create a rat radiation hybrid (RH) mapping database. RH panels allow for the mapping of both polymorphic and non-polymorphic markers, including SSLPs, sequence tagged sites (STSs), expressed sequence tags (ESTs), and genes. These maps act as an integration point between genetic linkage mapping and positional cloning of a gene.
- create a database to store the data produced by rat scientific community (the Rat Genome Database – RGD, (7)).

Comparative Genomics
One approach to the evaluation of gene function and regulation is to identify genetically conserved sequences and structures by interspecies comparisons. Conservation of function also occurs at higher levels. In many cases not only individual protein domains and proteins, but entire multi-subunit complexes and biochemical pathways are conserved. Often, the way in which these complexes and pathways are utilised are also conserved. Genome project data and transgenic experimentation data will determine the extent to which genes from different organisms can be swapped around, but still retain function. Thus will studies in rats benefit from the human and mouse genome projects. This approach also extends to lower organisms (8), particularly those with genomes amenable to analysis. Identifying regulatory elements in mammalian genomes is a tedious exercise because of the large amount of 'junk' DNA that is interspersed in the intergenic and intronic sequences. Nearly 90% of the 4000 Mbp of the rodent genome is comprised of repetitive elements, and this has slowed transgenic studies on the regulatory sequences contained within the VP and OT locus. In contrast, the pufferfish, *Fugu*, has a compact genome of 390 Mbp that contains very few repetitive elements, yet this teleost has a gene repertoire comparable to that of mammals. The majority of introns in the pufferfish are small (modal value 80 bp) and repetitive sequences account for less than 10% of the genome (9). Thus the pufferfish genome is thus an attractive model for identifying and characterising regulatory elements through comparative genomics.

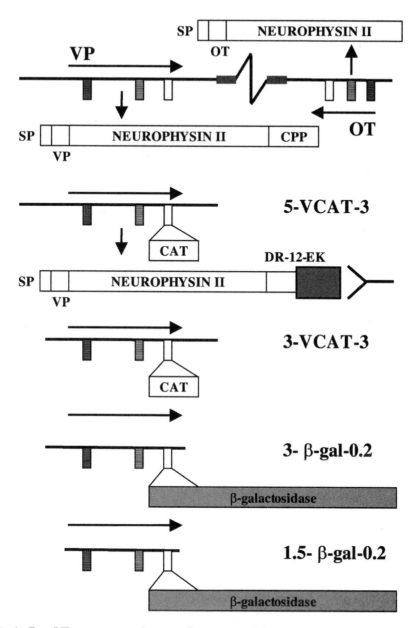

Fig 1. Rat VP transgenes in rats. Structure of the rat VP-OT gene locus and of rat transgene derivatives that have been studied in rats. The horizontal arrows indicate the direction of transcription. Boxes represent exons. The VP and OT genes are closely linked in the rat genome, being separated by 11 kbp, and are transcribed towards each other from opposite strands of the DNA duplex. The VP and OT mRNAs are translated into prepropeptides of similar structure. Following the signal peptide, the nine amino acid hormone moiety

(VP or OT) forms the N-terminal domain of its propeptide. This is followed by the disulphide-rich neurophysin (NP) moiety, a putative intracellular carrier molecule. The VP propeptide contains an additional C-terminal extension consisting of a 39 amino acid glycopeptide (copeptin; CPP) of unknown function. Prototype transgene 5-VCAT-3 consists of the rat VP structural gene containing a chloramphenicol acetyl transferase (CAT) reporter in exon III, flanked by 5kb of upstream and 3kb of downstream sequences. The transgene encodes a modified prepropeptide in which CPP is truncated, the C-terminal 26 amino acids being replaced with a novel 16 amino acid epitope (DR-12-EK), against which antisera have been raised. 3-VCAT-3 is derived from 5-VCAT-3 by the removal of upstream sequences such that the 5' flank is only 3 kbp long. 3-Vβgal-0.2 has the same 3 kbp 5' flank as 3-VCAT-3, but the 3' flank is reduced to 0.2 kbp, and the CAT reporter is replaced with β-galactosidase. In 1.5-Vβgal-0.2, the 5' flank is reduced to 1.5 kbp.

ENDOCRINE MODELS IN TRANSGENIC RATS - THE HYPOTHALAMO-NEUROHYPOPHYSEAL SYSTEM

The molecular genetics and physiology of the vasopressin (VP) and oxytocin (OT) expressing hypothalamo-neurohypophyseal system (HNS) have been extensively reviewed (8,10). The hypothalamic brain peptides VP and OT, the neurohypophysial peptides, play crucial roles in the regulation of salt and water balance in both land based (11) and aquatic vertebrates (12). In land mammals, the physiological challenge of dehydration results in a rise in plasma osmolality that is detected by an undefined osmoreceptor mechanism. Subsequent excitation of hypothalamic neurons leads to a release of VP into the general circulation. VP travels through the blood stream to its targets that exhibit specific receptors. Particularly, through an interaction with V2-type receptors located in the kidney, VP increases the permeability of the collecting ducts to water, promoting water conservation by decreasing the amount of water lost in urine. As well as its well-known roles in lactation (13) and, perhaps, parturition (14), OT is involved in the control of salt excretion from the kidney (15,16). Similar roles have been determined for the teleost homologues of VP and OT - vasotocin (VT) and isotocin (IT) respectively (12,17,18).

In all vertebrates, the neurohypophysial peptides are synthesised in, and secreted from the HNS, a specialised part of the brain which consists of specific neuronal cell bodies in the hypothalamus, and their axonal projections, which terminate in the posterior pituitary gland. VP and OT are synthesised in the cell bodies of anatomically defined groups (nuclei) of hypothalamic neurons each of which is involved in the maintenance of distinct physiological systems and is subject to functionally appropriate regulatory cues. VP neurons are found in the magnocellular supraoptic nucleus (SON) and in the paraventricular nucleus (PVN). OT is also expressed in magnocellular neurons of the SON and PVN, but VP and OT are rarely found in the same cell (19,20).

It is the magnocellular neurons that are involved in osmoregulation. VP, but not OT, is also found in the parvocellular neurons in the PVN and in the dorsomedial region of the suprachiasmatic nucleus (SCN), the circadian generator of the mammalian brain (21,22).

The neurohypophysial peptides genes are highly conserved, within and between vertebrate species, at both the structural and sequence level (23). The VP and OT genes are closely linked, tail-to-tail, in the genomes of all mammals studied, being separated by an intergenic region of 11 kbp in the rat (24, Figure 1), and 3 kbp in the mouse (25).

Within the magnocellular SON and PVN neurons, the VP gene is upregulated by the physiological stimuli (8,10). Osmotic stimuli such as dehydration (fluid deprivation for up to 3 days) or salt-loading (the normal tap-water diet is replaced with a solution of 2% w/v NaCl for up to 10 days), result in an increase in VP gene transcription (26) a concomitant increase in VP mRNA abundance (26-30) and an increase in the length of the VP RNA poly(A) tail (26, 31-33). The OT gene is similarly regulated by osmotic stimuli (29,33, 34,35) despite being expressed in distinct magnocellular neurons.

Cell-Specific Expression of the Vasopressin Gene in Transgenic Rats
A number of VP transgenes of rat origin have been introduced into rat hosts (Figure 1). The rationale behind these experiments has been to define chromosomal regions necessary for appropriate transgene expression in magnocellular neurons. It is then possible to further delineate the cis-acting sequences responsible by making further deletions until appropriate expression is lost.

A number of independently-derived transgenic rat lines bearing a derivative of the rat VP gene called 5-VCAT-3 (Figure1) show appropriate cell-specific and physiological expression in hypothalamic vasopressinergic magnocellular neurons (5). 5-VCAT-3 consists of the rat VP structural gene, containing reporter sequences in exon III derived from the bacterial chloramphenicol acetyl transferase (CAT) gene, flanked by 5 kbp of upstream and 3 kbp of downstream sequences. The CAT sequences provide a unique nucleic acid reporter of transgene RNA expression, which was used in Northern and *in situ* hybridisation analyses (5). 5-VCAT-3 is regulated by cell-specific cues in the hypothalami of transgenic rats, being confined to VP, but not OT, magnocellular neurons in the PVN and SON. However, the pattern of 5-VCAT-3 transgene expression was not identical to that of the endogenous VP gene. Firstly, the basal level of expression of the 5-VCAT-3 RNA was lower than the endogenous VP transcript in the hypothalami of physiologically unstimulated animals. Secondly, expression in the SCN was, at best, negligible. These data suggest that the transgene lacks enhancers that mediate high basal VP expression, particularly in the SCN.

Progress towards the fine-structure delineation of the cis-acting sequences required to direct expression to magnocellular neurons has been slow, but there is some evidence that distal sequences flanking the structural gene are important. Transgene 3-VCAT-3, with 3 kbp of 5' flanking sequences, has an expression pattern indistinguishable from that of 5-VCAT-3 (36). However, when 3' flanking sequences are deleted (transgenes 3-Vβgal-0.2 and 1.5-Vβgal-0.2, Figure 1) expression in the hypothalamus cannot be detected (37).

Comparative Genomics of the HNS
Studies on mice and rats bearing transgenes derived from *Fugu*, the Japanese pufferfish, have demonstrated that teleost genes can be expressed in mammalian system under the control of their own regulatory sequences. For example, the *Fugu* gene encoding the neuropeptide isotocin is expressed in rat brain in a cell-specific manner, and responds to physiological stimuli like the equivalent rat oxytocin gene (38).

Although the fish brain is anatomically not identical to that of mammals, there are distinct magnocellular neurons in the preoptic nucleus of the fish that express the isotocin (IT) and vasotocin (VT) genes - the teleost equivalents of OT and VP respectively. Immunohistochemical and *in situ* hybridisation studies have shown that the fish IT and VT are expressed in separate magnocellular neurons (39,40). *Fugu* VT and IT genomic sequences have been cloned and found to be closely linked on a single cosmid, along with at least six other genes (38). Comparison of the sequence of the *Fugu* VT-IT locus with that of the rat VP-OT locus (24) revealed numerous homologies that might correspond to conserved regulatory elements. In order to test the functions of these homologies empirically, transgenic rats were produced that contain a 40 kbp pufferfish genomic fragment derived from the IT-VT locus.

The IT gene was shown to be expressed only in rat OT neurons of the SON and PVN (38). Further, the expression of the IT gene is upregulated in response to the withdrawal of dietary water in parallel with the endogenous OT gene. These results are important for a number of reasons. They demonstrate conservation of regulatory mechanisms between two species that separated 400 million years ago. The cis-acting sequences and cognate trans-acting factors responsible for cell-type and physiological regulation have been conserved - this is all the more remarkable given that the two species, rat and *Fugu*, face profoundly different environmental challenges. Putative regulatory sequences can be readily identified by species comparison; non-significant sequences have had ample time to randomise by mutation and translocation, and conserved sequences in the non-coding sequences are likely to have a role in gene expression and regulation. High gene density in the *Fugu* and a dearth of repetitive sequences is an additional advantage in transgenic studies. A combination of comparative genomics using the compact genome of *Fugu* and transgenesis in rats is thus a useful tool for identifying and characterising conserved gene regulatory elements. Further, this approach will enable the identification of fundamental and conserved components of the limbic and brain stem systems that control body fluid homeostasis (8).

Physiological Regulation of Gene Expression

The physiological regulation in rat hosts of the rat VP gene-derived 5-VCAT-3 transgene has been well defined (5,41). Five days of salt-loading, an osmotic stimulus that increases the level of the endogenous VP RNA 2-fold, evokes an exaggerated effect on transgene expression, increasing the level of the transgene encoded RNA 20 fold. This is not a consequence of transgene copy number, as different lines, with different numbers of transgene copies, show the same exaggerated physiological response. It has been postulated that this exaggerated effect is due to the release of the transgene from the action of repressor sequences, absent from the transgene, but present in the normal context of the VP gene, that attenuate the physiological response.

For a transgene to participate in the normal (or abnormal) physiology of its host, it must, of course, be translated and processed into a biologically active peptide. The incorporation of the CAT reporter into exon III of the 5-VCAT-3 transgene (Figure 1) placed a unique hexadecapeptide (DRSAGYYGLFKDRKEK, abbreviated to DR-12-EK) at the C-terminus of a modified prepropeptide (41). Exons I and II of the transgene are normal, and thus the signal peptide, the VP nonapeptide and the regions of NP encoded by these exons are the same as in the wild-type prepropeptide. 72 bp of exon III are deleted from the transgene as compared to the normal VP gene, and these are replaced by sequences from the CAT gene (5). The C-terminal portion of NP encoded by transgene exon III is intact (thus the whole of NP is wild-type), as is the cleavage signal that separates NP from CPP. The first 13 residues of CPP are intact (including the unique glycosylation site) but the last 26 amino acids of CPP are replaced by a novel 16 amino acid peptide (DR-12-EK). An antibody was raised against the DR-12-EK 'tag' and, using immunohistochemistry, electron microscopy, RIA and HPLC, it was shown that the transgene RNA is translated into a protein product found, in a processed form, in secretory granules in the posterior pituitaries of transgenic rats (41). Replacement of the hydrophobic C-terminus of the VP precursor with the hydrophilic peptide 'tag' is thus well tolerated, and does not disrupt VP production or disturb salt and water balance. An osmotic stimulus was shown to increase hypothalamic DR-12-EK levels, in parallel with transgene RNA levels, but changes in posterior pituitary DR-12-EK levels were more complex. After 5 days salt-loading DR-12-EK levels fell, as would be expected if its release was co-ordinate with that of VP. However, after 10 days of salt-loading, posterior pituitary DR-12-EK levels increased, despite the lower level of VP. This probably reflects the greater response of the transgene to osmotic challenge at the RNA level, increasing the proportion of DR-12-EK-containing translation products transported to the posterior pituitary relative to those derived from the endogenous gene. These observations are further evidence in support of models of neurohypophyseal homeostasis that suggest that pituitary VP peptide levels passively reflect changes in hormone release and synthesis (42), and that the availability of mRNA is the primary determinant of pituitary VP content in the basal state (43).

Physiological Engineering

As well as aspects of gene regulation, transgenic experiments can be used to address problems in gene function, and the role of gene products in the overall physiology of the organism (2). Thus far, little of functional consequence has emerged from such transgenic studies on the HNS. Attempts have been made to develop expression systems based on the 5-VCAT-3 transgene that would target expression of any protein or RNA to magnocellular hypothalamic VP neurons. Two properties of the 5-VCAT-3 transgene are very useful for functional studies. Firstly, expression in magnocellular cells is separable from expression in parvocellular neurons. Secondly, the exaggerated response of 5-VCAT-3 to osmotic challenge, compared to the endogenous VP gene, means that by simple manipulation of the drinking diet of the transgenic rats we can predictably regulate the level of expression of these molecules. The facile but dramatic regulation of the 5-VCAT-3 transgene in precisely defined neurons of the rat brain is in marked contrast to previously described neuronal promoters that, when expressed in mice, are active in diverse cell types and cannot be controlled. However, as yet an effective expression system for the HNS has not been reported, despite attempts to achieve this. The first vector built, 5-VCAT-3(Sal I), was derived from 5-VCAT-3 (Figure 1) by the site-directed deletion of the VP prepropeptide translation start sites in exon I, and their replacement by a unique Sal I restriction endonuclease site, into which any gene sequence can be inserted. Two cDNAs and one genomic sequence were cloned into this site:

- PKIα - a cDNA encoding an inhibitor of protein kinase A (44)
- PKCI-1/HINT - a cDNA encoding a putative inhibitor of protein kinase C (45)
- bVP - a genomic clone encoding the bovine VP gene (clone VP-B; 46)

No transgene-derived RNA could be detected in any tissue of numerous independently derived rat lines bearing the cDNA constructs (47). The construct containing the VP-B gene was, however, expressed in the SON and PVN (36). These data suggest that disruption of exon I with cDNA has a deleterious effect on VP gene expression. This is possibly due to recognition of a premature termination codon into exon I, and the subsequent degradation of the message by nonsense-mediated decay (48).

ENDOCRINE MODELS IN TRANSGENIC RATS - GROWTH

The major co-ordinator of post-natal growth is the pituitary-derived peptide, growth hormone (GH). The release of GH from somatotroph cells of the anterior lobe of the pituitary gland is controlled by two hypothalamic peptides; growth-hormone-releasing hormone (GHRH) and somatostatin. GHRH and somatostatin are synthesised in the hypothalamic neurones of the arcuate and periventricular nuclei respectively, from which they are released into the hypophyseal-portal venous system (49). This capillary network extends through the external zone of the median eminence to the anterior lobe (50) where GHRH stimulates the expression and release of GH (51-53), whereas somatostatin has an inhibitory effect (54). GH autoregulates its own secretion by mediating the levels of these neuropeptides via a negative feedback system. GH excess causes decreased hypothalamic GHRH content (55,56) and increases somatostatin levels (57). Conversely GH deficiency increases GHRH and reduces somatostatin mRNA levels (58).

GH is the cause of much clinical interest as an excess of circulating peptide leads to giantism and a deficiency to dwarfism. As with the neurohypophysis, a greater understanding of the control and regulation of such an integrated system has resulted from the study of rodent models. The *little* mouse and Snell and Jackson dwarf mice, which contain spontaneous mutations, have been widely used in studies of GH deficiency (59,60). With the development of transgenic technology, both dwarf (61,62) and giant mice (63,64) have been generated. Although valuable for cellular and expression studies, the physiological data which can be collected from these animals is minimum. The size of a mouse (even a giant one!) greatly limits the number of canulation techniques which can be performed. For these type of studies the rat is an excellent model which has been used extensively in the growth hormone field.

Rats carrying spontaneous mutations leading to GH deficiencies have also been classified. The most prominent examples are the dwarf *dr/dr* rat, which carries a mutation in the rat GH gene (65) and the spontaneously-deficient *dw/dw* rat, with an unknown recessive mutation (66). Useful as these dwarf models have been, rat transgenesis lends as opportunity to introduced defined genetic modifications into a species where complex and wide ranging physiological manipulations can be accomplished. A Transgenic growth-retarded rat (Tgr) has been generated for this purpose (67).

The Tgr rat utilises the negative feedback system which regulates the GH axis. Using the rat GHRH gene locus, the human GH gene was targeted to the GHRH neurones in the hypothalamus so as to induce dwarfism by local feedback of the GH system. In order to generate these rats, a DNA fragment spanning 38kbp of the rat GHRH locus including 16kbp of 5' and 14kbp of 3' flanking sequence, was introduced into fertilised rat oocytes. It was hoped that, by virtue of its size, this fragment contains sufficient sequence to direct cell-specific expression of the transgene to GHRH neurones. The structural gene for human GH, which is capable of activating rat GH receptors, was inserted into the first exon of the rat GHRH gene. One line, bearing a single copy of a GHRH-hGH transgene, has been characterized in detail. Expression of the hGH transgene was detected by RT-PCR in the hypothalami of these transgenic rats. Tgr rats display a dominant, sexually-dimorphic dwarfism due to a retardation of their linear growth. The body weight of a 12 week old male Tgr rats is 75% that of a wild-type male and as this would suggest the pituitary content of endogenous rat growth hormone and the related hormone prolactin are also reduced in this animal.

During the extensive physiological analysis performed on Tgr rats, serial blood sampling studies revealed that the 3 hour episodes in which male rats secrete GH is conserved in these transgenic animals. It has also been shown that the application of an exogenous GH secretagogue can produce a large release of endogenous GH (68) and stimulate growth. These data indicate that the endogenous GH axis of these transgenic animals, although down-regulated, is intact. These rats, therefore, provide a much more suitable model for the testing of GH secretagogues to be used to treat human dwarfism than rodents with impaired GH responses.

PROBLEMS WITH RAT GERMLINE TRANSGENESIS

Knockouts in rats?
Whilst gene knockouts are routine in mice, it has not been possible to make knockouts in rats. This is because, although rat embryonal stem cell-like cultures have been described (69), it has not been possible to derive chimeric rats from these cells (70).

Nuclear transfer technologies, proven in both sheep (71,72) and mouse (73) will in the future be successfully adapted to the rat. This will enable knockout manipulations which, until now, were not possible in rats, to be carried out on cultured embryonic or foetal somatic cells, followed by the generation of the genetically modified animal by nuclear transfer into enucleated rat oocytes.

Epigenetic consequences of germline transgenesis
In a germline transgenic animal, the genetic change represented by the transgene is manifested throughout development, from conception onwards. The lack of temporal or spatial specificity can complicate the interpretation of the phenotypes resulting from a transgenic experiment (74) for three reasons:

- the overall phenotype observed may be a summation of transgene effects in different tissues at different times. The parts of the overall effect might be very difficult to dissect.
- an early or severe effect of a transgene might preclude the study of subsequent or downstream processes involving the same gene.
- transgenic animals are genetic "reactionisms". Epigenetic responses to a genetic lesion can have physiological effects (75). Thus, a phenotype may be due to endogenous genes being switched on or off in response to the transgene or knockout, rather than being a direct effect of the primary genetic lesion. In contrast, many investigators have invested effort and resources into the generation of a knockout mouse only to find that the

homozygous null mutant animal has no overt or obvious phenotype. The knocked-out gene may be "redundant" - other genes, possibly of the same family, may take over the function of the knocked-out gene (75-78).

ALTERNATIVES TO KNOCKOUTS

Dominant Effectors
Studies on intracellular signalling pathways have resulted in the development of dominant acting mutants of signalling molecules that are antagonists or agonists of their wild-type counterparts. The expression of the latter molecules in transgenic animals is a useful alternative to knockouts. These mutants can be expressed under the control of cell specific regulatory sequences resulting in the inhibition or the constitutive activation of particular pathways. This approach has been proven in studies in transgenic, on a number of systems, including the cAMP regulation of somatotroph development and function; its application to the rat is greatly anticipated as the rewards, in terms of physiologically important data, will be all the more valuable. cAMP, generated by adenylate cyclases as a consequence of G-protein-coupled receptor activation, activates protein kinase A (PKA), which phosphorylates the cAMP response element binding protein (CREB) transcription factor, thus promoting association with the CREB-binding protein (CBP), the assembly of the transcriptional machinery, and the activation of cAMP-responsive genes. The cAMP pathway can be constitutively activated by the intracellular A1 chain of cholera toxin (Ctx), which is non-cytotoxic, but irreversibly activates $G_{s\alpha}$ by ADP-ribosylation, resulting in a chronic activation of adenylate cyclase and an increase in the level of intracellular cAMP (79). When expressed in transgenic mice under the control of growth hormone regulatory sequences, Ctx caused somatotroph proliferation, pituitary hyperplasia and gigantism (80). In contrast, pituitary gland overexpression of a transcriptionally inactive CREB mutant, which cannot be phosphorylated by PKA, effectively competes with wild-type CREB activity, blocking the transcriptional response to cAMP (81). These mice exhibit a dwarf phenotype with atrophied pituitary glands markedly deficient in somatotroph but not other cell types, suggesting that transcriptional activation of CREB is necessary for the normal somatotroph development.

Antisense RNA Expression
Watson and Crick base pairing of an antisense RNA corresponding to a messenger RNA can potentially inhibit gene expression. The antisense RNA could interfere at one of a number of levels in the overall process of gene expression:

- transcription - the generation of the primary transcript
- processing of the primary transcript
- transport of the RNA from nucleus to cytoplasm
- translation

Alternatively, antisense RNA could promote mRNA degradation by forming a substrate for nucleases specific for double stranded RNA.

A number of groups have reported that specific expression of an antisense RNA can reduce the expression of the corresponding gene in transgenic rats, with interesting physiological consequences. Angiotensin II produced in the brain plays important roles in the regulation of blood pressure, heart rate, vasopressin (VP) release, drinking behaviour and, possibly, in the development of hypertension. Transgenic rats specifically expressing an antisense RNA in the brain that is complementary to the mRNA encoding the angiotensin II precursor protein angiotensinogen have a 90% reduction in brain angiotensinogen (82) and lower blood pressure. Angiotensinogen is converted to the active peptide angiotensin II by renin, and intracerebroventricular infusions of renin evoke a characteristic drinking response; this is markedly reduced in the angiotensinogen antisense transgenic rats. Interestingly, the angiotensinogen antisense transgenic rats exhibit a diabetes insipidus-like syndrome, producing an increased amount of urine with decreased osmolarity, coincident with a 35% reduction in plasma vasopressin.

Antisense in rats has also been used to investigate the endocrinology of growth. Transgenic rats bearing a construct consisting of the rat growth hormone (GH) promoter containing four copies of a thyroid hormone response element driving the expression antisense cDNA sequences for rat GH exhibited dwarfism at as early as 3-4 weeks of age. Plasma rat GH levels were approximately 40-50% lower in transgenic rats compared to their nontransgenic littermates, results in a 70-85% reduction in growth rate (83,84).

Antisense oligonucleotides
The application of antisense oligonucleotides to somatic cells has the potential to be an alternative to germline gene-knockout technologies. Attempts have been made to block gene expression in rat magnocellular HNS neurons by the application of large quantities of antisense single-stranded DNA oligonucleotides corresponding to the sequence of a target gene or its mRNA. Some striking results have been obtained. Injection into the brain of

oligonucleotides complementary to the VP mRNA, chemically modified to increase their stability *in vivo*, resulted in a rapid reduction in VP biosynthesis, and induced diabetes insipidus-like symptoms (85-87). Similarly, chemically modified antisense oligonucleotides corresponding to the OT mRNA had rapid effects on the level of systemic OT (88,89), and blocked lactation and suckling (89,90). However, these effects were seen without any changes in OT biosynthesis (89,90). Further, the OT antisense oligonucleotides elicited effects that were clearly unrelated to OT biosynthesis - including reduced electrophysiological excitability, reduced cholecystokinin (CCK)- and electrically-stimulated OT release, and inhibition of CCK-induced c-fos expression (91,92).

Antisense technology is highly controversial, and remains unproven in most circumstances. Whilst it is possible that antisense molecules are able to specifically interfere, through Watson and Crick base-pairing, with their intended targets, it has never been proven that an oligonucleotide can knock-out just one gene product, and that all of the other expressed genes in the target cell remain unaltered. Antisense molecules, particularly oligonucleotides which have been chemically modified to increase their stability and efficacy, can have profound non-sequence specific effects on cells. Modified oligonucleotides can bind avidly to proteins, and breakdown products can inhibit cell proliferation. Central administration of chemically-modified oligonucleotides into the brain have been shown to elevate body temperature, suppress food and water intake, and inhibit night-time activity; pyrogenic effects accompanied by elevated concentrations of circulating corticosterone, and an increase in the synthesis of interleukin-6 mRNA in the brain and spleen (93). A biological effect seen as a result of applying an antisense molecule might therefore be the result of (94):

- a specific Watson and Crick interaction with its intended target.
- a specific Watson and Crick interaction with an unintended, but unidentified, target.
- a non-antisense interaction with another RNA or RNAs.
- an interaction with protein.
- a non-specific effect on cell proliferation or metabolism.

Whilst antisense technology may one day mature into a valuable tool in both basic research and gene therapy, most of the results obtained to date must be viewed with caution.

SUMMARY

The rat has historically been the species of choice for experimental endocrinologists. The extension of gene transfer technologies to this species will further enhance its utility. Whilst these technologies, and the level of their exploitation, remain primitive, the forthcoming diversion of resources from gene sequencing to the determination of gene function will fuel their development. Commercial and medical researchers interested in high quality physiological data gleaned from appropriate models will increasingly exploit the rat. Endocrinologists, rather than having to rely on less-than-ideal murine models, will enjoy the benefits of working on their favourite species.

Acknowledgements

We wish to thank the Wellcome Trust for generous support.

REFERENCES

1. Dunham I, Shimizu N, Roe BA, Chissoe S, Hunt AR et al. The DNA sequence of human chromosome 22. Nature 1999;402:489-495.
2. Murphy D, Carter DA. Transgenic approaches to modifying cell and tissue function. Curr Opp Cell Biol 1992;4:274-279.
3. Murphy D, Carter DA. Transgenesis Techniques: Principles and Protocols. Methods in Molecular Biology, Volume 18. New Jersey: Humana Press, 1993.
4. Flavell DM, Wells T, Wells SE, Carmignac DF, Thomas GB, Robinson IC. Dominant dwarfism in transgenic rats by targeting human growth hormone (GH) expression to hypothalamic GH-releasing factor neurons. EMBO J 1996;15:3871-3879.
5. Zeng Q, Carter DA, Murphy D. Cell specific expression of a vasopressin transgene in rats. J Neuroendocrinol 1994;6:469-477.
6. http://ratmap.gen.gu.se/
7. http://www.rgd.mcw.edu/
8. Murphy D, Si-Hoe S-L, Brenner S, Venkatesh B. Something fishy in the rat brain: molecular genetics of the hypothalamo-neurohypophyseal system. BioEssays 1998;20:741-749.
9. Brenner S, Elgar G, Sandford R, Macrae A, Venkatesh B, Aparicio S. Characterisation of the pufferfish (Fugu) genome as a compact model vertebrate genome. Nature 1993;366:265-268.
10. Burbach JPH, Luckman, SM, Murphy D, Gainer H (1999) Gene Regulation in the magnocellular hypothalamo-neurohypophysial system. Physiol Rev 2000; (in press)
11. Reeves WB, Andreoli TE. The posterior pituitary and water metabolism. In Williams Textbook of Endocrinology (eds. Wilson JD, and Foster DW), pp 311-356. Philadelphia: WB Saunders, 1992.
12. Pierson PM, Guibbolini ME, Mayer-Gostan N, Lahlou B. ELISA measurements of vasotocin and isotocin in plasma and pituitary of the rainbow trout: effect of salinity. Peptides 1995;16:859-865.

13. Young WS 3rd, Shepard E, DeVries AC, Zimmer A, LaMarca ME, Ginns EI, Amico J, Nelson RJ, Hennighausen L, Wagner KU. Targeted reduction of oxytocin expression provides insights into its physiological roles. Adv Exp Biol Med 1998;449:231-240.

14. Russell JA, Leng G. Sex, parturition and motherhood without oxytocin. J Endocrinol 1998; 157:343-359.

15. Verbalis JG, Mangione MP, Stricker EM. Oxytocin produces natriuresis in rats at physiological plasma concentrations. Endocrinology 1991;128:1317-1322.

16. Huang W, Lee SL, Arnason SS, Sjoquist M. Dehydration natriuresis in male rats is mediated by oxytocin. Am J Physiol 1996;270: R427-R433.

17. Perks AM. The neurohypophysis. In: Fish Physiology Vol II (eds W.S. Hoar WS, Randall DJ) pp 111-205. New York: Academy Press, 1969.

18. Urano A, Kubokawa K, Hiraoka S. Expression of the vasotocin and isotocin gene family in fish. In: Fish Physiology, Vol XIII (eds Sherwood NM, Hew CL). New York: Academy Press, 1994.

19. Kiyama H, Emson PC. Evidence for the co-expression of oxytocin and vasopressin messenger ribonucleic acids in magnocellular neurosecretory cells: Simultaneous demonstration of two neurophysin messenger ribonucleic acids by hybridisation histochemistry. J Neuroendocrinol 1990;2:257-260.

20. Mohr E, Bahnsen U, Kiessling C, Richter D. Expression of the vasopressin and oxytocin genes occurs in mutually exclusive sets of hypothalamic neurons. FEBS Lett 1988;242:144-148.

21. Moore RY. Organisation and function of a central nervous system circadian oscillator: the suprachiasmatic nucleus. Fed Proc 1983;42:2783-2789.

22. Carter DA, Murphy D. Circadian rhythms and autoregulatory transcription loops - going around in circles? Mol Cell Endocrinol 1996:124:1-5.

23. Young WS. Expression of the oxytocin and vasopressin genes. J Neuroendocrinol 1992;4:527-540.

24. Schmitz E, Mohr E, Richter D. Rat vasopressin and oxytocin genes are linked by a long interspersed repeated DNA element (LINE): sequence and transcriptional analysis of LINE. DNA and Cell Biol 1991;10:81-91.

25. Ratty AK, Jeong S-W, Nagle JW, Chin H, Gainer H, Murphy D, Venkatesh B. A systematic survey of the intergenic region between the murine oxytocin and vasopressin genes. Gene 1996;174:71-78.

26. Murphy D, Carter DA. Vasopressin gene expression in the rodent hypothalamus: transcriptional and post-transcriptional responses to physiological stimulation. Mol Endocrinol 1990;4:1051-1059.

27. Burbach JPH, De Hoop MJ, Schmale H, Richter D, De Kloet ER, Ten Haaf JA, De Wied D (1984) Differential responses to osmotic stress of vasopressin-neurophysin mRNA in hypothalamic nuclei. Neuroendocrinol 1984;39:582-584.

28. Zingg HH, Lefebvre D, Almazan G. Regulation of vasopressin gene expression in hypothalamic neurons. J Biol Chem 1986;261:12956-12959.

29. Lightman SL, Young WS. Vasopressin, oxytocin, dynorphin, enkephalin, and corticotrophin releasing factor mRNA stimulation in the rat. J Physiol (London) 1987;394:23-39.

30. Sherman TG, Day R, Civelli O, Douglas J, Herbert E, Akil H, Watson SJ. Regulation of hypothalamic magnocellular neuropeptides and their mRNAs in the Brattleboro rat: coordinate responses to further osmotic challenge. J Neurosci 1988;8:3785-3796.

31. Carrazana EJ, Pasieka KB, Majzoub JA. The vasopressin poly (A) tail is unusually long and increases during stimulation of vasopressin gene expression in vivo. Mol Cell Biol 1988;8:2267-2274.

32. Carter DA, Pardy K, Murphy D Regulation of vasopressin gene expression: changes in level, but not size, of VP mRNA following endocrine manipulations. Cell Mol Neurobiol 1993;13:87-95.
33. Carter DA, Murphy D. Rapid changes in poly(A) tail length of vasopressin and oxytocin mRNAs form a common early component of neurohypophysial peptide gene activation following physiological stimulation. Neuroendocrinol 1991;53:1-6.
34. Van Tol HHM, Voorhuis TAM, Burbach JPH. Oxytocin gene expression in discrete hypothalamic magnocellular groups is stimulated by prolonged salt loading. Endocrinology 1987;120:71-76.
35. Carter DA, Murphy D. Independent regulation of neuropeptide mRNA level and poly(A) tail length. J Biol Chem 1989;264:6601-6603.
36. Si-Hoe SL, Murphy D. Physiological regulation of vasopressin mRNA abundance and poly(A) tail length: tissue and species differences revealed by comparative analysis in transgenic rodents. (Submitted)
37. Zeng Q, Foo N-C, Funkhouser JM, Carter DA, Murphy D. Expression of a rat vasopressin transgene in rat testis. J Reprod Fert 1994;102:471-481
38. Venkatesh V, Si-Hoe S-L, Murphy D, Brenner S. Transgenic rats reveal functional conservation of regulatory controls between the Fugu isotocin and rat oxytocin genes. Proc Natl Acad Sci USA 1997;94:12462-12466.
39. Goossens N, Diericks K, Vandesande F. Immunocytochemical localization of vasotocin and isotocin in the preopticohypophysial neurosecretory system of teleosts. Gen Comp Endocrinol 1977;32:371-375.
40. Hyodo S, Urano A. Changes in expression of provasotocin and proisotocin genes during adaptation to hyper- and hypo-osmotic environments in rainbow trout. J Comp Physiol B 1991;161:549-556.
41. Waller S, Fairhall KM, J Xu, Robinson ICAF, Murphy D. Neurohypophyseal and Fluid Homeostasis in Transgenic Rats Expressing a Tagged Rat Vasopressin Prepropeptide in Vasopressinergic Magnocellular Neurons. Endocrinology 1996;137:5068-5077.
42. Fitzsimmons MD, Roberts MM, Sherman TG, Robinson AG. Models of neurohypophyseal homeostasis. Am J Physiol 1992;262:R1121-R1130.
43. Fitzsimmons MD, Roberts MM, Robinson AG. Control of posterior pituitary vasopressin content: implications for the control of the vasopressin gene. Endocrinology 1994;134:1874-1878.
44. Olsen SR, Uhler MD. Inhibition of protein kinase A by overexpression of the cloned human protein kinase inhibitor. Mol Endocrinol 1991;5:1246-1256.
45. Pearson JD, DeWald DB, Mathews WR, Mozier NM, Zurcher-Neely HA, Heinrikson RL, Morris MA, McCubbin WD, McDonald JR, Fraser ED, Vogel HJ, Kay CM, Walsh MP. Amino acid sequence and characterisation of a protein inhibitor of protein kinase C. J Biol Chem 1990;265:4583-4591
46. Ang H-L, Carter DA, Murphy D. Neuron-specific and physiological regulation of bovine vasopressin transgenes in mice. EMBO J 1993;12:2397-2409.
47. Waller SJ, Murphy D. Expression of rat vasopressin transgenes in rats. In, The 1st Joint World Congress of Neurohypophysis and Vasopressin. Excerpta Medica International Congress Series 1098 (Eds, Saito et al.) pp89-98. Amsterdam: Elsevier Science, 1995.
48. Hentze MW, Kulozik AE. A perfect message: RNA surveillance and nonsense-mediated decay. Cell 1999;96:307-310.
49. Harvey S. Growth hormone release. In: Growth Hormone (eds Harvey S, Scanes CG, Daughaday WH) pp 97-130. 1995.

50. Clarke I J. What can we learn from sampling hypophysial portal blood? In: Functional anatomy of the neuroendocrine hypothalamus. Ciba Foundation symposium. 1992;168:87-103.

51. Barinaga M, Yamonoto G, Rivier G, Vale W, Evans R, Rosenfeld MG. Transcriptional regulation of growth hormone gene expression by growth hormone-releasing factor. Nature 1987;306:84-5.

52. Guillemin R, Brazeau P, Böhlen P, Esch F, Ling N, Wehrenberg, WB. Growth hormone-releasing factor from a human pancreatic tumor that caused agromegaly. Science 1982;218:585-587.

53. Rivier J, Speiss J, Thorner M, Vale W. Characterization of growth hormone-releasing factor from a pancreatic islet tumor. Nature 1982;300:276-278.

54. Brazeau P, Vale W, Burgus R, Ling L, Butcher M, Rivier J, Guillemin, R. Hypothalamic polypeptide that inhibits the secretion of immunoreactive pituitary growth hormone. Science 1973;179:77-79.

55. de Gennaro CV, Cattaneo E, Cocchi D, Müller EE, Maggi A. Growth hormone regulation of growth hormone-releasing hormone gene expression. Peptides 1988;9:985-988.

56. Miki N, Ono M, Miyoshi H, Tsushima T, Shizume K. Hypothalamic growth hormone-releasing factor (GRF) participates in the negative feeback regulation of growth hormone secretion. Life Sciences 1989;44:469-476.

57. Berelowitz M, Firestone SL, Frohman LA. Effects of growth hormone excess and deficiency on hypothalamic somatostatin content and release and on tissue somatostatin distribution. Endocrinology 1981;109:714-9.

58. Levy A, Matovelle MC, Lightman SL, Young III WS. The effects of pituitary stalk transection, hypophysectomy and thyroid hormone status on insulin-like growth factor 2-, growth hormone releasing hormone-, and somatostatin mRNA prevalence in rat brain. Brain Res 1992;579:1-7.

59. Godfrey P, Rahal JO, Beamer WG, Copeland NG, Jenkins NA, Mayo KE. GHRH receptor of little mice contains a missense mutation in the extracellular domain that disrupts receptor function. Nature Gen 1993;4:227-231.

60. Li S, Crenshaw III EB, Rawson E.J, Simmons M, Swanson LW, Rosenfeld MG. Dwarf locus mutants lacking three pituitary cell types　　　　result from mutations in the POU-domain gene pit-1. Nature 1990;347:528-533.

61. Stewart TA, Clift S, Pitts-Meek S, Martin L, Terrell TG, Liggitt D, Oakley H. An evaluation of the functions of the 22-kilodalton (kDa), the 20 kDa, and the N-terminal polypeptide forms of human growth hormone using transgenic mice. Endocrinology 1992;130:405-414.

62. Banerji J, Olson L, Schaffner W. A lymphocyte-specific cellular enhancer is located downstream of the joining region in immunoglobulin heavy chain genes. Cell 1983;33:729-740.

63. Palmiter R.D, Brinster RL, Hammer RE, Trumbauer ME, Rosenfeld MG, Birnberg NC, Evans RM. Dramatic growth of mice that develop from eggs microinjected with metallothionein-growth hormone fusion genes. Nature 1982;300:611-615.

64. Palmiter RD, Norstedt G, Gelinas E, Hammer RE, Brinster RL. Metallothionein-human GH fusion genes stimulate growth of mice. Science 1983;222:809-814.

65. Takeuchi T, Suzuki H, Sakurai S, Nogami H, Okuma S, Ishikawa H. Molecular mechanism of growth hormone (GH) deficiency in the spontaneous dwarf rat: detection of abnormal splicing of GH messenger ribonucleic acid by polymerase chain reaction. Endocrinology 1990;126:31-38.

66. Charlton HM, Clark RG, Robinson ICAF, Porter Goff AE, Cox BS, Bugnon C, Bloch BA. Growth hormone-deficient dwarfism in the rat: a new mutation. J Endocrinol 1988;119:51-58.

67. Flavell DM, Wells T, Wells SE, Carmignac DF, Thomas, GB, Robinson ICAF. Dominant dwarfism in transgenic rats by targeting human growth hormone (GH) expression to hypothalamic GH-releasing factor neurons. EMBO J 1996;15:3871-3879.

68. Wells T, Flavell DM, Wells SE, Carmignac DF, Robinson ICAF. Effects of growth hormone secretagogues in the transgenic growth-retarded (Tgr) rat. Endocrinology 1997;138:580-587.

69. Iannaccone PM, Taborn GU, Garton RL, Caplice MD, Brenin DR. Pluripotent embryonic stem cells from the rat are capable of producing chimeras. Dev Biol 1994;163:288-292.

70. Iannaccone PM, Taborn GU, Garton RL, Caplice MD, Brenin DR Pluripotent embryonic stem cells from the rat are capable of producing chimeras (erratum). Dev Biol 1997;185:124-125.

71. Schnieke AE, Kind AJ, Ritchie WA, Mycock K, Scott AR, Ritchie M, Wilmut I, Colman A, Campbell KH. Human factor IX transgenic sheep produced by transfer of nuclei from transfected fetal fibroblasts. Science 1997;278:2130-2133.

72. Wilmut I, Schnieke AE, McWhir J, Kind AJ, Campbell KHS. Viable offspring from fetal and adult mammalian cells. Nature 1997;385:810-813.

73. Wakayama T, Perry AC, Zuccotti M, Johnson KR, Yanagimachi R. Full-term development of mice from enucleated oocytes injected with cumulus cell nuclei. Nature 1998;394:369-374.

74. Nelson RJ. The use of genetic "knockout" mice in behavioural endocrinology research. Horm Behav 1997;31:188-96.

75. Routtenberg A. Knockout mouse fault lines. Nature 1995;374:314-315.

76. Hochgeschwender U, Brennan MB. Mouse knockouts rule OK. Nature 1995;374:543.

77. Thomas JH. Thinking about genetic redundancy. TIG 1993;9:395-399.

78. Nowak MA, Boerlijst MC, Cooke J, Smith JM. Evolution of genetic redundancy. Nature 1997;388:167-171.

79. Gilman AG. G proteins: transducers of receptor-generated signals. Ann Rev Biochem 1987;56:615-649.

80. Burton FH, Hasel KW, Bloom FE, Sutcliffe JG. Pituitary hyperplasia and gigantism in mice caused by a cholera toxin transgene. Nature 1991;350:74-77.

81. Struthers RS, Vale WW, Arias C, Sawchenko PE, Montminy MR. Somatotroph hypoplasia and dwarfism in transgenic mice expressing a non-phosphorylatable CREB mutant. Nature 1991;350:622-624.

82. Schinke M, Baltatu O, Bohm M, Peters J, Rascher W, Bricca G, Lippoldt A, Ganten D, Bader M. Blood pressure reduction and diabetes insipidus in transgenic rats deficient in brain angiotensinogen. Proc Natl Acad Sci U S A 1999;96:3975-3980.

83. Matsumoto K, Kakidani H, Anzai M, Nakagata N, Takahashi A, Takahashi Y, Miyata K. Evaluation of an antisense RNA transgene for   inhibiting   growth   hormone gene expression in transgenic rats. Dev Genet 1995;16:273-277.

84. Matsumoto K,   Kakidani H,   Takahashi A,   Nakagata N,   Anzai M,   Matsuzaki Y, Takahashi Y, Miyata K, Utsumi K, Iritani A. Growth retardation in rats whose growth hormone gene expression was suppressed by antisense RNA transgene. Mol Reprod Dev 1993;36:53-58.

85. Flanagan LM, McCarthy MM, Brooks PJ, Ptaff DW, McEwan BS. Arginine vasopressin levels after daily infusions of antisense oligonucleotides into the supraoptic nucleus. Ann NY Acad Sci 1993;689:520-521.

86. Skutella T, Probst JC, Engelmann CT, Wotjak CT, Landgraf R, Jirikowski GF. Vasopressin antisense oligonucleotide induces temporary diabetes insipidus in rats. J Neuroendocrinol 1994;6:121-125.

87. Meeker R, LeGrand G, Ramirez J, Smith T, Shih YH. Antisense vasopressin oligonucleotides: uptake, turnover, distributioin, toxicity and behavioural effects. J Neuroendocrinol 1995;7:419-428.

88. Neumann I, Porter DW, Landgraf R, Pittman QJ. Rapid effect on suckling of an antisense oligonucleotide administered into the rat supraoptic nucleus. Am J Physiol 1994;267:R852-858.

89. Jirikowski GF, Celeda D, Jantz M, Prufer K, Lee JS. Sense- and antisense-targeting of oxytocinergic systems in rat hypothalamus. Adv Exp Med Biol 1995;395:59-65.

90. Skutella T, Probst JC, Caldwell JD, Pederson CA, Jirikowski GF. Antisense oligodeoxynucleotide complementary to oxytocin mRNA blocks lactaion in rats. Exp Clin Endocrinol Diabetes 1995;103:191-195.

91. Neumann I, Kremarik P, Pittman QJ. Acute, sequence-specific effects of oxytocin and vasopressin antisense oligonucleotides on neuronal responses. Neuroscience 1995;69:997-1003.

92. Neumann I, Pittman QJ, Landgraf R. Release of oxytocin within the supraoptic nucleus. Mechanisms, physiological significnce and antisense targeting. Adv Exp Med Biol 1995;395:173-183.

93. Schobitz B, Pezeshki G, Probst JC, Reul JM, Skutella T, Stohr T, Holsboer F, Spanagel R. Centrally administered oligonucleotides in rats: occurance of non-specific effects. Eur J Pharmacol 1997;331:97-107.

94. Branch AD. A good antisense molecule is hard to find. TIBS 1998;23:45-50.

2
TRANSGENIC MODELS FOR STUDIES OF OXYTOCIN AND VASOPRESSIN

Harold Gainer and W. Scott Young, III

Laboratory of Neurochemistry, National Institute of Neurological Disorders and Stroke, and the Section on Neural Gene Expression, National Institute of Mental Health, National Institutes of Health, Bethesda, MD 20892, USA.

INTRODUCTION

The number of studies of the nervous system using transgenic mice has grown explosively over the past decade, presenting a wide spectrum of approaches. Transgenic mice have been used to model human disease, to understand physiologic roles of genes, and to understand the regulation of genes (1). For example, transgenes composed of a gene's promoter directing expression of a reporter gene such as beta-galactosidase may be used to follow expression of a gene during development (2), after various mutations are made in the promoter region to look for cell or regulatory specificity (3), and to study various physiological states (4). Regulatory studies may also use reporters (e.g., green fluorescent protein) that allow for real-time measurement of activity, either *in vivo* or in various tissue preparations, that is precluded by difficulties in assaying the promoter's natural gene product. Transgenic mice may also be used to perturb a particular system by overexpressing a gene or by reducing the gene's expression and/or effectiveness through antisense or dominant negative expression (5). Transgenic expression of certain products, such as tumor promoters or fluorescent substances, may allow for the isolation of immortalized and homogeneous cells for further study (6,7). Finally, transgenic mice may be used in the attempt to correct defects in mutant mice, either those found accidentally or those produced through homologous recombination or random mutagenesis (8).

Transgenesis in mice has also been accomplished by homologous recombination, usually to produce a mouse whose targeted gene is rendered nonfunctional. As most brain genes seem to be expressed in more than one cell type at one time or another during the life-span of a mouse or are widespread in their expression, many "knocked out" mice may alter gene expression within the magnocellular neurons through indirect influences. Targeting genes for knockout specifically in the magnocellular vasopressin (VP) and oxytocin (OT) neurons has not been accomplished yet. However, knockouts of a few genes have produced results that are likely to reflect direct importance of those genes within the magnocellular neurons. These studies will be presented below. Finally, since most endocrinological and physiological data have been obtained using rats it would be highly desirable to perform such experiments on transgenic rats. In fact, significant efforts towards this goal have been made in studies (9,10) on the rat hypothalamo-neurohypophysial system (HNS). However, at this time, studies using transgenic rats still lag behind those with transgenic mice, and we are not aware of any studies using homologous recombination to study the rat nervous system.

The HNS as a Model System

The magnocellular neurons in the HNS have been among the most intensively studied peptidergic neurons in the CNS (11-19). The HNS neurons have served as excellent model systems for the study of peptide neurosecretion mechanisms *in vivo* in large part due to their compact nuclear organization in the CNS, e.g., most of the approximately 12,000 cells in the rodent hypothalamus are located in two bilateral nuclei, the paraventricular (PVN) and supraoptic (SON) nuclei. The HNS neurons project via well-defined axonal tracts to the posterior pituitary (i.e., neural lobe), where each axon is estimated to branch into hundreds of nerve terminals (13,16), and where these axonal branches and terminals represent about fifty percent of the total tissue mass of the neural lobe. The relatively easy access to the HNS cellular components, the cell bodies in the PVN and SON and axons in the median eminence by both stereotaxic and micropunch assay methods, and the nerve terminals by their presence outside of the blood-brain-barrier in the posterior pituitary, have made these magnocellular neurons favorite objects of many biochemical and physiological studies. Consequently, a substantial database about these peptidergic neurons exists in the literature.

Based on biochemical, morphological and physiological criteria, the HNS neuronal population has been divided into two distinct phenotypes, the OT and VP neurons (20). This historic classification has been reinforced by many immunocytochemical (ICC) and *in situ* hybridization histochemical (ISHH) experiments that have been performed on this system, and it had been the generally accepted view that the expression of the OT and VP genes were mutually exclusive and occurred in separate cells in the HNS (21). The two neuronal phenotypes in rats have also been discriminated by their distinct electrical activity patterns in response to sustained physiological stimuli, i.e., OT cells increase their action potential frequencies with regular, continuous firing, whereas VP cells generate phasic bursting patterns (11-14) in response to steady depolarization. These different properties have been attributed, in part, to higher levels of calcium-binding proteins (e.g., calbindin, calretinin) in oxytocin neurons which prevent depolarizing after potentials considered critical for bursting (11-14), and to a novel non-activating outward potassium current found only in oxytocin cells (22). The latter may underlie the milk-ejection high-frequency bursts observed only in OT cells. In addition to the principal peptides, OT and VP, these neurons (and their LDCVs) also contain smaller amounts of other coexisting peptides (e.g., galanin, cholecystokinin, CRH, dynorphin, enkephalin, TRH) which can vary between cells depending on functional conditions (19,23), thereby revealing additional heterogeneity within the OT and VP neuronal phenotypes.

Several laboratories have reported coexistence between OT and VP in some HNS neurons (i.e., about 1-3% of the total population under normal conditions) (24,25), and that this can increase to a maximum of 17% after 2 days of lactation (25,26). In addition, while the segregation of OT and VP gene expression in separate cells in the HNS is the rule, recent studies using a sensitive RT-PCR analysis of single magnocellular neurons (MCNs) showed that virtually all of the MCNs of the OT-phenotype contain some VP mRNA, and those of the VP-phenotype also contain OT mRNA at low levels (26,27). Quantitative RT-PCR analyses of MCNs in normal female rats showed that in the OT and VP MCN phenotypes, the major nonapeptide mRNA species was more than one hundred-fold greater than the minor peptide mRNA. In contrast, in the OT- and VP-coexpressing MCN phenotypes the ratio of the two peptide mRNAs is around two (28). Hence, it is clear that expression of these peptide genes is not mutually exclusive in the MCNs, although the physiological significance of this and its underlying mechanisms are unclear and require further study.

TRANSGENIC ANALYSIS OF OT AND VP GENE EXPRESSION IN THE HNS

Studies of the mechanisms that are responsible for cell-specific gene expression of OT and VP in the magnocellular neurons of the HNS have been limited by the absence of relevant experimental models in which to examine these issues. The lack of homologous cell lines that express these genes has led most investigators to study their regulation either in heterologous culture systems or in transgenic mice (9,27,29-31). The experimental work using heterologous systems has identified a number of putative regulatory elements in the OT and VP genes. These include various nuclear hormone receptors, class III POU proteins, and fos/jun/ATF family members as candidate transcriptional activating factors involved in the regulation of the OT and VP genes (32,33). However, interpretation of the data from heterologous cell lines used for this purpose has been complicated by the observations that the behavior of the identified *cis*-motifs in the 5' untranslated regions of these genes often differs depending upon which cell line is used (32,33). Consequently, the most relevant data to date has come from studies done in transgenic mice with the assays being performed *in vivo* in the magnocellular neurons themselves (27,29,33).

The OT and VP Genes
An additional complexity for cell-specific gene expression studies is the possibility of interactions between the OT and VP genes. Each gene contains three exons and two introns, and both genes are on the same chromosomal locus but are transcribed in opposite directions (Figure 1). The domain separating the OT and VP genes has been called the "intergenic region" (IGR) and is relatively short. The IGR region in the rat is about 11 kbp in length (34), whereas in the mouse it is only 3.6 kbp (35,36). The rat and mouse IGRs have been sequenced and more than half of the rat IGR is known to be represented by a long interspersed repeated DNA element (37), which is completely missing in the mouse IGR (36). Except for the LINE element and a few hundred bp of random, non-homologous sequences, most of the rest of the two IGR sequences were highly conserved. The principal value of this information is that the high sequence conservation found both upstream and downstream of the genes (in the IGR) suggested that both of these domains contained regulatory DNA sequences and that the LINE element found only in the rat IGR was not critical for gene function.

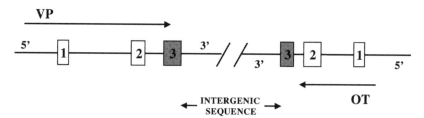

Fig 1. Organization of the intron-exon structure of the VP and OT genes. Both genes are composed of three exons shown as numbered rectangles, separated by two introns (shown as lines between the exons). The genes are present on the same chromosome but in opposite transcriptional directions (shown by arrows). Similar gene structures but with variable lengths of intergenic sequences have been found in various mammalian species. Upstream (5') and downsteam (3') flanking sequence domains are indicated.

Initial Studies: The Age of Innocence

The earliest studies of HNS gene expression in transgenic mice made use of the 5' flanking regions of heterologous VP genes attached to exogenous reporters (38-41). Murphy et al. (38) used 1.25 kbp of the bovine promoter attached to the early region of the tumor virus SV40 encoding the large T-antigen. This transgene construct (termed AVP.SVER 1.25, see Table 1) was not expressed in the HNS, but was in the anterior pituitary, where it produced tumors (38,39). The same 1.25 kbp bovine promoter when attached to a chloroamphenicol acetyl transferase (CAT) reporter (VP-A) produced ubiquitous expression of the construct in two lines of transgenic mouse lines, but still no notable expression in the HNS (40). Russo et al. (41) studied 14 different fusion genes in transgenic mice. One of these contained 2 kbp of the human AVP promoter linked to the human growth hormone gene (AVP-GH, in Table 1). They found ubiquitous (ectopic) expression of this transgene throughout the mouse CNS, particularly in the cerebral cortex, and also in the hypothalamus and HNS neurons. However, this construct was no more specific or robust in its expression in the HNS than a metallothionin promoter that had been connected to the growth hormone reporter as a transgene. Interestingly, the CNS expression of the latter construct could be eliminated by removing all the introns from the growth hormone reporter gene (41), and only expression in tissues normally characteristic of metallothionin promoter expression (e.g., liver, pancreas, intestine, and kidney) was found with the latter construct.

While the above observations focus on inappropriate (ectopic) expression of these transgenes, they also alert one to the general caveats in the interpretation of transgenic data. Expression in a given tissue can reflect so-called "position effects" of the transgenes' integration (42), enhancing influences of heterologous sequences (43), or silencing effects (44) that may derive from exogenous sequences present in the constructs. In some cases, the same reporter gene (e.g., Lac-Z) can act as an enhancer (e.g., see VP-BGL-pA construct in Table 1) or as a repressor (e.g., see the 3.VP-Gal-0.55 construct in Table 1, and OT-BGL-pA in Table 2), depending upon the specific configurations of the DNAs in the construct. Habener et al. (45) produced a transgenic mouse with a complex construct containing 800 bp of the mouse metallothionin I (MT) 5' flanking region (promoter) and 35 bp of its exon 1 connected through a 14 bp artificial DNA linker to the entire rat prepro AVP NP II gene. The authors reported the expression of the fusion gene in appropriate tissues (corresponding to the metallothionin promoter's endogenous expression), but also in the HNS which normally does not express this gene. The authors referred to the HNS expression as "eutropic expression", and attributed this to a "synergy" between the MT promoter sequence and the AVP gene sequence. It is clear from the above studies that the heterologous DNA and the reporter's DNA sequences in the transgene can interact with its endogenous sequences to produce ectopic expression, and sometimes even reproducible expression in relevant tissue, thereby confounding the analysis regarding cell specificity.

The "Minilocus" Construct and the IGR

In view of the aforementioned complexity produced by exogenous sequences in the constructs, and since it was unclear what part of the genes (including introns and exons) contained the elements that were critical for cell-specific expression, many of the subsequent studies used intact bovine or rat OT and VP genes in transgenic mice. This decision was based on the fact that while there was great conservation in the gene sequences between the species, there was enough divergence to allow for the production of specific probes to differentiate the expression of the transgene from the endogenous mouse gene. The first report of robust cell-specific gene expression in the HNS of transgenic mice was made using a combined rat VP and OT ("minilocus") construct (46-48). The construct used was called V1 (see Tables 1 and 2). The results were surprising in that although this construct contained about four times more VP 5'-upstream sequence than OT 5'-upstream sequence (the latter being only 0.36 kbp), the rat OT transgene was robustly expressed (at between 10-30% of endogenous OT gene expression levels) in mouse OT cells only whereas the VP gene was not expressed at all (Figure 2).

Subsequent studies on the VP gene (10,49,50) showed that by extending the 3'-downstream region of the VP gene to 3 kbp, it was possible to get cell-specific expression of these transgenes in VP neurons only. Some of the successful constructs used in these studies (see Table 1) contained an exogenous reporter (chloramphenicol acetyltransferase, CAT) inserted into exon III. This indicated that increasing the distance of the putative element in the IGR from the promoter region in the VP by an additional 660 bp was not deleterious for the cell-specific expression of this gene and also suggested that the cell-specific elements for both the OT and VP genes were in the IGR within 3 kbp of the 3'-downstream region of the rat VP gene (15,51). These studies of various oxytocin and vasopressin constructs in transgenic mice indicated that constructs containing genomic DNA from 0.5 to 9 kbp 5' upstream of the OT and VP genes but with no endogenous 3'-downstream sequences do not produce significant expression in the hypothalamic magnocellular neurons, and that the IGR contained *cis*-elements that are essential for their cell-specific gene expression in the magnocellular neurons. This viewpoint has been called the IGR hypothesis (15,51).

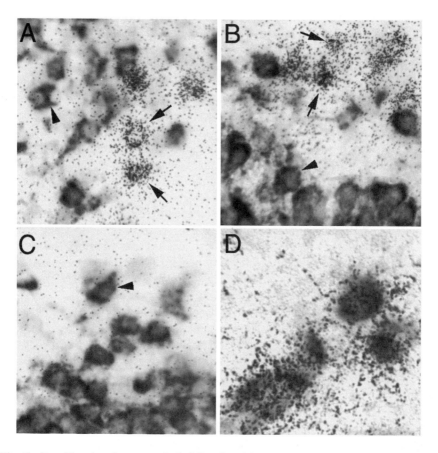

Fig 2. Double simultaneous hybridization histochemistry is used to examine expression of mouse OT (D) or VP (A-C) genes and the rat OT transgene (A-D). Expression of the mouse genes was detected using digoxigenin-labeled oligodeoxynucleotide probes (cells indicated by arrowheads) and the rat OT transgene using a 35-S-labeled oligodeoxynucleotide probe (cells indicated by arrows). Mouse VP and rat transgene expression was located in exclusive populations of cells in the PVN (A) and SON (B). No transgene expression was seen in a control mouse SON (C). In D, label for both the mouse OT and rat transgene OT was found in the same SON neurons.

Given the fact that a CAT reporter could be inserted into exon III of the rat transgene (10,50), we set out to further examine the IGR hypothesis using CAT-bearing mouse OT and VP gene constructs extended downstream by 2.1 or 3.6 kbp of IGR sequence. The results of these studies were that of five separate founder lines containing the VP-III-CAT-2.1 construct (Table 1) all showed equivalent cell-specific expression of CAT in VP magnocellular neurons but no expression at all in OT neurons. All five lines also had some ectopic expression of CAT in the brains, consistently in the large neurons of the reticular thalamic nucleus. Similar data were obtained using a VP-III-CAT-3.6 construct. We also found using a mouse OT gene-CAT construct and the 3.6-kbp IGR domain (Table 2) that this OT construct also produced cell-specific expression of CAT, but in this case in OT neurons only and not in VP cells nor in ectopic sites in the brain. Validation of the cell-specific expression of the CAT protein was done by double-label immunofluorescence studies using polyclonal CAT and monoclonal OT- and VP-associated neurophysin antibodies. Double-label experiments using immunogold labels were also performed at the electron microscopic level and both constructs were also successful in targeting the CAT protein to the regulated secretory pathway, i.e., the immunoreactive CAT was sorted to RER, Golgi and LDCVs (in preparation, HG lab). The above data lend further support to the IGR hypothesis, and indicate that for VP gene expression only 2.1 kbp downstream of the 3' flanking region was necessary for cell specificity. Clearly, more experimentation is necessary to identify and characterize the elements in the IGR that are necessary for cell-specific expression of the OT and VP genes.

Some of the transgenes were studied for their ability to be regulated by physiological stimuli *in vivo*. These studies are denoted in Tables 1 and 2 by asterisks. To our knowledge, few of the putative regulatory elements that were identified in the *in vitro* experiments have been re-evaluated *in vivo* by transgenic studies. In one case, the removal of the putative ERE in the rat OT promoter (-169/-157) in the V1 minilocus construct (V18 construct), which reduced by 80-90% its responsiveness to estrogen, thyroid hormone and retinoic acid *in vitro* (WSY and J.P.H. Burbach labs, unpublished), also eliminated the transgene's basal expression in the HNS (unpublished, WSY lab). Mutation of bases -168 and -167 from GG to TT (V17 construct) reduced responsiveness to the three hormones by approximately 50% *in vitro* (52), but had no significant impact on basal transgenic expression (WSY lab, unpublished).

TABLE 1. VASOPRESSIN TRANSGENES

Selective DNA Source	Host Species	Transgene Name	5'UT (kbp)	Structural Gene	Reporter Gene	3'UT (kbp)	Expression in HNS	Ectopic Expression	Ref.
bovine	mouse	AVP.SV.ER.1.25	1.25	no	SV40T-Ag	———	no	anterior pituitary	(38)
bovine	mouse	VP-A	1.25	no	CAT	———	no	ubiquitous	(54)
human	mouse	AVP-GH	2	no	human growth hormone	———	yes*	ubiquitous in CNS	(41)
bovine	mouse	——	1.5-3	yes	Lac-Z	0.2	no	testis, germ cells	(9)
mouse	mouse	VP-BGL-pA	1.4	no	Lac-Z	———	no	ubiquitous	(d)
bovine	mouse	VP-B	1.25	yes	———	0.2	yes*	ubiquitous in CNS, adrenal medulla	(54)
rat	mouse	V1-minilocus	1.4	yes[a]	no	0.17	no	n.d.	(46)
rat	rat	3.VP-GLO.2	3	yes	Lac-Z	0.2	no	no	(50)
rat	rat	3-VP-Gal-0.55	3	yes	Lac-Z	0.55	no	no	(9)
bovine	mouse	VP-C	9	yes	no	3	yes*	pituitary, ovary	(54)
rat	mouse	8.2.rVP	3	yes	———	3	yes*	lung, pancreas	(49)
rat	rat	5-VCAT-3	5	yes	CAT[b]	3	yes	low	(10, 50)
mouse	mouse	VP-3-CAT-2.1	3	yes	CAT[c]	2.1	yes	little in CNS	(d)

a   Fused to OT structural gene in "minilocus" configuration.
b   CAT inserted in exon 3.
c   CAT inserted after exon 3.
d   Unpublished data, HG lab.
*   Physiological regulation of transgene expression reported.
    n.d. – not determined.

TABLE 2. OXYTOCIN TRANSGENES

DNA Source	Host Species	Transgene Name	5'flank (kbp)	Structural Gene	Reporter Gene	3'flank (kbp)	Expression in HNS	Selective Ectopic Expression	Ref.
rat	mouse	ROT-1.63	0.36	yes	no	0.5	*	*	(i)
bovine	mouse	b.BOT 6.4	3.0	yes	no	2.6	*	*	(53)
bovine	mouse	b.OT	0.6	yes	no	2.5	no	testes, lungs	(40, 54)
mouse	mouse	OT-BGL-pA	1.3	no	Lac-Z	1.8[a]	no	no	(j)
bovine	mouse	b.OT.3.5	0.5	yes	no	1.8[a]	yes**	testes, lung	(53)
mouse	mouse	AI-02	1.05	yes	EGFP[b]	>3.5[c]	no	no	(55)
mouse	mouse	AI-01	1.05	yes	EGFP[d]	>3.5[c]	very low	no	(55)
rat	mouse	V1-minilocus	0.36	yes	no	0.43	yes**	no	(46)
mouse	mouse	OT-3-CAT-3.5	0.5	yes	CAT[e]	3.5	yes	no	(i)
mouse	mouse	A1-03	1.05	yes	EGFP[f]	>3.5[c]	yes	no	(55)
rat	mouse	JL-01	1.05	yes	IRES-EGFP[g]	0.43[h]	yes	no	(55)

a Fused to bovine VP structural gene with 1.25 kbp VP-5' flanking and 0.2 kbp VP-3' flanking regions.

b EGFP reporter in place of OT-exon 1.

c Includes entire IGR plus VP structural gene minus VP-exon 1.

d EGFP reporter is immediately after OT-exon-1, plus VP structural gene minus VP-exon 1.

e CAT inserted after exon 3, and 3.5 kbp represents entire IGR in mouse.

f EGFP inserted in middle of OT-exon 3, fused to entire VP structural gene minus VP-exon 1.

g IRES-EGFP inserted into middle of OT-exon 3.

h Fused to entire VP-structural gene containing 0.17 kbp of VP 3' flanking region (IGR).

i Unpublished data, WSY lab.

j Unpublished data, HG lab.

* No transgenic mice were produced.

** Physiological regulation of expression reported.

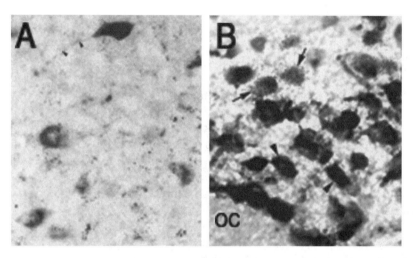

Fig 3. Green fluorescent protein expression in neurons of the PVN (A) and SON (B). Panel A shows EGFP fluorescence in a number of cells and fibers (arrowheads). Panel B shows EGFP-expressing cells through the use of a 35S-labeled riboprobe (arrows). Darkly staining VP cells (arrowheads) are labeled using a digoxigenin-labeled riboprobe. No coexpression of EGFP and VP is seen.

Using the OT-VP minilocus construct (Table 2) as a starting point, four types of transgenic mice, in which the OT genes contained green fluorescent protein (GFP) as the reporter gene (57,58), were generated (55). Placement of the GFP within the first OT exon either before or after the signal peptide, yielded little to no expression. However, like the CAT constructs, placement of the GFP in the third exon (as an in-frame fusion with the carboxyl terminus of the OT preprohormone) resulted in cell-specific expression of GFP in OT neurons (Figure 3). This expression was seen only within the OT neurons as well as their axons and nerve endings in the posterior pituitary (55). Furthermore, placement of the GFP sequence downstream of a picornavirus internal ribosomal entry site (IRES), also in the third exon, allowed expression of the GFP as a separate protein. As expected, the GFP in this construct, synthesized separately from the signal sequence in the pro-OT, was not packaged into vesicles for transport to the pituitary, but was most prominent in the cytoplasm and proximal processes (55,56). Therefore, constructs with inserts in the third exon could target expression specifically to OT neurons, whereas inserts in the first exon did not. One curious observation was that even in the most robustly expressing line (AI-03), the transgene was only expressed in a minority (10-30%) of the OT cells (unpublished, WSY lab). In contrast, the original V1 construct, which didn't contain foreign coding sequence as a reporter, was expressed in over 90% of the OT neurons. Is the reduced expression that was observed with the GFP constructs due to the influence of the reporter? In this regard, it is interesting that a similar cell-specific gene expression pattern with a low penetrance was also observed when CAT was used as the reporter in the OT-III-CAT-3.6 construct (unpublished, HG lab).

The Use of the OT-EGFP Mouse to Study Neurosecretion
The existence of this OT-EGFP transgenic mouse in which the fluorescent reporter is targeted to secretory granules in the OT magnocellular neurons offers a unique opportunity to study neurosecretion from the nerve terminals of individual neurons. Previous measurements of OT and VP peptide secretion in the HNS, from nerve terminals in the neural lobe (59-61) or from dendrites in the hypothalamus (62), have been made by radioimmunoassay (RIA) of samples, collected either by superfusion *in vitro* from neural lobe or by microdialysis *in vivo* in hypothalamus. While the RIA method is extremely sensitive and has provided much valuable information about OT and VP secretion, the topographic (dendrite versus nerve terminal) and kinetic aspects of the secretion of these peptides cannot be studied by these procedures. One solution to this limitation would be the development of HNS cell systems (derived from transgenic mice) in which their large dense core vesicles (LDCVs) would be fluorescent (e.g., by EGFP targeting), and the use of imaging methods by which the secretion processes could be directly visualized (63-65). Given such transgenic mice, and the HNS cell culture models that could be derived from them (66,67), it should be possible to visualize secretion directly in HNS axons and terminals.

We have done preliminary studies using the AI-03 transgenic mouse line in which the OT neurons contain EGFP in their LDCVs. Electron microscopic immunochemistry has shown that the expressed EGFP-fusion protein in the HNS of these mice is restricted to LDCVs containing OT peptides (unpublished, HG and WSY labs). The main assay of secretion in these studies is the quantitative imaging of decreased fluorescence (due to exocytosis of EGFP from the LDCVs) in either individual dendritic or nerve terminal processes (see 63-65 for methods). At present, we are utilizing a Zeiss Atto digital fluorescence imaging system, with a photometry capacity for quantitative measurements and fast kinetics for this assay. In preliminary studies on the HNS we have been able to demonstrate calcium-dependent secretion from individual OT nerve endings or neurosecretosomes (Figure 4) isolated from the neural lobes of AI-03 mice.

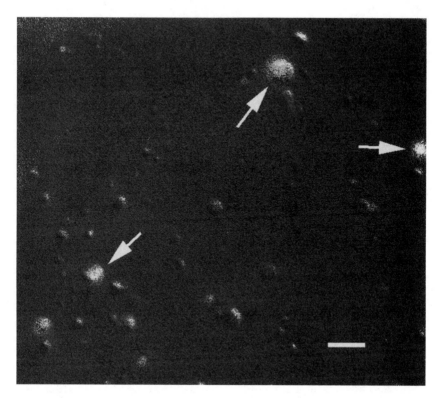

Fig 4. Dispersed, live posterior lobe nerve terminals from an AI-03 transgenic mouse are immobilized on to a poly-L-lysine-coated slide. Arrows indicate EGFP fluorescent terminals (bar = 5μm).

OTHER APPROACHES TO THE STUDY OF GENE FUNCTION IN THE HNS

Knockout Mice

As noted above, we are not aware of homologous recombination targeted specifically to the OT or VP magnocellular neurons. However, some relatively specific knockouts have been produced that are relevant. Two transcription factors that have been knocked out have profound effects on the OT and VP magnocellular neurons. Null mutations of the basic helix-loop-helix-PAS *Sim1* (68) or the POU protein *Brn-2* (69,70) genes lead to failure of development of those magnocellular neurons. In addition, the mice die shortly after birth and also lack parvocellular neurons of the PVN encoding thyrotropin-releasing hormone, corticotropin-releasing hormone and somatostatin. Absence of Sim1 also leads to lack of Brn-2 suggesting that Sim1 functions upstream of Brn-2 (68).

To our knowledge, no report of a knockout of the mouse VP gene has been published. However, there is an extensive literature about the Brattleboro rat that contains a single nucleotide mutation in the second exon of the VP gene leading to central diabetes insipidus in homozygous mutants. A recent review of how the CNS is affected in the Brattleboro rat is available (71). Knockout mice have been produced that lack OT (72,73). The mice are fertile and able to deliver their litters, but the pups die because they are not able to obtain milk from the dams. Milk is present in the mammary glands and can be released by injecting them with OT, at which point the pups can successfully suckle. Administration of 22.4 U of OT intraperitoneally, which does not change noticeably the mother's nursing behavior, rescues the pups who then have productive suckling within 30min to 1hr. The failure of the newborn pups to obtain milk is not just due to ineffective suckling, as six day old WT (wild type) pups placed with the HO (homozygous) dams (that are within 24hrs of parturition) are also unable to obtain milk, despite presumably more forceful suckling. In addition, whole-mount examination of the mammary glands from HO dams shows milk accumulation prior to OT administration. Further studies revealed that alveolar density and mammary epithelial-cell differentiation at parturition are similar in WT and OT-deficient dams. However, within 12 hr after parturition, about 2% of the alveolar cells in the WT dams incorporate DNA and proliferate, but virtually no proliferation is detected in the HO dams. Continuous suckling of pups leads to the expansion of the lobulo-alveolar units in WT but not HO dams. Despite suckling and the presence of systemic lactogenic hormones, mammary tissue in the HO dams partially involutes. These results demonstrate that post-partum alveolar proliferation requires not only systemic lactogenic hormones, such as prolactin, but also the presence of OT in conjunction with continued milk removal (74,75).

The distributions of OT and VP receptors in the brains of various species have been correlated with various behaviors (76-79). Interestingly, the absence of OT has no affect on the distribution of OT receptors in those mice (72). Furthermore, a number of behaviors, including sexual and maternal, are minimally, if at all, affected (72,79). In fact, a centrally administered OT receptor antagonist does not affect maternal behavior indicating that VP is not compensating for OT (80). In our strain (mixed 129SV-C57BL/6J), HO mice exhibit reduced aggression as compared to their WT or HE (heterozygous) mice littermates, particularly in agonistic bouts within a neutral arena. WT and HE mice do not differ in aggressive behavior. Although the frequency of aggressive encounters between the WT and HO mice are similar, the OT-/- mice spend significantly less of the test time in aggressive encounters than the WT mice, and the average duration of each aggressive encounter is reduced.

In other words, the HO mice attack as frequently as the WT mice, but the aggressive encounters end quickly. HO mice do not display reduced defensive aggression. A battery of sensorimotor skills failed to reveal any detectable sensorimotor deficits. Young et al. (80) found in their strain reduced olfactory investigation and increased aggression. The different findings on aggression likely reflect strain differences. They also found that the KO pups made decreased ultrasonic calls upon separation from the mothers, perhaps due to deficient social attachments (80). These authors cautioned that significant species and, even, strain differences exist with respect to behaviors making interpretations of mouse behaviors in OT KOs problematic.

The OT-deficient mice were also crossed with 2 transgenic lines, demonstrating that an experimentally created null background may be used to examine the effectiveness of particular transgenic constructs related to the knockout gene. The first line, V7A, was similar to the rat transgene with which we first obtained successful cell-specific and physiologic expression of oxytocin (46), except that the OT and VP transcriptional units were in the normal 3'-to-3' orientation. These double homozygous lines had successful lactation restored and reproduced successfully (81). The other line, V20, used the rat transgene with replacement of the sequence coding for amino acids 41-107 of the rat neurophysin with the human counterpart (7 differences from mouse). This enabled us to determine if this highly conserved neurophysin sequence between OT and VP within the same species is important for proper protein expression and processing. OT knock-out animals, into whose genome this transgene was crossed, are rescued (e.g., they lactate successfully) indicating that the conservation of both nucleotide and amino acid sequences in neurophysin is not necessary for proper expression (56).

In an interesting investigation on the interaction between oxytocin and prostaglandins in regulating mouse parturition, Gross and coworkers (82) crossed a cyclooxygenase-1 (COX-1) deficient mouse with an OT deficient line. The cross corrects the delayed parturition found in the COX-1 knockout mice. The COX-1 deficient mice show impaired luteolysis (and, consequently, elevated serum progesterone levels) and delayed induction of uterine OT receptors. The absence of the luteotrophic OT in the double knockouts allows luteolysis to occur and therefore promotes parturition.

The Comparative Genomics Approach
In a bold experiment, Venkatesh et al. (83) made transgenic rats containing a 40 kbp cosmid sequence derived from the isotocin (IT)-vasotocin (VT) gene locus in the pufferfish, Fugu. Remarkably, they found cell-specific expression of the fish IT gene only in the magnocellular OT neurons in the rat HNS, and that this transgene could be physiologically regulated by dehydration in a

manner similar to the endogenous gene. This experiment showed that there was sufficient conservation of the cell-specific expression and physiological regulatory mechanisms despite an evolutionary distance of 400 million years between these species. They further suggest that the pufferfish genome, because of it's high information density [390 Mbp genome with very little "junk" DNA, and small (about 80bp) introns (84)], when used in this rodent transgenic assay, might provide a unique opportunity to efficiently dissect the key *cis* elements in the IT-VT locus that are involved in the cell-specific expression and physiological regulation (83,85).

Gene Transfer Using Viral Vectors

An interesting alternative to transgenesis for the transfer of foreign DNA into differentiated neurons *in vivo* is the use of viral vectors. Several types of modified virus vectors have been used as efficient vehicles to transfer genes into differentiated neurons in the central nervous system *in vitro* and *in vivo* (86-88). These include herpes-related viruses (89,90), attenuated adenoviruses (91,92), adeno-associated viruses (93,94), and lentiviruses (95). To date, only adenovirus has been used in the HNS *in vivo* (91,92,96,97) and the adeno-associated virus has been used to transfect oxytocin cells *in vitro* (98). A comprehensive discussion of this approach to gene transfer in the CNS can be found in Chapter 10 in this volume.

FUTURE DIRECTIONS

There are a number of issues in the study of VP and OT gene expression that could benefit from the use of transgenic animals. For example, little has been done so far to validate *in vivo* the *in vitro* work on specific elements within the VP or OT promoters that regulate expression. Targeting of specific sequences that express dominant negatives or other proteins to affect magnocellular physiology are just becoming possible, at least in the OT neurons. In addition to targeting proteins to directly affect neuronal functions, it should be possible to express cre (cyclization recombination) site-specific DNA recombinase gene of bacteriophage P1 in magnocellular neurons. It is likely that a large number of lines will be generated in which genes are flanked by loxP sites ("floxed") so that cre recombinase can be used to excise the intervening sequence, thus inducing a knockout (99). By crossing with mice expressing the cre recombinase specifically in magnocellular cells with the floxed lines, one should be able to eliminate expression specifically within those cells.

The transgenic expression and elimination of the various receptors is also in its infancy. Recent studies that have examined the influence of receptor distribution on behavior have demonstrated increased affiliative behavior in transgenic mice expressing the prairie vole VP V1a receptor (100,101). These mice have increased V1a receptor expression within areas that also have high levels in the prairie vole (cingulate cortex and laterodorsal and ventroposterior thalamus). These exciting data are consistent with affiliative behavior being defined, at least in part, by the distribution of the V1a receptor. Finally, simply targeting GFP to specific neuronal phenotypes can provide a major tool for physiological studies. For example, Suter and co-workers (102) produced transgenic mice in which the gonadotropin-releasing hormone (GnRH) promoter targeted GFP to GnRH neurons of the hypothalamus. Therefore, they were able to identify individual GnRH neurons in slices by their fluorescence *in vitro*, thereby enabling them to perform efficient patch-clamp recordings and subsequent morphological analysis of this neuronal subtype. The OT-EGFP mouse line represents a similar model for physiological studies of this HNS neuronal subtype.

REFERENCES

1. Herrup K. Transgenic and ES cell chimeric mice as tools for the study of the nervous system. Discuss Neurosci 1995;10:41-64.
2. Kapur RP, Hoyle GW, Mercer EH, Brinster RL, Palmiter RD. Some neuronal cell populations express human dopamine beta-hydroxylase-lacZ transgenes transiently during embryonic development. Neuron 1991; 7:717-727.
3. Fromont-Racine M, Bucchini D, Madsen O, Desbois P, Linde S, Nielsen JH, Saulnier C, Ripoche MA, Jami J, Pictet R. Effect of 5'-flanking sequence deletions on expression of the human insulin gene in transgenic mice. Mol Endocrinol 1990;4:669-677.
4. Smeyne RJ, Schilling K, Robertson L, Luk D, Oberdick J, Curran T, Morgan J.I. Fos-lacZ transgenic mice: mapping sites of gene induction in the central nervous system. Neuron 1992;8:13-23.
5. Moitra J, Mason MM, Olive M, Krylov D, Gavrilova O, Marcus-Samuels B, Feigenbaum L, Lee E, Aoyama T, Eckhaus, M, Reitman M, Vinson C. Life without white fat: a transgenic mouse. Gen Dev 1998;12:3168-3181.
6. Windle JJ, Weiner RI, Mellon PL. Cell lines of the pituitary gonadotrope lineage derived by targeted oncogenesis in transgenic mice. Mol Endocrinol 1990;4:597-603.
7. Radovick S, Wray S, Lee E, Nicols DK, Nakayama Y, Weintraub BD, Westphal H, Cutler GBJ, Wondisford FE. Migratory arrest of gonadotropin-releasing hormone neurons in transgenic mice. Proc Natl Acad Sci USA 1991;88:3402-3406.
8. Mason AJ, Pitts SL, Nikolics K, Szonyi E, Wilcox JN, Seeburg PH, Stewart TA. The hypogonadal mouse: reproductive functions restored by gene therapy. Science 1986;234:1372-1378.
9. Waller SJ, Murphy D. Expression of rat vasopressin genes in rats. In: Neurohypophysis: Recent Progress of Vasopressin and Oxytocin Research. (Eds. T. Saito, K. Kurokawa, & S. Yoshida). Amsterdam : Elsevier Press, 1995.
10. Waller S, Fairhall KM, Xu J, Robinson ICAF, Murphy D. Neurohypophyseal and fluid homeostasis in transgenic rats expressing a tagged rat vasopressin prepropeptide in hypothalamic neurons. Endocrinology 1996;137:5068-5077.
11. Armstrong WE. Hypothalamic supraoptic and paraventricular nuclei. In: The Rat Nervous System, 2nd Ed, pp377-390. New York: Academic Press, 1995
12. Armstrong WE. Morphological and electrophysiological classification of hypothalamic supraoptic nuclei. Prog Neurobiol 1995;47:291-339.

13. Hatton GI. Emerging concepts of structure-function dynamics in adult brain: the hypothalamo-neurohypophysial system. Prog Neurobiol 1990;34:437-504.

14. Hatton GI. Function-related plasticity in hypothalamus. Annu Rev Neurosci 1997;20:375-397.

15. Gainer H, Wray S. Cellular and molecular biology of oxytocin and vasopressin. In: The physiology of reproduction pp1099-1129. (Eds E Knobil & JD Neill). New York: Raven Press, 1994.

16. Morris JF, Nordmann JJ, Dyball REJ. Structure-function correlation in mammalian neurosecretion. Int Rev Exp Pathol 1978;18:1-95.

17. Swanson LW, Sawchenko PE. Hypothalamic integration of the paraventricular and supraoptic nuclei. Annu Rev Neurosci 1983;6:269-324.

18. Silverman A, Zimmerman EA. Magnocellular neurosecretory system. Annu Rev Neurosci 1983;6:357-380.

19. Meister B. Gene expression and chemical diversity in hypothalamic neurosecretory neurons. Mol Neurobiol 1993;7:87-110.

20. Hatton GI. Oxytocin and vasopressin neurones: vive la difference! J Physiol. 1997;500:284.

21. Mohr E, Bahnsen U, Kiessling C, Richter D. Expression of the vasopressin and oxytocin genes in rats occurs in mutually exclusive sets of hypothalamic neurons. FEBS Lett 1988;242:144-148.

22. Stern JE, Armstrong WE. Sustained outward rectification of oxytocinergic neurons in the rat supraoptic nucleus: ionic dependence and pharmacology. J Physiol 1997;500:497-508.

23. Bondy CA, Whitnall MH, Brady LS, Gainer H. Coexisting peptides in hypothalamic neuroendocrine systems: some functional implications. Cell. Molec Neurobiol 1989; 9:427-446.

24. Kiyama H, Emson PD. Evidence for the coexpression of oxytocin and vasopressin messenger ribonucleic acids in magnocellular neurosecretory cells: simultaneous demonstration of two neurophysin messenger ribonucleic acids by hybridization histochemistry. J Neuroendocrinol 1990;2:257-259.

25. Mezey E, Kiss J. Coexpression of vasopressin and oxytocin in hypothalamic supraoptic neurons of lactating rats. Endocrinology 1991;129:1814-1820.

26. Glasgow E, Kusano K, Chin H, Mezey E, Young WS, III, Gainer H. Single cell RT-PCR analysis of rat supraoptic magnocellular neurons: Neuropeptide phenotypes and high voltage-gated calcium channel subtypes. Endocrinology 1999;140:5391-5401.

27. Gainer H. Cell-specific gene expression in magnocellular oxytocin and vasopressin neurons. Adv Exp Med Biol 1998;449:15-27

28. Xi D, Kusano K, Gainer H. Quantitative analysis of oxytocin and vasopressin mRNAs in single magnocellular neurons isolated from supraoptic nucleus of rat hypothalamus. Endocrinology 1999;140: 4677-4682.

29. Young WS III. Expression of the oxytocin and vasopressin genes. J Neuroendocrinol 1992;4:529-540.

30. Murphy D, Carter DA. Transgenic approaches to modifying cell and tissue function. Curr Opin Cell Biol 1992;4 274-279.

31. Murphy D, Carter DA. Transgenesis techniques: Principles and protocols. In: Methods in Molecular Biology, Volume 18. New Jersey: Humana Press, 1993.

32. Burbach JPH, van Shaick HSA, deBree FM, Lopes da Silva S, Adan RAH. Functional domains in the oxytocin gene for regulation of expression and biosynthesis of gene products. In: Oxytocin (Eds: Ivell R, Russell J), pp. 9-21. New York: Plenum Press, 1995.

33. Waller SJ, Ratty, A, Burbach JPH, Murphy D. Transgenic and transcriptional studies on neurosecretory cell gene expression. Cell Mol Neurobiol 1998;18:149-172.

34. Mohr, E., Schmitz, E., Richter D. A single rat genomic DNA fragment encodes both the oxytocin and vasopressin genes separated by 11 kilobases and oriented in opposite

transcriptional directions. Biochimie 1988;70:649-654.

35. Hara Y, Battey J, Gainer H. Structure of mouse vasopressin and oxytocin genes. Mol. Brain Res. 1990;8:319-324.

36. Ratty A, Jeong S-W, Nagle JW, Chin H, Gainer H, Murphy D, Venkatesh B. A systematic survey of the intergenic region between the murine oxytocin and vasopressin encoding genes. Gene 1996;174:71-78.

37. Schmitz E, Mohr E, Richter D. Rat vasopressin and oxytocin genes are linked by a long interspersed repeated DNA element (LINE): sequence and transcriptional analysis of LINE. DNA Cell Biol 1991;10:81-91.

38. Murphy D, Bishop A, Rindi G, Murphy MN, Stamp G, Hanson J, Polak J, Hogan BLM Mice transgenic for a vasopressin-SV40 hybrid oncogene develop tumors of the endocrine pancreas and the anterior pituitary: a possible model for human multiple endocrine neoplasia. Am J Pathol 1987;129:552-566.

39. Stefaneanu L, Rindi G, Horvath E, Murphy D, Polak JM, Kovacs K. Morphology of adenohypophyseal tumors in mice transgenic for vasopressin-SV40 hybrid oncogene. Endocrinology 1992;130:1780-1788.

40. Ang H-L, Ungerfroren H, de Bree F, Foo N-C, Carter DA, Burbach JP, Ivell R, Murphy D. Testicular oxytocin gene expression in seminiferous tubules of cattle and transgenic mice. Endocrinology 1991;128:2110-2117.

41. Russo AF, Crenshaw III EB, Lira SA, Simmons DM, Swanson LW, Rosenfeld MG. Neuronal expression of chimeric genes in transgenic mice. Neuron 1988;1:311-320.

42. Wilson C, Bellen HJ, Gehring WJ. Position effects on eukaryotic gene expression. Ann Rev Cell Biol 1990;6:679-714.

43. Palmiter, RD, Sandgren EP, Avrbock MR, Allen DD, Brinster RL. Heterologous introns can enhance expression of transgenes in mice. Proc Natl Acad Sci USA 1991;88:478-482.

44. Schoenherr CJ, Anderson DJ. Silencing is golden: negative regulation in the control of neuronal gene transcription. Curr Opin Neurobiol 1995;5:566-571.

45. Habener JF, Cwikel BJ, Hermann H, Hammer RE, Palmiter RD, Brinster RL. Metallothionin-vasopressin fusion gene expression in transgenic mice. J Biol Chem 1989;264:18844-18852.

46. Young WS III, Reynolds K, Shepard EA, Gainer H, Castel M. Cell specific expression of the rat oxytocin gene in transgenic mice. J Neuroendocrinol 1990;2:917-925.

47. Young WS III, Reynolds K, Shepard EA. Tissue- and stimulus-specific expression of the rat oxytocin gene in transgenic mice. Society for Neuroscience Abstracts 1990;16:157.

48. Belenky M, Castel M, Young WS III, Gainer, Cohen S. Ultrastructural immunolocalization of rat oxytocin-neurophysin in transgenic mice expressing the rat oxytocin gene. Brain Res 1992;583:279-286.

49. Grant FD, Reventos J, Gordon JW, Kawabata S, Miller M, Majzoub JA. Expression of rat arginine vasopressin gene in transgenic mice. Mol Endocrinol 1993;7:659-667.

50. Zeng Q, Carter DA, Murphy D. Cell specific expression of a vasopressin transgene in rats. J Neuroendocrinol 1994;6:469-477.

51. Gainer H, Jeong SW, Witt DM, Chin H. Strategies for cell biological studies in oxytocinergic neurons. Adv Exp Med Biol 1995;395:1-8.

52. Adan RA, Cox JJ, Beischlag TV, Burbach JP. A composite hormone response element mediates the transactivation receptors. Mol Endocrinol 1993;7:47-57.

53. Ho M-Y, Carter DA, Ang H-L, Murphy D. Bovine oxytocin transgenes in mice: Hypothalamic expresion, physiological regulation and interactions with the vasopressin gene. J Biol Chem 1995;270:27199-27205.

54. Ang H-L, Carter DA, Murphy D. Neuron-specific and physiological regulation of bovine vasopressin transgenes in mice. EMBO J 1993;12:2397-2409.

55. Young WS III, Iacangelo A, Luo X-ZJ, King C, Duncan K., Ginns EI. Transgenic expression of green fluorescent protein in mouse oxytocin neurons. J

Neuroendocrinology 1999;11:935-939.

56. Young WS III, Iacangelo I, Luo, X-ZJ, King C, Duncan K, Ginns EI. Transgenic expression of green fluorescent protein and human oxytocin neurophysin in mouse oxytocin neurons. Society for Neuroscience Abstracts 1999;795.7.

57. Chalfie M. Green fluorescent protein. Photochem Photobiol 1995; 62:651-656.

58. Tsien RY. The green fluorescent protein. Ann Rev of Biochem 1998;67:509-544.

59. Gainer H, Chin H. Molecular diversity in neurosecretion: reflections on the hypothalamo-neurohypophysial system. Cell Mol Neurobiol 1998;18:211-230.

60. Sabatier N, Richard P, Dayanithi G. L-, N- and T- but neither P- nor Q-type Ca2+ channels control vasopressin-induced Ca2+ influx in magnocellular vasopressin neurones isolated from rat supraoptic nucleus. J Physiol 1997;503:253-268.

61. Wang G, Dayanithi G, Kim S, Horn D, Nadaschi L, Kristipati R, Ramachandran J, Stuenkel EL, Nordmann JJ, Newcomb R, Lemos JR. Role of Q-type Ca2+ channels in vasopressin secretion from neurohypophysial terminals of the rat. J Physiol 1997;502:351-363.

62. Neumann I, Russell JA, Landgraf R. Oxytocin and vasopressin release within the supraoptic and paraventricular nuclei of pregnant, parturient, and lactating rats: a microdialysis study. Neurosci 1993;53:65-75.

63. Burke NV, Han W, Danqing L, Takimoto K, Watkins SC, Levitan ES. Neuronal peptide release is limited by secretory granule mobility. Neuron 1997;19:1095-1102.

64. Kaether C, Gerdes H-H. Visualization of protein transport along the secretory pathway using green fluorescent protein. FEBS Lett 1995;369:267-271.

65. Lang T, Wacker I, Steyer J, Kaether C, Wunderlich I, Soldati T, Geddes H-H, Almers W. Ca2+-triggered peptide secretion in single cells imaged with green fluorescent protein and evanescent-wave microscopy. Neuron 1997;18:857-863.

66. House SB, Thomas A, Kusano K, Gainer H. Stationary organotypic cultures of oxytocin and vasopressin magnocellular neurons from rat hypothalamus. J Neuroendocrinol 1998;10:849-861.

67. Kusano K, House S.B, Gainer H. Effects of osmotic pressure and brain-derived neurotrophic factor on the survival of postnatal hypothalamic oxytocinergic and vasopressinergic neurons in dissociated culture. J Neuroendocrinol 1999;11:145-152.

68. Michaud JL, Rosenquest T, May NR, Fan C-M. Development of neuroendocrine lineages requires the bHLH-PAS transcription factor SIM1. Gen Devel 1998;12:3264-3275.

69. Nakai S, Kawano H, Yudate T, Nishi M, Kuno J, Nagata A, Jishage K, Hamada H, Fujii H, Kawamura K, Shiba K, Noda T. The POU domain transcription factor Brn-2 is required for the determination of specific neuronal lineages in the hypothalamus of the mouse. Gen Devel 1995;9:3109-3121.

70. Schonemann MD, Ryan AK, McEvilly RJ, O'Connell SM, Arias CA, Kalla KA, Li P, Sawchenko PE, Rosenfeld MG. Development and survival of the endocrine hypothalamus and posterior pituitary gland requires the neuronal POU domain factor Brn-2. Gen Devel 1995;9:3122-3135.

71. Bohus B, de Wied D. The vasopressin deficient Brattleboro rats: a natural knockout model used in the search for CNS effects of vasopressin. Prog Brain Res 1998;119:555-573.

72. Nishimori K, Young LJ, Guo Q, Wang Z, Insel TR, Matzuk M.M. Oxytocin is required for nursing but is not essential for parturition or reproductive behavior. Proc Natl Acad Sci USA 1996;93:11699-11704.

73. Young WS III, Shepard E, Amico J, Hennighausen L, Wagner K-U, LaMarca ME, McKinney C, Ginns EI. Deficiency in mouse oxytocin prevents milk ejection, but not fertility or parturition. Neuroendocrinol 1996;8:847-853.

74. Li M, Liu X, Robinson G, Bar-Peled U, Wagner K-U, Young WS, Hennighausen L, Furth PA. Mammary-derived signals activate programmed cell death during the first stage of mammary gland involution. Proc Natl Acad Sci USA 1997;94:3425-3430.

75. Wagner K-U, Young WS III, Liu X, Ginns EI, Li M, Furth PA, Hennighausen L. Oxytocin and milk removal are required for post-partum mammary gland development. Gen Func 1997;1:233-244.
76. Insel TR, Winslow JT, Williams JR, Hastings N, Shapiro LE, Carter CS. The role of neurohypophyseal peptides in the central mediation of complex social processes-evidence from comparative studies. Reg Pept 1993;45:127-131.
77. Insel TR, Young L, Wang Z. Central oxytocin and reproductive behaviors. Rev Reprod 1997;2:28-37.
78. Young LJ. Oxytocin and vasopressin receptors and species -typical social behaviors. Horm Behav 1999;36:212-221.
79. DeVries AC, Young WS III, Nelson RJ. Reduced aggressive behaviour in mice with targeted disruption of the oxytocin gene. Journal of Neuroendocrinol 1997;9:363-368.
80. Young LJ, Winslow JT, Wang Z, Gingrich B, Guo Q, Matzuk M.M, Insel TR. Gene targeting approaches to neuroendocrinology: oxytocin, maternal behavior, and affiliation. Horm Behav 1997;31:221-231.
81. Young WS III, Shepard E, Amico J, DeVries AC, Nelson RJ, Hennighausen L, Wagner K-U, Zimmer A, LaMarca ME, Ginns EI. Targeted reduction of oxytocin expression provides insights into its physiological roles. Adv Exp Mede Biol 1998;449:231-240.
82. Gross GA, Imamura T, Luedke C, Vogt SK, Olson LM, Nelson DM, Sadovsky Y, Muglia LJ. Opposing actions of prostaglandins and oxytocin determine the onset of murine labor. Proc Natl Acad Sci USA 1998;95:11875-11879.
83. Venkatesh B, Si-Ho S-L, Murphy D, Brenner S. Transgenic rats reveal remarkable functional conservation of regulatory controls between the fish isotocin and rat oxytocin genes. Proc Natl Acad Sci USA 1997;94:12462-12466.
84. Brenner S, Elgar G, Sandford R, Macrae A, Venkatesh B, Aparicio S. Characterisation of the pufferfish (Fugu) genome as a compact model vertebrate genome. Nature 1993;366:265-268.
85. Murphy D, Si-Hoe S-L, Brenner S, Venkatesh B. Something fishy in the rat brain: molecular genetics of the hypothalamo-neurohypophyseal system. BioEssays 1998;20:741-749.
86. Karpati G, Lochmuller H, Nalbantoglu J, Durham H. The principles of gene therapy for the nervous system. Trends Neurosci 1996;19:49-54.
87. Slack RS, Miller FD. Viral vectors for modulating gene expression in neurons. Curr Opin Neurobiol 1996;6:576-583.
88. Lowenstein PR, Enquist LW (eds). Protocols for Gene transfer in Neuroscience. Chichester: John Wiley & Sons, 1996.
89. Fink DJ, DeLuca NA, Goins WF, Glorioso JC. Gene transfer to neurons using herpes simplex virus-based vectors. Ann Rev Neuroscience. 1996;19:265-287.
90. Federoff HJ. Novel Targets for Gene Therapy. Gene Ther 1999;6:1907-1908.
91. Geddes B J, Harding TC, Hughes DS, Byrnes AP, Lightman SL, Conde G, Uney JB. Persistent transgene expression in the hypothalamus following stereotaxic delivery of a recombinant adenovirus: suppression of the immune response with cyclosporin. Endocrinology 1996;137:5166-5169.
92. Geddes BJ, Harding TC, Lightman S, Uney JB. Assessing viral gene therapy in neuroendocrine models. Front Neuroendocrinol 1999;20:296-316.
93. Du BP, Wu DM, Boldt-Houle, Terwilligier EF. Efficient transduction of human neurons with an adeno-associated virus vector. Gene Ther 1996;3:254-261.
94. Kaplitt MG, Leone P, Samulski RJ, Xiao X, Pfaff DW, O'Malley KL, During MJ. Long term gene expression and phenotypic correction using adeno-associated virus vectors in the mammalian brain. Nature Gen 1994;8:148-154.
95. Federico M. Lentiviruses as gene delivery vectors. Curr Opin in Biotechnol. 1999;10:448-453.
96. Geddes BJ, Harding TC, Lightman SL, Uney JB. Long-term gene therapy in the CNS:

reversal of hypothalamic diabetes insipidus in the Brattleboro rat by using an adenovirus expressing arginine vasopressin. Nat Med 1997;3:1402-1404.

97. Vascquez EC, Johnson RF, Beltz TG, Haskell RE, Davidson BL, Johnson AK. Replication-deficient adenovirus vector transfer of gfp reporter gene into supraoptic nucleus and subfornical organ neurons. Exp Neurol 1998;154:353-365.

98. Keir SD, House SB, Li J, Xiao X, Gainer H. Gene transfer into hypothalamic organotypic cultures using an adeno-associated virus vector. Exp Neurol 1999;160:313-316.

99. Sauer B. Inducible gene targeting in mice using the Cre/lox system. Methods: A Companion to Methods in Enzymology 1998;14:381-392.

100. Young LJ, Waymire KG, Nilsen R, Macgregor GR, Wang Z, Insel TR. The 5' flanking region of the monogamous prairie vole oxytocin receptor gene directs tissue-specific expression in transgenic mice. Ann New York Acad Sci 1997;807:514-517.

101. Young LJ, Nilsen R, Waymire KG, MacGregor GR, Insel TR. Increased affiliative response to vasopressin in mice expressing the V1a receptor from a monogamous vole. Nature 1999;400:766-768.

102. Suter KJ, Song WJ, Sampson TL, Wuarin JP, Saunders JT, Dudek FE, Moenter SM. Genetic targeting of green fluorescent protein to gonadotropin-releasing hormone neurons: characterization of whole-cell electrophysiological properties and morphology. Endocrinology 2000;141:412-419.

3

GnRH TRANSGENIC MODELS

Allan E. Herbison

Laboratory of Neuroendocrinology, The Babraham Institute, Cambridge CB2 4AT, UK

INTRODUCTION

The gonadotropin-releasing hormone (GnRH) neurons have represented something of an enigma ever since the isolation of the decapeptide in 1971 (1,2) and the first immunocytochemical mappings of their location within the mammalian forebrain shortly thereafter (3,4). Unusually, they were not found to exist within any defined nucleus or subregion of the brain but, instead, were identified as a scattered continuum of neurons within the basal forebrain. This was found to occur in all mammalian brains, although the relative distribution of GnRH neurons along the "pathway" from the olfactory lobes to the basal hypothalamus differs (4). It now seems most likely that this peculiar distribution arises from the extraordinary migration of the GnRH neurons from the olfactory placode into the brain during embryogenesis (4,5,6). However, it remains that the scattered nature of GnRH phenotype has made the detailed molecular and cellular investigation of these neurons extremely difficult. Given the critical role of these cells in the neural regulation of fertility, this has been unfortunate. While many strategies have been attempted to facilitate the characterisation of the GnRH phenotype, the most successful to date has been through the use of transgenics. I intend here to review these various transgenic strategies and suggest what we might expect from them in the near future.

IN THE BEGINNING...

The first report using GnRH transgenics was over a decade ago in 1986 and involved the recovery of fertility to the hypogonadal (*hpg*) mouse through one of the first examples of "gene therapy" (7). The infertile *hpg* mouse has a spontaneous 33kb deletion involving the distal half of the GnRH gene and results in an absence of GnRH peptide. Mason and colleagues (7) replaced the defective gene by injecting a 13.5kb murine GnRH gene fragment (Figure 1A) into the pronucleus of fertilized mouse oocytes. From this, they derived two founder lines in which the GnRH transgene was expressed, and bred both lines to homozygosity. This strategy resulted in the expression of GnRH peptide within GnRH neurons of the hypothalamus and restored normal fertility to both males and females. While all reproductive parameters were returned to near normal in these mice, aberrant GnRH expression was noted in the liver as well as within the paraventricular nucleus and bed nucleus of the stria terminalis (BNST). It was suggested at the time that this may have resulted from the absence of negative regulatory elements on the 13.5kb transgene. Overall, this striking study demonstrated that this 13.5kb of the GnRH gene was sufficient to direct expression to the GnRH phenotype in the mouse and provided an elegant start to GnRH transgenics.

GnRH IMMORTALISATION THROUGH TRANSGENICS

Approximately five years after the seminal paper by Mason et al. (7), two groups reported the use of GnRH transgenics to create immortalized GnRH-expressing cell lines (8,9). These cell lines have probably had the single biggest impact upon GnRH neurobiology in the last decade (10,11). Both groups attempted targeted oncogenesis of the GnRH phenotype using constructs consisting of 1-2kb of rat and human GnRH promoter linked to the SV40 T antigen oncogene (Figure 1B). Oocyte pronucleus injection of these constructs resulted in founder mice exhibiting different degrees of fertility. In the case of the rat 2kb GnRH promoter transgene, Mellon and colleagues (10) derived 9 infertile founders which exhibited tumors in a variety of brain regions. Two of these founders exhibited anterior hypothalamic tumors and one expressed high levels of GnRH from which the immortalised GT1 cell lines were derived (10). In the case of the 1kb human GnRH promoter transgene, Radovick and colleagues derived 3 founders, two of which were infertile with no, or few, GnRH neurons while the other gave offspring with varying degrees of fertility and tumors. One of these mice exhibited a GnRH-expressing tumor in the nasal cavity from which two different cell lines (NLT and Gn11) were derived (11,12).

Together, these various GnRH-expressing cell lines provided evidence that even relatively small regions of heterologous GnRH proximal promoter could direct expression, albeit imperfectly, to the GnRH phenotype. They have also provided interesting models in which the differences between nasal (NLT/Gn11) and hypothalamic (GT1) GnRH neurons may be explored. Most importantly, however, these GnRH cell lines provided the first simple experimental model for the investigation of GnRH-expressing neurons and they have now been used extensively in the analysis of GnRH gene expression, peptide processing, secretion and GnRH neuron electrical activity (10-15). Although these immortalized cell lines have undoubtedly represented a wonderful tool for those interested in GnRH neurons, their relationship to real GnRH neurons *in vivo* is controversial (16) and difficult to assess. Leaving aside concerns over the effects of tumorgenesis on these cells, it is not clear what developmental stage they might each represent and how their isolation from glial and other neuronal phenotypes may affect their properties. While it seems reasonable to think that these cells lines are likely to be useful models for the investigation of GnRH gene structure and regulation (14,15), their relevance to GnRH neuron physiology *in vivo* awaits the accumulation of data from real GnRH neurons in their native environment. Led by the current wave of GnRH transgenic methodology, this data is now starting to appear.

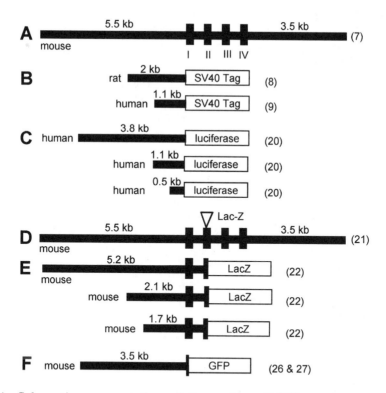

Fig 1. Schematic representation of the various GnRH constructs used in published GnRH transgenics. Numbers in parenthesis give reference.

TARGETING GnRH NEURONS *IN VIVO*

While the GnRH gene therapy and immortalisation approaches were important in their own right, neither addressed our ability to target GnRH neurons *in vivo* for physiological studies. Following on from the success of these studies in targeting the GnRH phenotype, however, it seemed likely that this approach could be used to drive the expression of various reporters in GnRH neurons. This has now been achieved and several laboratories are using this strategy to examine GnRH gene expression *in vivo* and facilitate the identification of the GnRH phenotype in the acute brain slice procedure.

Spatiotemporal Targeting of the GnRH Phenotype with Promoter-driven Transgenics

The targeting of neuropeptide genes with promoter-driven transgenics has proven to be extremely difficult in the past (17). Whereas, some endocrine-related genes such as growth hormone have been successfully targeted to pituitary cells with as little as 200bp of promoter (18), the appropriate expression of transgenes in specific neuropeptide-expressing phenotypes has often required much larger constructs. For example, 15kb of 5' and 3' somatostatin gene sequence results in expression of a LacZ reporter within the mouse brain but fails to target somatostatinergic neurons (19). Thus, the first major obstacle to overcome in using transgenics to evaluate the GnRH phenotype was to establish how much GnRH-GAP sequence was sufficient to drive transgene expression in a selective manner within the GnRH neuronal population.

The first attempt at defining this issue was undertaken by Wolfe and colleagues (20) using a construct comprised of 3.8 to 0.5kb of human GnRH 5' flanking sequence fused to the luciferase reporter gene (Figure 1C). With this approach, luciferase was detected within the hypothalamus of both the 3.8 and 1.1kb GnRH-luciferase mice but was absent in the 0.5kb mice. The temporal profile of luciferase expression was also determined postnatally with a robust increase observed prior to puberty (20). Although it is difficult to accurately assess the luciferase expression at a cellular level, dual *in situ* hybridisation experiments revealed that GnRH mRNA-expressing neurons in the rostral preoptic area, in particular, expressed luciferase mRNA. Only a small amount of luciferase expression was detected elsewhere in the brain. Thus, these studies clearly indicated that more than 0.5kb of human GnRH 5' flanking sequence is required to target GnRH neurons in the mouse and, ignoring issues of luciferase sensitivity at a cellular level, that this was likely to target at least some of the GnRH neurons in a spatiotemporal-specific manner.

In order to assess the elements of the murine GnRH gene required to direct expression of reporters to the GnRH population in the mouse, we examined the expression patterns of a variety of GnRH gene fragments coupled to the LacZ reporter (21,22). Following on from the work of Mason and colleagues (7), we began our studies by using the same murine 13.5kb GnRH fragment in which we positioned the LacZ cassette into the coding sequence for GnRH (to prevent over-expression of GnRH in transgenic mice, Figure 1D). Due to X-gal histochemistry and the availability of excellent antibodies directed against β-galactosidase (the product of the lacZ gene) we were able to determine the precise cellular expression patterns of the transgene in the 3 separate lines of so-called GNZ mice bearing this construct (21). As expected (7), we found that approximately 90% of the GnRH population was targeted by this construct and, additionally, that transgene expression followed the correct developmental pattern in migrating GnRH neurons during embryogenesis. It is not entirely clear why 10% of the GnRH phenotype do not appear to express the reporter but we presently favour the hypothesis that they do in fact express the transgene, but do so at a level below our threshold for detection.

In a further series of experiments (22) we examined the effects of both 3' and 5' deletions of the 13.5kb GnRH-LacZ construct upon reporter expression in the GnRH neurons (Figure 1E). These studies showed that deletion of sequence 3' to exon 2 had no effect upon transgene expression in GnRH neurons but did result in the presence of substantial "ectopic" reporter expression throughout the brain (22). Subsequent deletion of 5' GnRH promoter sequence from 5.2 to 2.1kb resulted in a small but significant drop in transgene expression within the GnRH population while mice bearing constructs with only 1.7kb of 5' sequence showed no transgene expression in the GnRH population (22). Together, these studies revealed (a) that elements between exon 2 and 3.5kb of 3' sequence are important in helping to restrict GnRH gene expression to the GnRH phenotype and (b) that an important 400bp enhancer region exists between -2.1 and -1.7kb in the GnRH promoter. Interestingly, the analogous 400bp region of the rat GnRH promoter had not been defined previously as a critical enhancer in GT1 cells *in vitro*.

New Populations of GnRH Neurons Identified Through Transgenics
The big surprise in studies with GNZ mice was that neuronal populations located in the lateral septum and BNST were also targeted by the transgene, albeit at much reduced levels (21). This phenomenon was consistent in all three lines and, as we found out later, also present in an independent murine

GnRH-LacZ line (23). Furthermore, the BNST expression was highly reminiscent of the spurious GnRH expression detected by Mason and colleagues earlier (7). At first this appeared to us to be the rather "typical" ectopic expression seen in many neuron-specific transgenic lines (17). However, the subsequent evaluation of transgene and GnRH in the developing mouse brain revealed otherwise.

Through the use of an enhanced immunocytochemical procedure we were able to show that authentic GnRH is, in fact, expressed by both lateral septal and BNST cell populations in addition to the classic GnRH neurons during embryogenesis and early postnatal life (Figure 2). Transgene expression throughout the brain was found to parallel exactly that of GnRH during development but then persisted in the postnatal lateral septum and BNST whilst GnRH immunoreactivity declined (Figure 2). A further GnRH-expressing population located in the tectum, which had been discovered previously in the mouse brain (24), was also detected to express the transgene and GnRH transiently (Figure 2). We were able to demonstrate that, unlike the classic GnRH neurons, these novel GnRH-expressing populations did not originate from the olfactory placode (21). Although the roles of these new GnRH-expressing neurons are presently unknown, their discovery through GnRH reporter transgenics provides a good illustration of the unexpected benefits of this type of strategy.

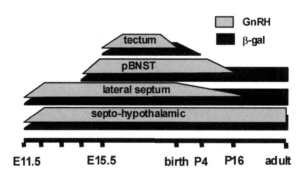

Fig 2. Schematic developmental timeline showing the expression of GnRH and β-galactosidase (β-gal) immunoreactivity in 4 different brain regions of GNZ mice. Note that GnRH and β-gal appear transiently in the tectum, while transgene expression persists in the BNST and lateral septum as GnRH expression declines after birth.

The presence of transgene expression in the lateral septum and BNST of adult mice does, however, provide a serious drawback to the use of GnRH promoter transgenics in targeting of the GnRH phenotype. At present we believe that lacZ expression in the lateral septum and BNST of adult mice results from the lack of a critical repressor element in the 13.5kb GnRH fragment. This element is probably used in late development to switch off GnRH gene expression in the lateral septum and BNST. How much more GnRH sequence might be required in a construct to enable this inactivation is unknown but could conceivably require many thousands of bases (25). Nevertheless, in terms of the visual identification of GnRH neurons in brain slices this is not too much of a problem as BNST neurons expressing the reporter will not be in the same coronal brain section and the lateral septal cells can be differentiated by their more lateral position to the midline GnRH neurons of the medial septum. Even more troublesome, however, will be the appearance of true "ectopic" reporter-expressing cells within the GnRH distribution if the GnRH transgene construct does not contain sufficient 3' GnRH sequence (22). Thus, although imperfect, the strategy of using murine GnRH gene sequence to drive the expression of transgenes in the majority of mouse GnRH neurons is feasible with due care.

Visualizing Living GnRH Neurons Using Transgenics
One of the most important goals of current GnRH transgenic experiments has been to enable living GnRH neurons to be visualized and investigated *in situ*. The first reports of success (26,27), in this regard, involved the production of transgenic mice bearing constructs comprised of approximately 3.5kb of murine 5' GnRH sequence linked to the endogenously fluorescent green fluorescent protein (GFP, Figure 1F). Spergel and colleagues (26) demonstrated that approximately 65% of the GnRH population was sufficiently fluorescent to be detected in the acute brain slice procedure, and using this fluorescent marker went on to carry out the first series of whole cell and nucleated patch-clamp studies on living GnRH neurons. A second line of GnRH-GFP mice was made by Suter and colleagues (27) and similarly used to undertake whole cell electrophysiological recordings. However, in the case of the latter mice, approximately 95% of the GnRH neurons were reported to be endogenously fluorescent and this was attributed to the use of a cassette encoding an enhanced GFP molecule (27). Thus, in both cases, GFP has proved to be a useful fluorescent marker with which to identify the scattered GnRH neurons.

While these GnRH-GFP mice likely represent as big a boost to the field of GnRH neurobiology as the production of the GnRH-expressing cell lines, important questions regarding the effects of fluorescence and GFP expression on cell physiology remain to be established. In light of the need for 3' GnRH sequence to prohibit ectopic transgene expression in the region of the GnRH neurons (22), it is also interesting that ectopic GFP expression has not been reported in either of these transgenic lines. One likely explanation is that the level of GFP expression in lateral septal and BNST, as well as true "ectopic" cells, is below the threshold level for detection with conventional fluorescent microscopes. Certainly, the level of transgene expression in lateral septal cells is known to be at least 5-fold lower than in the GnRH neurons (21).

Another approach for identifying living GnRH neurons has been to use fluorescent β-galactosidase (β-gal) substrates in conjunction with GnRH-LacZ transgenic mice. In this procedure, membrane permeant derivatives of molecules such as fluoroscein-di-β-galactopyranosides (28) enter LacZ-targeted cells whereupon they are cleaved by β-galactosidase to fluorescent moieties. This approach works equally well in detecting GnRH neurons and can be used to isolate GnRH neurons for morphological dye filling and single cell RT-PCR experiments. Furthermore, by directing the transgene to the nucleus of the cell, it is possible to place a very convenient nuclear located tag within the GnRH phenotype. Thus, dual labelling immunocytochemical and *in situ* hybridisation experiments can now be undertaken using either βgal immunocytochemistry (Figure 3) or X-gal histochemistry to reveal the GnRH neurons, while leaving the cytoplasm free for protein or mRNA detection.

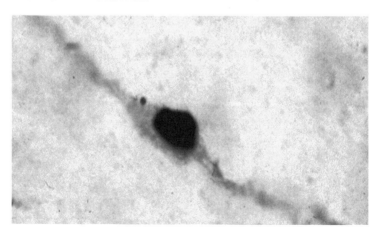

Fig 3. Nuclear localised β-gal transgene (black) in a GnRH-immunoreactive neuron

In the course of developing this methodology, we somewhat ironically realized that bipolar GnRH neurons could actually be identified in the acute brain slice preparation by their neuroanatomical position, morphology, and orientation without the need for a fluorescent marker. Using post-patching single cell RT-PCR to identify the presence of GnRH transcripts, we found that over half of such morphologically-identified cells could be shown to be genuine GnRH neurons and have now undertaken the first molecular and electrophysiological investigations of unmodified living GnRH neurons (29,30). While this approach has its own caveats, such as the investigation of exclusively bipolar GnRH neurons, it does avoid any potential confounding factors introduced by transgenes. It will be of great interest to compare the results obtained from GFP-tagged and unmodified GnRH neurons in the near future.

Using Transgenics to Modify GnRH Neuronal Physiology *in vivo*
As noted above, large GnRH promoter-driven transgenes can be used to very effectively target the GnRH phenotype. Thus, it should be possible to use this strategy to over express molecules within living GnRH neurons *in vivo*. The use of dominant-negative strategies will enable the activity of specific molecules within GnRH neurons to be reduced or ablated. While a number of laboratories are involved in such experiments, the only published datum to date is the preliminary report by Postigo and colleagues (31) where GnRH sequences were used to drive the expression of a dominant-negative fibroblast growth factor receptor in mouse GnRH neurons. This was reported to reduce the numbers of GnRH neurons present in the adult brain and, thus, suggested that the strategy is likely to work. As with the live visualization approaches, care will be necessary in determining the specificity of the targeting in terms of the numbers of GnRH neurons targeted as well as the non-GnRH neurons effected. There is, nevertheless, a very promising future for this strategy in defining physiologically relevant molecules within the GnRH phenotype.

A FUTURE FOR GnRH TRANSGENICS?

In the past, transgenic methodologies have played a very considerable role in the investigation of the GnRH phenotype and they seem likely to continue. As soon as any doubts over unwanted effects of transgene reporters have been dispelled, the GnRH-GFP and GnRH-LacZ mice should become standard *in vitro* tools in GnRH neurobiology laboratories. While we have shown that the GnRH neurons can be accessed without the need for transgenics, it is undoubtedly easier in the first instance to use fluorescence-based identification. Used with care and appreciation of the possible pitfalls, thin brain slices maintained *in vitro* will allow any neuroscientist access to living GnRH neurons in their native, albeit differentiated, environment. In addition, the use of nuclear-localised transgene markers of the GnRH phenotype will greatly facilitate conventional dual labelling procedures on these neurons. Together, this will be of enormous benefit and we should expect to see a rapid characterization of the membrane and molecular properties of these cells. The current advances being made in the production of transgenes capable of signalling intracellular events will also open up new avenues for examining such events in living GnRH neurons in real time.

With respect to "whole animal physiology" of the GnRH neuron, the use of GnRH transgenics to modulate the expression of specific molecules within the phenotype (as above) will be of great benefit. However, great care will be required in defining exactly where the various transgenes have been expressed. The use of GnRH promoter-driven transgenes already suggests that it will be difficult to achieve perfect targeting of the GnRH neuron. Furthermore, because of the random nature of transgene insertion into DNA, a minimum of two independent lines will be needed to be sure there are no integration or "position" effects on the transgene. One option, although more laborious than the production of promoter transgenics, will be to target the GnRH gene itself with homologous recombination techniques. This should provide "perfect" transgene expression in the GnRH phenotype. By employing conditional strategies, such as the Cre/loxP-recombinase system (32), this approach will be very attractive in the analysis of molecules critical for GnRH functioning in the intact brain. While it is possible to undertake multiple blood sampling procedures on mice (33), it is clearly easier in the rat. For this reason alone, we may soon have to return to the rat and the more arduous task of making GnRH transgenic rats.

REFERENCES

1. Amoss M, Burges R, Blackwell R, Vale W, Fellows R, Guillemin R. Purification, amino acid composition and N-terminus of the hypothalamic luteinizing hormone releasing hormone factor (LRF) of ovine origin. Biochem Biophys Res Comm 1971;44: 205-210.
2. Schally AV, Arimura A, Baker Y. Isolation and properties of the FSH and LH-releasing hormone. Biochem Biophys Res Comm 1971;43:393-399.
3. Barry J, DuBois MP, Poulain P. LRF producing cells of the mammalian hypothalamus. Z. Zellforsch 1973;146:351-366.
4. Silverman A, Livne I, Witkin JW. The gonadotrophin-releasing hormone (GnRH) neuronal systems: hybridization. In: The Physiology of Reproduction (Eds: Knobil E & Neill JD) pp1683-1706. New York: Raven 1994.
5. Schwanzel-Fukuda M, Pfaff DW. Origin of luteinizing hormone-releasing neurons. Nature 1989;338:161-164.

6. Wray S, Grant P, Gainer H. Evidence that cells expressing luteinizing hormone-releasing hormone mRNA in the mouse are derived from progenitor cells in the olfactory placode. Proc Natl Acad Sci USA 1989;86:8132-8136.

7. Mason AJ, Pitts SL, Nikolics K, Szonyi E, Wilcox JN, Seeburg PH, Stewart, TA. The hypogonadal mouse: Reproductive functions restored by gene therapy. Science 1986;234:1372-1378.

8. Mellon PL, Windle JJ, Goldsmith PC, Padula CA, Roberts JL,Weiner RI. Immortalization of hypothalamic GnRH neurons by genetically targeted tumorigenesis. Neuron 1990;5:1-10.

9. Radovick S, Wray S, Lee E, Nichols DK, Nakayama Y, Weintraub BD, Westphal H, Cutler GB, Wondisford FE. Migratory arrest of gonadotropin-releasing hormone neurons in transgenic mice. Proc Natl Acad Sci USA 1991;88:3402-3406.

10. Weiner RI, Wetsel W, Goldsmith P, Escalera de la Martinez G, Windle J, Padula C, Choi A, Negro-Vilar A, Mellon P. Gonadotropin-releasing hormone neuronal cell lines. Front Neuroendocrinol 1992;13:119.

11. Turgeon JL. Gonadotropin-releasing hormone neuron cell biology. TEM 1996;7:55-56.

12. Zhen S., Dunn IC., Wray S., Liu Y., Chappell PE., Levine JE., Radovick S. An alternative gonadotropin-releasing hormone (GnRH) RNA splicing product found in cultured GnRH neurons and mouse hypothalamus. J Biol Chem 1997;272:12620-12625

13. Krsmanovic LZ, Stojilkovic SS, Catt KJ. Pulsatile gonadotropin-releasing hormone release and its regulation. TEM 1996;7:56-59.

14. Wierman ME, Fang Z, Kepa JK. GnRH gene expression in neuronal cell lines. TEM 1996;7:60-65.

15. Nelson SB, Eraly SA, Mellon PL. The GnRH promoter: target of transcription factors, hormones, and signaling pathways. Mol Cell Endocrinol 1998;140:51-155.

16. Selmanoff M. Commentary on the use of immortalized neuroendocrine cell lines for physiological research. Endocrine 1997; 6:1-3.

17. Waschek JA. Transgenic targeting of neuroendocrine peptide genes in the hypothalamic-pituitary axis. Mol Neurobiol 1995;10:217.

18. Lira SA, Crenshaw III EB, Glass CK, Swanson LW, Rosenfeld MG. Identification of rat growth hormone genomic sequences targeting pituitary expression in transgenic mice. Proc Natl Acad Sci USA 1988;85:4755-4759.

19. Rubinstein M, Liu B, Goodman RH, Low MJ. Targeted expression of somatostatin in vasopressinergic magnocellular hypothalamic neurons of transgenic mice. Mol Cell Neurosci 1992;3:152-161.

20. Wolfe AM., Wray S, Westphal H, Radovick S. Cell-specific expression of the human gonadotropin-releasing hormone gene in transgenic animals. J Biol Chem 1996;271:20018-20023.

21. Skynner MJ, Slater R, Sim JS, Allen ND, Herbison AE. Promoter transgenics reveal multiple gonadotropin-releasing hormone-1-expressing cell populations of different embryological origin in mouse brain. J Neurosci 1999;19:5955-5966.

22. Pape J-R, Skynner MJ, Allen ND, Herbison AE. Transgenics identify distal 5'- and 3' sequences specifying gonadotropin-releasing hormone expression in adult mice. Mol Endocrinol 1999;13:2203-2211.

23. Spergel DJ, Krueth U, Hanley DF, Sprengel R, Seeburg PH. GNRH/LACZ mice exhibit two major populations of beta-galactosidase-stained neurons. Soc Neurosci Abst 1998;24:237.9.

24. Wu TJ, Gibson MJ, Silverman AJ. Gonadotropin-releasing hormone (GnRH) neurons of the developing tectum of the mouse. J Neuroendocrinol 1995;7:899-902.

25. Quinn JP. Neuronal-specific gene expression-The interaction of both positive and negative transcriptional regulators. Prog Neurobiol 1996;50:363-379.

26. Spergel DJ, Kruth U, Hanley DF. Sprengel R, Seeburg PH. GABA-and glutamate-activated channels in green fluorescent protein-tagged gonadotropin-releasing hormone neurone in transgenic mice. J Neurosci 1999;19:2037-2050.

27. Suter KJ, Song WJ, Sampson TL, Wuarin J-P, Saunders JT, Dudek FE, Moenter SM.. Genetic targeting of green fluorescent protein to gonadotropin-releasing hormone neurons: characterization of whole-cell electrophysiological properties and morphology. Endocrinology 1999;141:412-419.

28. Zhang Y-Z, Naleway JJ, Larison KD, Huang Z, Haugland RP. Detecting lacZ gene expression in living cells with new lipophilic, fluorogenic β-galactosidase substrates. FASEB J 1991; 5:3108-3113.

29. Skynner MJ, Sim JS, Herbison AE. Detection of estrogen receptor α and β messenger ribonucleic acids in adult gonadotropin-releasing hormone neurons. Endocrinology 1999;140:115195-5201.

30. Sim JA, Skynner MJ, Dyer RG, Herbison AE. Patch-clamp studies on GnRH neurons in acute slice preparations. Soc Neurosci Abst 1998;24:827.4.

31. Postigo H, Weiner R, Moenter S, Zhang L, Tsai P. Characterization of transgenic mice overexpressing a dominant negative FGF receptor in GnRH neurons. Soc Neurosci Abst 1999;25:777.4.

32. Tsien JZ, Chen DF, Gerber D, Tom C, Mercer EH, Anderson DJ, Mayford M, Kandel ER, Tonegawa S. Subregion-and cell type-restricted gene knockout in mouse brain. Cell 1996;87:1317-1326.

33. Gibson MJ, Miller GM, Silverman AJ. Pulsatile luteinizing hormone secretion in normal female mice and in hypogonadal female mice with preoptic area implants. Endocrinology 1991;128:965-971.

4

LH HYPERSECRETING MICE: A MODEL FOR OVARIAN GRANULOSA CELL TUMORS

Gabe E. Owens, Ruth A. Keri, and John H. Nilson

Case Western Reserve University School of Medicine, Dept. of Pharmacology, Cleveland OH

INTRODUCTION AND OVERVIEW

Proper function of the ovary requires carefully orchestrated proliferation and differentiation of granulosa cells that respond dynamically to signals from LH and FSH, and a host of other paracrine and autocrine factors. These numerous signals regulate a complex array of responsive genes. Any perturbation of these pathways, especially those that change the ratios of LH and FSH, compromises ovarian function and results in a number of outcomes including infertility, disruption of pregnancy, and even ovarian malignancy such as granulosa cell (GC) tumors.

Several years ago, we developed a line of transgenic mice that chronically hypersecrete LH and as a consequence display a number of ovarian pathological responses including development of GC tumors (1). A unique feature of this transgenic model became apparent when we discovered that oligogenic differences between transgenic mouse strains that hypersecrete LH dictates formation of one of two phenotypes; a non-neoplastic luteoma of pregnancy that occurs in most strains and a GC tumor that appears to be unique to CF-1 mice (2). Both of these phenotypes are found in women with ovarian sex cord stromal tumors (3-5). Furthermore, oligogenic differences in women may explain why ovarian sex cord stromal tumors only represent approximately 10% of all ovarian tumors (3-6). Finally there is a growing body of data suggesting that in addition to genetic modifiers and protooncogenes, development of ovarian malignancies may result from disruption of the extensive cross-talk between endocrine (particularly gonadotropins), paracrine, and autocrine signals that normally regulate GC growth and differentiation (6). Yet, despite extensive effort, the molecular pathogenesis of GC tumors remains poorly understood (4,6). The same holds for the molecular pathways that mediate LH or FSH action. Given the molecular and genetic complexity that underlies both GC tumor formation and the action of gonadotropins, understanding their relationship will require development of a number of animal models that isolate specific facets of these pathways.

In this review, we will discuss pertinent features of our transgenic mouse model with an emphasis on events that may underlie formation of GC tumors. This requires a brief overview of the hypothalamic-pituitary-ovarian axis and critical elements associated with murine reproductive physiology. We will also briefly discuss the strategy underlying our transgenic approach. In addition, we will compare our mouse model of LH hypersecretion to other mouse models of GC tumors to highlight important differences. In the end, we are convinced that by employing genomic approaches that analyze the action of LH in animals with defined genetic backgrounds, we will learn much about the genetic network that underlies both its normal and pathological action, including the genesis of GC tumors. As genetic and molecular networks are more highly conserved across mammalian species than overall phenotypes, we believe that critical insights into human biology can be gained from the continued study of molecular pathways in pertinent mouse models.

THE HYPOTHALAMIC-PITUITARY-GONADAL AXIS

LH and FSH are glycoprotein hormones secreted by the gonadotropes of the anterior pituitary (7,8). These hormones are both heterodimeric, each consisting of a common alpha subunit (αGSU) and a unique beta subunit which confers biologic specificity (7,8). The expression and secretion of these hormones are tightly regulated as illustrated in Figure 1 above. GnRH from the hypothalamus stimulates the secretion of both hormones (9). LH, in females, stimulates theca cells to produce androstenedione, which is converted to estrogen by granulosa cells (10). FSH acts on granulosa cells to regulate LH receptors and the activity of aromatase, the enzyme responsible for the conversion of androstenedione to estrogens (11). The steroids produced feedback at the level of both the pituitary and hypothalamus to inhibit secretion of LH while also repressing the expression of both the alpha and beta subunits (12,13). When regulation of LH is altered, pathological consequences occur, thus disrupting reproductive physiology in both humans and mice (1,14-16). These consequences include infertility, ovarian hyperstimulation and tumor development as discussed below.

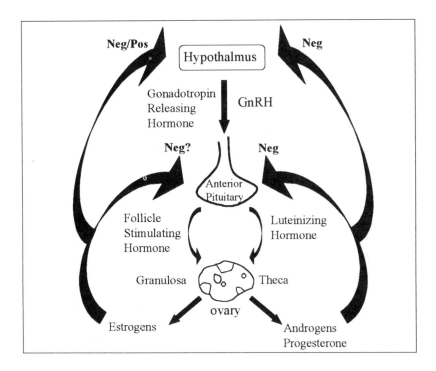

Fig 1. The hypothalamic-Pituitary-Axis.

PHYSIOLOGY OF THE MOUSE OVARY

For reproductive function, a delicate balance between differentiation and proliferation must occur in the ovary. Before puberty, follicular growth remains independent of gonadotropins (LH and FSH) and follicles remain at the primordial or primary stage (17). When puberty begins (usually around day 28 in wild-type mice), GnRH is released from the hypothalamus, stimulating secretion of FSH that acts on granulosa cell to stimulate proliferation and maturation of follicles (17,18). In addition, FSH induces the production of estrogen by granulosa cells by increasing the transcription of P450 aromatase (11). The combined action of estrogen and FSH induce granulosa cell proliferation by up-regulating cell cycle molecules such as cyclin D2 (19,20). In addition, these hormones also stimulate the formation of LH receptors on granulosa cells (18). Proliferation ends at ovulation when a surge of GnRH from the hypothalamus induces a surge of LH from the pituitary (9). This LH surge induces granulosa cells to stop proliferating and signals them to differentiate into a corpus luteum (17,21). The corpus luteum then begins producing progesterone. Granulosa cell differentiation is stimulated by a down-regulation of cyclin D2 expression and the increase of anti-mitotic proteins such as p27 and p21, all of which are induced by this surge of LH (20). When any of these events are disrupted, dire reproductive consequences occur (1,19,22). Several mouse models with significant reproductive phenotypes have been generated and alterations of either LH or FSH are commonly associated with mice that develop GC tumors. Therefore, the remainder of this review will focus on the role of gonadotropins, especially LH, in development of GC tumors.

A TRANSGENIC MODEL OF LH HYPERSECRETION

Other models or options of producing high serum levels of LH in mice utilize pharmacological dosing strategies that are laborious and non-physiologic. Thus, we sought to develop a mouse model with elevated LH by altering the hypothalamic-pituitary-gonadal axis through transgenic technology, without having to resort to pharmacological injection paradigms. As illustrated above in Figure 2, we achieved this by constructing a transgene that contained a bovine αGSU promoter (315bp) fused to the coding region of the LHβ subunit. The bovine αGSU promoter directed expression of the transgene specifically to gonadotropes (23-26). From previous studies, this promoter was also shown to retain responsiveness to hormones involved in the hypothalamic-pituitary-gonadal axis such as GnRH (23), estrogens (27), and androgens (28). Additionally, the carboxyl terminus of the LHβ subunit was fused in frame to the carboxyl terminal peptide (CTP) from the human chorionic gonadotropin β subunit (1). This addition of CTP lengthened the half-life of serum LH containing a chimeric LHβ subunit dimerized with the endogenous αGSU. This transgene was designated αLHβCTP.

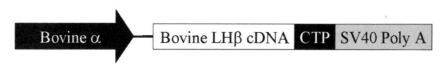

Fig 2. Construction of transgene for LH hypersecretion.

We predicted that this construct would produce elevated serum levels of LH due to the use of a strong, cell-specific promoter and through the addition of the CTP which would increase the proteins' half-life. Since the chimeric LHβ subunit is regulated by the αGSU promoter, we anticipated some escape from negative feedback since this promoter is less tightly regulated than the cognate LHβ promoter (29). Indeed, subsequent work from our laboratory has validated this expectation (30).

REPRODUCTIVE PHENOTYPES IN αLHβCTP MICE

As summarized in Table 1, male transgenic mice have normal levels of LH and no obvious reproductive females. In striking contrast, female transgenic mice hypersecrete LH and have a number of profound reproductive phenotypes. In intact αLHβCTP females, serum LH is elevated 5-10 fold in comparison to non-transgenic littermates. Levels of testosterone (T) and estradiol (E2) are also elevated with an overall increase in the ratio T to E2 (1). While the elevated steroids completely suppress the endogenous mouse LHβ gene, the transgene and endogenous α subunit have become refractory to negative feedback by E2 (30). The elevated steroids also suppress GnRH, indicating that activity of the promoters for the transgene and endogenous α subunit genes are only partially dependent on GnRH. In contrast, the promoter for the mouse LHβ gene has no detectable activity in the absence of the hypothalamic-releasing hormone (30). Consequently, levels of serum LH in intact female mice derive entirely from expression of the transgene and the endogenous α subunit gene.

Table 1. Reproductive Phenotypes in αLHβCTP Mice

- Males:
  Fertile, Normal LH
  No Obvious Phenotype

- Females:
  Elevated LH, Androgens, Estrogens, A/E Ratio
  Precocious Puberty
  Polycystic Ovaries
  Chronic Anovulation
  Defects in Uterine Receptivity
  Midgestation Pregnancy Failure
  Hydronephrosis/Pyelonephritis Due to Bladder Atony
  Granulosa Cell Tumors
  Pituitary Adenomas
  Mammary Hyperplasia

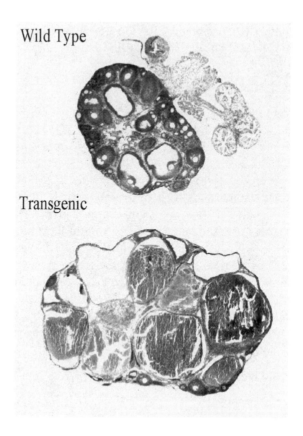

Fig 3. Histological sections of ovaries from both 6 week old wild type and 6 week old LH hypersecreting, transgenic mice. Transgenic mice display large fluid and blood filled cysts.

Female αLHβCTP mice display early vaginal opening and advanced folliculogenesis but are anovulatory and display a prolonged luteal phase (1,16). Anovulation reflects the absence of GnRH (30) and the inability of the animal to produce a preovulatory surge of LH. Their ovaries are enlarged with reduced numbers of primordial follicles (31) and numerous giant hemorrhagic follicles as shown in Figure 3 (16). Despite the pathological appearance of the ovary, females can be superovulated and mated (32). Although pregnancy occurs, implantation is compromised due to defects in uterine receptivity (32). In addition, pregnancy fails at midgestation, reflecting a maternal defect such as estrogen toxicity (32). Oocytes in these animals are healthy and can develop into normal embryos when transplanted into pseudopregnant wild-type females. Furthermore, when the transgene is in a CF-1 background, all females develop GC tumors by five months of age (1,30). They die shortly thereafter due to bladder atony and subsequent kidney failure (1). In addition to these pathologies, elevated LH effects the pituitary and mammary gland. Both develop hyperplasia and eventually form adenomas. In all, elevated LH drastically effects many tissues with its most penetrant and dramatic effect being GC tumor formation.

OLIGOGENIC DIFFERENCES PREDISPOSE LH HYPERSECRETING MICE TO GC TUMORS

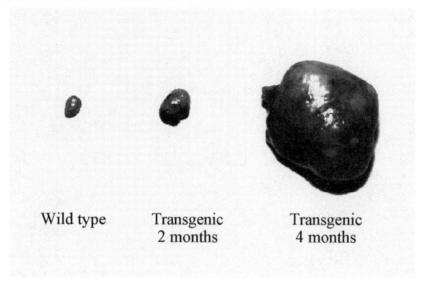

Wild type   Transgenic   Transgenic
             2 months   4 months

Fig 4. Ovaries from wild type and transgenic mice. By 4-5 months LHCTP mice develop large GC tumors.

Initial studies analyzing the impact of LH hypersecretion involved the use of mice in a combined genetic background with contributions from CF-1, C57BL/6, and SJL. In these studies, some older mice developed GC tumors (1). However, upon repetitive multi-generational breeding into the outbred CF-1 strain, the frequency of the GC tumor phenotype progressively increased to 100% at five months of age. This suggested that the development of GC tumors might be dependent on genetic background.

Genetic background effects were evaluated by breeding transgenic CF-1 males with females from one of four different strains: CF-1, CD-1, C57BL/6, and SJL. Transgenic, male mice were bred with females from each of these strains. After breeding, the progeny were screened for the phenotypes observed in CF-1 transgenic females. Vaginal opening was evaluated and transgenic females in both hybrids analyzed [(C57BL/6♀ x CF-1, Tg ♂) and (CD-1♀ x CF-1, Tg ♂) displayed vaginal opening five days earlier than their wild-type controls and comparable to CF-1 transgenic females (2). This indicated that the αLHβCTP transgene was active in different strains of mice and that precocious vaginal opening occurred independently of the mouse strain.

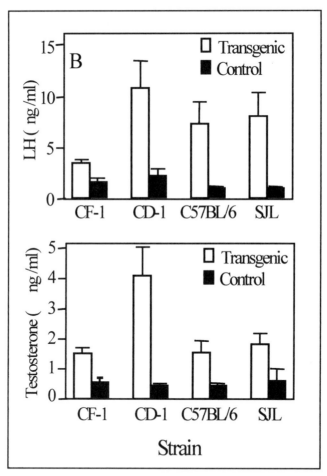

Fig 5. LH and Testosterone elevations are independent of genetic background and do not correlate with the tumor phenotype.

Next, serum levels of LH and androgens were determined. LH and androgens were significantly elevated in all stains of mice when compared to littermate controls as shown in Figure 5 (2). Remarkably, 5-month-old CF-1 transgenic mice did not have the same magnitude of LH hypersecretion as was observed in younger animals. This may be attributed to the fact that CF-1 animals at this age bear a GC tumor that may be secreting compounds such as inhibin (33) that may dampen the hypersecretion of LH from the gonadotropes.

When the ovaries from F1 hybrids and CF-1 animals were examined grossly it was observed that CF-1 females had large GC tumors as expected but the F1 hybrids did not. This was confirmed by histological analysis and the data is reviewed in Table 2. Hybrids had ovaries that were reminiscent of a luteoma of pregnancy, a phenomenon commonly found in women (34). These ovaries are completely luteinized and look like giant copora luteum (2). Thus, it was confirmed that GC tumors caused by LH hypersecretion are dependent on genetic background. Therefore, measures were taken to determine the exact genetic requirement predisposing CF-1 animals and protecting other strains from the development of these tumors.

Table 2. Results of backcross analyses assaying for GC tumor formation.

	Granulosa Cell Tumor	Luteoma/ Thecoma
CF-1	8/8	0/8
CD-1 x CF-1	0/13	13/13
C57BL/6 xCF-1	0/6	6/6
SJL x CF-1	0/4	4/4

To determine the number of genes responsible for the formation of GC tumors in CF-1 females, transgenic F1 males (C57BL/6♀ x CF-1, Tg ♂) were backcrossed to CF-1 females. At five months of age the transgenic female progeny were scored for the formation of GC tumors. Out of 138 transgenic female progeny, 19 developed tumors (2). To determine the number of genes underlying tumor development we tested the fit between the predicted and observed frequency of tumor appearance. It was assumed that each gene acted in an equal, additive, and independent manner. 100% penetrance of tumor appearance by 5 months of age was also assumed for the backcross progeny as was observed for CF-1 animals. After chi-square goodness-of-fit analysis with one degree of freedom, it was determined that the best fit represented a 3 gene model ($\chi^2 = 0.20$, p > 0.66), with the 2- and 4-gene models being rejected ($\chi^2 = 9.28$ p > 0.003, $\chi^2 = 13.31$ p > 0.0003 respectively). Thus, this data suggested that LH induced tumorigenesis is predisposed and controlled by 3 unlinked genes (2). Therefore, GC tumor formation is a complex genetic trait that may underscore the rarity of GC tumors in the human population. However, the αLHβCTP mouse provides a unique model to elucidate the biology and genetics of GC tumors that may aid in preventing or treating the tumors that do form in the human population.

OTHER MOUSE MODELS OF GC TUMORS

The SWR inbred strain of mice is a natural mouse model of spontaneous GC tumor formation (33,35-42). Tumors arise by 8 weeks of age in 1-2% of pubertal females with malignancy occurring frequently (35). This incidence can be increased by treatment with DHEA or testosterone, whereas DHT has no effect and estradiol exerts a suppressive effect (38). In tumor bearing females, serum levels of LH and FSH tend to be low whereas serum levels of androstenedione, DHEA, and estradiol are normal (36). Although serum levels of LH and FSH were uninformative subsequent grafting studies, where ovaries from tumor susceptible donors were placed into the kidney capsule of normal or gonadotropin-deficient hosts, indicated that tumors form only in hosts replete with gonadotropins (43). This provided strong evidence that tumor formation requires gonadotropins. In addition, the low incidence of tumor formation in the SWR strain is indicative of a complex genetic trait. Indeed, Beamer and colleagues have now identified at least four chromosomal loci that underlie tumor susceptibility (40). Taken together, the developmental age of onset, endocrine profile, propensity for malignancy, and low frequency of occurrence of GC tumors in the SWR model has a striking parallel with the spontaneous and rare occurrence of juvenile ovarian GC tumors in young girls (39,44,45). In contrast, the SWR mouse model is much less applicable to GC tumors that occur in postmenopausal women that constitute the largest cadre of patients with this cancer (3).

GC tumors can also be induced in mice through a variety of exogenous paradigms including ectopic ovarian transplantation (43,46), chemical carcinogens (47-51), irradiation (52), neonatal thymectomy (53-56), and pharmacological treatment with metabolites of amino acids (57) and steroids (38). When taken together, these studies support the notion that gonadotropins play a critical role in development of GC tumors. For example, treatment with the chemical carcinogen DMBA induced the formation of GC tumors in 5 of 15 mice by 360 days (51). GC tumor formation, however, was significantly enhanced (14 of 18) during the same period if females were first masculinized by high dose administration of estradiol within 3 days after birth and then treated with the DMBA (51). The estradiol-treated mice had significantly higher levels of LH and FSH than their non-treated littermates. High levels of LH and FSH also occur when ovaries are ectopically transplanted to the spleen of an ovariectomized host. This transplantation results in the formation of GC tumors, which can be blocked by hypophysectomy and anti-gonadotropin agents (49).

Additional supporting evidence for the importance of gonadotropins in GC tumorigenesis comes from gamma-irradiation studies carried out with mice heterozygous or homozygous for GnRH deficiency (*hpg*) (52). In heterozygotes (+/*hpg*), all irradiated females develop GC tumors whereas no tumors develop in irradiated homozygotes deficient for GnRH (*hpg/hpg*). This difference underscores the importance of LH and FSH in GC tumor development. Studies with mice that undergo neonatal thymectomy offer a third layer of support for the role of gonadotropins as well as provide an interesting parallel to our αLHβCTP model. Mice that undergo this procedure undergo premature ovarian failure caused by the development of autoimmunity; they also have a high incidence of GC tumors (53). Ovarian dysgenesis and GC tumor formation in this model is accompanied by high levels of gonadotropins that precede appearance of the tumor and then wane subsequently (54-56,58), leading the authors to conclude that prolonged stimulation with gonadotropins induces GC tumors.

Transgenic mouse models have also evaluated the role of gonadotropin in tumorigenesis and are illustrated in Table 3. Each of these animals presents with gonadotropin levels that range from low to high. For convenience, the animals were ranked primarily by their ability to form a GC tumor and secondarily by their levels of LH (from low to high). Although several of the transgenic animals have elevated LH (ERβ-, ERα-/ERβ-, and P450$_{arom}$-) and substantial ovarian pathology, they are resistant to formation of GC tumors. This is particularly noteworthy for the P450$_{arom}$ transgenic mice, as they have elevated levels of both LH and FSH. Animals deficient in ERα rarely form tumors even though they have elevated LH. Together, this grouping of transgenic mice suggests that elevations of LH alone cannot account for GC tumor formation at high frequency.

Studies with transgenic mice deficient for inhibin provide additional evidence that while essential, elevated gonadotropins are not sufficient for formation of GC tumors. Homozygous females with a disrupted inhibin α subunit gene develop multiple sex cord-stromal tumors with 100% penetrance (62). They also have high levels of serum FSH and estrogen but

Transgenic	Serum Hormone Levels			GC Tumor	Fertile	Follicular development
	LH	FSH	E2			
MT-hFSHβ (65)	+	+++	+++	No	No	Accelerated-cystic
p27- (22)	++	++	++	No	No	Arrested (no CL)
ERβ- (66)	++	++	++	No	Yes	Normal
Cyclin D2- (19)	++	++	++	No	No	Arrested (1)
ERα-/ERβ- (59)	+++	++	+++	No	No	Sex-reversed
ERα- (60)	+++	++	+++	Rare*	No	Accelerated-cystic
P450$_{arom}$- (61)	+++	+++	+	No	No	Arrested (1)
FSHβ- (64)	+++	—	++	No	No	Arrested (1)
Inhα-/GnRH- (63)	—	—	—	No	No	Arrested (1)
Inhα- (62)	++	+++	+++	Yes	No	N/A
Inhα-/FSHβ- (65)	+++	—	++	Yes	No	Arrested (1)
αLHβCTP (1)	+++	++	+++	Yes	No	Accelerated-cystic

+++ = high   *only sporadically in older animals
++   = normal
+   = low
—   = absent

normal levels of LH (62). Nevertheless, even though inhibin may be regarded as a tumor suppressor, when gonadotropins are also lacking, as they are in mice deficient for GnRH (63), GC tumors no longer form. GC tumors also fail to form in female mice that are homozygous for a targeted disruption of the FSHβ gene. Although these mice have elevated levels of LH, reductions in inhibin were probably insufficient to unmask the tumorigenic potential of the gonadotropins (64). Matzuk and colleagues have recently tested this assumption further by generating a new strain of mice that carried homozygous mutations in the genes encoding both inhibin α and FSHβ (65). These mice continue to form GC tumors but at a slower rate and with reduced penetrance when compared to mice deficient only in inhibin. In addition, these mice have estrogen levels near the limit of detection and presumably elevated levels of LH. This suggests that, in an inhibin deficient environment, LH can act independently of FSH to cause GC tumors and that estradiol is not required.

While the above transgenic data implicate gonadotropins as playing an essential role in GC tumorigenesis, it is clear that other factors are required, such as reductions in other hormones like inhibin that appear to function as tumor suppressors. This is consistent with the proposal that some apparently spontaneous carcinomas, like GC tumors, can be explained through a polygenic model of inheritance (40). As noted above, this concept clearly applies to the SWR mouse model of juvenile GC tumors. Our αLHβCTP transgenic mice now expand the polygenic concept to the formation of GC tumors in adult female mice since it appears as though three unlinked allelic variants act as essential modifiers of the action of chronically elevated LH (67). Our model differs significantly from the studies of Matzuk and colleagues in that the tumors are forming in the presence of normal to high levels of inhibin. Thus, the three unlinked allelic variants must be counteracting the tumor suppressive effect of inhibin.

OVARIAN GC TUMORS IN HUMANS

Ovarian cancer is the fourth most common cancer in women and the leading cause of fatality from gynecological cancer (6). Most of the cancers (90%) are epithelial with the remainder consisting of sex cord-stromal tumors (5). Although GC tumors are rare, the low survival rate of women with these tumors makes them clinically important (5,68,69). Despite their clinical importance, the molecular events underlying their formation in women remain poorly understood (4,6). This applies especially to the role of gonadotropins (70).

Because studies with rodents, especially mice, point to the potential role of gonadotropins in GC tumorigenesis, it is tempting to conclude from the biochemical features of GC tumors in women that gonadotropins play a similar role. GC tumors bind FSH and hCG (70-72), respond to FSH by producing inhibin (3,73-75) and MIS (76), contain progesterone receptors (77), and have elevated levels of cyclin D2 (19). In fact, the functional integration between FSH-R and cyclin D2 has stimulated a search for activating mutations in the FSH-R that might be associated with at least some GC tumors in women (4). Unfortunately, the outcome thus far has been negative (4).

The relatively rare occurrence of GC tumors in women fits well with the possibility that their spontaneous appearance has polygenic determinants as it clearly does in mice. At this point, however, most investigations have focused on single genetic determinants. For example, patients with Peutz-Jeghers syndrome can acquire sporadic GC tumors with relatively high incidence (78,79). Unfortunately, somatic mutations in the LKB1/Peutz-Jeghers syndrome gene have not been detected as they have in other sporadic adenocarcinomas of the cervix and lung (78). Interestingly, loss of heterozygosity (LOH) has been reported on the locus that contains the LKB1 gene (19p13.3) in approximately 50% of ovarian cancers that form in a subset of these patients (79). Similar to the aforementioned study, however, the LOH almost certainly targets a gene other than LKB1 since most of the samples were devoid of mutations in this gene. In another study, altered expression of the p53 tumor suppressor gene, c-erbB2 oncogene, and proliferation antigen Ki-67 were not detected in ovaries removed prophylactically from women with a family history of ovarian cancer (80). Therefore, if these genes do play a role, it must be late during the onset of the disease. Finally, monosomy 22 and trisomy 14 have been detected in an adult onset granulosa-thecoma cell tumor (81). Although the authors speculate that these genetic abnormalities may play an early role in GC tumorigenesis, it is not possible to rule out that these disruptions occur as a consequence of the tumor. The same concern holds for two cases that report trisomy 12 in juvenile GC tumors (82).

With the lack of genetic evidence that indict either gonadotropins, or down-stream participants of its signal transduction pathway, as critical agents of GC tumorigenesis, it is necessary to rely on clinical studies. Although the findings are only suggestive and often incomplete, the role for gonadotropins in promoting GC tumorigenesis remains viable and likely. The strongest indirect evidence for a role for gonadotropins comes from post-menopausal women who also represent the largest cadre of patients with CG tumors (6). With decline in ovarian function, due presumably to an aging hypothalamus (83,84), estradiol levels fall, which in turn causes a rise in LH and FSH. It is tempting to speculate that the post-menopausal rise LH and FSH increase the risk for development of GC tumors. But, this is probably too simplistic, especially if the appearance of this carcinoma has a polygenic basis. On the other hand, should women accrue an abnormal gonadotropin status during their youth, then long-term exposure may impact their risk for acquiring a GC tumor after the onset of menopause.

There is general agreement that nulliparity resulting from infertility associates with an increased risk in acquiring GC tumors (85,86). This has raised the question of whether gonadotropin treatment for infertility further increases the incidence of GC tumors in this patient cohort. While some studies support this notion (87,88), others reach the opposite conclusion (85,86,89). These latter studies have recently been reinforced by a Finnish study indicating that the rate of GC tumors is actually falling concomitantly with increasing use of ovulation inducers (90). Thus, although gonadotropins are strongly suspected as culprits in human GC tumor development, more research needs to be done to elucidate the mechanisms behind their tumorigenic action.

SUMMARY

Despite extensive effort, the molecular pathogenesis of GC tumors in women remains unexplained. Much of this reflects the inherent difficulty of performing carefully controlled prospective studies that extend over time and examine many different molecular and genetic parameters. In contrast, studies with mice are much more complete and provide compelling evidence that elevated levels of gonadotropins play a significant role in GC tumorigenesis. Nevertheless, it is becoming increasingly clear, through studies in both mice and women, that gonadotropins cannot act alone but instead in partnership with genetic modifiers that predispose females to this carcinoma. We are particularly excited because the weight of current evidence suggests that both juvenile and adult forms of spontaneous GC tumors in mice are mediated by a polygenic mechanism that probably involves similar as well as different allelic variants. While continued studies with the SWR mouse model will likely lead to identification of the allelic variants that underlie the juvenile form of GC tumors, these mice do not mimic the adult form. Our αLHβCTP mice represent a model to study adult onset GC tumors that are also polygenic. Thus, its continued study represents the only known approach for identifying allelic variants that unmask the tumorigenic potential of LH in adult female mice. Identification of these variants will open door for probing whether similar homologs are found in post-menopausal women with GC tumors.

REFERENCES

1.   Risma KA, Clay CM, Nett TM, Wagner T, Yun J, Nilson JH. Targeted overexpression of luteinizing hormone in transgenic mice leads to infertility, polycystic ovaries, and ovarian tumors. Proc Natl Acad Sci USA 1995;92:1322-1326.

2.   Keri RA, Lozada KL, Abdul-Karim FW, Nadeau JH, Nilson JH. Luteinizing hormone induction of ovarian tumors: oligogenic differences between mouse strains dictates tumor disposition. Proc Natl Acad Sci USA 2000;97:383-387.

3.   Lappohn RE, Burger HG, Bouma J, Bangah M , Krans M, de Bruijn HW. Inhibin as a marker for granulosa-cell tumors. N Engl J Med 1989;321:790-793.

4.   Fuller PJ, Verity K, Shen Y, Mamers P, Jobling T, Burger HG. No evidence of a role for mutations or polymorphisms of the follicle- stimulating hormone receptor in ovarian granulosa cell tumors. J Clin Endocrinol Metab 1998;83:274-279.

5.   Cooke I, O'Brien M, Charnock FM, Groome N, Ganesan TS. Inhibin as a marker for ovarian cancer. British J Cancer 1995;71:1046-1050.

6.   Amsterdam A, Selvaraj N. Control of differentiation, transformation, and apoptosis in granulosa cells by oncogenes, oncoviruses, and tumor suppressor genes. Endocr Rev 1997;18:435-461.

7.   Bousfield GR, Perry WM, Ward DN. Gonadotropins: Chemistry and Biosynthesis. In: The Physiology of Reproduction, Second Edition (Eds: Knobil E, Neill JD), pp1749-1791. New York: Raven Press, Ltd., 1994.

8.   Pierce J, Parsons TF. Glycoprotein hormones: structure, function. Ann Rev Biochem 1981;50:465-495.

9.   Marshall JC, Dalkin AC, Haisenleder DJ, Paul SJ, Ortolano GA, Kelch RP. Gonadotrophin-releasing hormone pulses: Regulators of gonadotropin synthesis and ovulatory cycles. Recent Prog Horm Res 1991;47:155-189.

10.   Falck B. Site of production of oestrogen in rat ovary as studied in micro-transplants. Acta Physiol 1959;193:1-101.

11.   Steinkampf MP, Mendelson CR, Simpson ER. Regulation by follicle-stimulating hormone of the synthesis of aromatase cytochrome P-450 in human granulosa cells. Mol Endocrinol 1987;1:465-471.

12.   Haisenleder DJ, Dalkin AC, Marshall JC. Regulation of gonadotropin gene expression. In: The Physiology of Reproduction (Eds: Knobil E, Neill JD), pp1793-1813. New York: Raven Press, Ltd., 1994.

13.   Gharib SD, Wierman ME, Shupnik MA, Chin WW. Molecular biology of the pituitary gonadotropins. Endocr Rev 1990;11:177-199.

14.   Balen AH, Tan SL, Jacobs HS. Hypersecretion of luteinizing hormone: a significant cause of infertility and miscarriage . Br J Obst Gyn 1993;100:1082-1089.

15.   Shoham Z, Jacobs HS, Insler V. Luteinizing hormone: its role, mechanism of action, and detrimental effects when hypersecreted during the follicular phase. Fertil Steril 1993;59:1153-1161.

16.   Risma KA, Hirshfield AH, Nilson JH. Elevated LH in prepubertal transgenic mice causes hyperandrogenemia, precocious puberty, and substantial ovarian pathology. Endocrinology 1997;138:3540-3547.

17.   Richards JS. Maturation of ovarian follicles: actions and interactions of pituitary and ovarian hormones on follicular cell differentiation. Physiol Rev 1980;60:51-89.

18.   Richards JS, Ireland JJ, Rao MC, Bernath GA, Midgley AR, Jr., Reichert LE, Jr. Ovarian follicular development in the rat: hormone receptor regulation by

estradiol, follicle stimulating hormone and luteinizing hormone. Endocrinology 1976;99:1562-1570.

19. Sicinski P, Donaher JL, Geng Y, Parker SB, Gardner H, Park MY, Robker RL, Richards JS, McGinnis LK, Biggers JD, Eppig JJ, Bronson RT, Elledge SJ, Weinberg RA. Cyclin D2 is an FSH-responsive gene involved in gonadal cell proliferation and oncogenesis. Nature 1996;384:471-474.

20. Robker RL, Richards JS. Hormone-induced proliferation and differentiation of granulosa cells: a coordinated balance of the cell cycle regulators cyclin D2 and p27Kip1. Mol Endocrinol 1998;12:924-940.

21. Richards JS, Hedin L, Caston L. Differentiation of rat ovarian thecal cells: evidence for functional luteinization. Endocrinology 1986;118:1660-1668.

22. Fero ML, Rivkin M, Tasch M, Porter P, Carow CE, Firpo E, Polyak K, Tsai LH, Broudy V, Perlmutter RM, Kaushansky K, Roberts JM. A syndrome of multiorgan hyperplasia with features of gigantism, tumorigenesis, and female sterility in p27(Kip1)-deficient mice. Cell 1996;85:733-744.

23. Hamernik DL, Keri RA, Clay CM, Clay JN, Sherman GB, Sawyer Jr. HR, Nett TM, Nilson JH. Gonadotrope- and thyrotrope-specific expression of the human and bovine glycoprotein hormone alpha-subunit genes is regulated by distinct cis-acting elements. Mol Endocrinol 1992;6:1745-1755.

24. Kendall SK, Saunders TL, Jin L, Lloyd RV, Glode LM, Nett TM, Keri RA, Nilson JH, Camper SA. Targeted ablation of pituitary gonadotropes in transgenic mice. Mol Endocrinol 1991;5:2025-2036.

25. Nilson JH, Bokar JA, Clay CM, Farmerie TA, Fenstermaker RA, Hamernik DL, Keri RA. Different combinations of regulatory elements may explain why placenta-specific expression of the glycoprotein hormone □- subunit gene occurs only in primates and horses. Biol Reprod 1991;44:231-237.

26. Clay CM, Keri RA, Finicle AB, Heckert LL, Hamernik DL, Marschke KM, Wilson EM, French FS, Nilson JH. Transcriptional repression of the glycoprotein hormone alpha subunit gene by androgen may involve direct binding of androgen receptor to the proximal promoter. J Biol Chem 1993;268:13556-13564.

27. Keri RA, Andersen B, Kennedy GC, Hamernik DL, Clay CM, Brace AD, Nett TM, Notides AC, Nilson JH. Estradiol inhibits transcription of the human glycoprotein hormone □-subunit gene despite the absence of a high affinity binding site for estrogen receptor. Mol Endocrinol 1991;5:725-733.

28. Heckert LL, Wilson EM, Nilson JH. Transcriptional repression of the alpha-subunit gene by androgen receptor occurs independently of DNA binding but requires the DNA- binding and ligand-binding domains of the receptor. Mol Endocrinol 1997;11:1497-1506.

29. Keri RA, Wolfe MW, Saunders TL, Anderson I, Kendall S, Wagner T, Yeung J, Gorski J, Nett TM, Camper SA, Nilson JH. The proximal promoter of the bovine luteinizing hormone β-subunit gene confers gonadotrope-specific expression and regulation by gonadotropin-releasing hormone, testosterone, and 17β-estradiol in transgenic mice. Mol Endocrinol 1994;8:1807-1816.

30. Abbud RA, Ameduri RK, Rao JS, Nett TM, Nilson JH. Chronic hypersecretion of luteinizing hormone in transgenic mice selectively alters responsiveness of the alpha-subunit gene to gonadotropin-releasing hormone and estrogens. Mol Endocrinol 1999;13:1449-1459.

31. Flaws JA, Abbud R, Mann RJ, Nilson JH, Hirshfield AN. Chronically elevated luteinizing hormone depletes primordial follicles in the mouse ovary. Biol Reprod 1997;57:1233-1237.

32. Mann RJ, Keri RA, Nilson JH. Transgenic mice with chronically elevated luteinizing hormone are infertile due to anovulation, defects in uterine receptivity, and midgestation pregnancy failure. Endocrinology 1999;140:2592-2601.

33. Gocze PM, Beamer WG, de Jong FH, Freeman DA. Hormone synthesis and responsiveness of spontaneous granulosa cell tumors in (SWR x SWXJ-9) F1 mice. Gynecol Oncol 1997;65:143-148.

34. Joshi R, Dunaif A. Ovarian disorders of pregnancy. Endocrinol Metab Clin North Am 1995;24:153-169.

35. Beamer WG, Hoppe PC, Whitten WK. Spontaneous malignant granulosa cell tumors in ovaries of young SWR mice. Cancer Res 1985;45:5575-5581.

36. Beamer WG. Gonadotropin, steroid, and thyroid hormone milieu of young SWR mice bearing spontaneous granulosa cell tumors. J Natl Cancer Inst 1986;77:1117-1123.

37. Beamer WG, Tennent BJ, Shultz KL, Nadeau JH, Shultz LD, Skow LC. Gene for ovarian granulosa cell tumor susceptibility, Gct, in SWXJ recombinant inbred strains of mice revealed by dehydroepiandrosterone. Cancer Res 1988;48:5092-5095.

38. Beamer WG, Shultz KL, Tennent BJ. Induction of ovarian granulosa cell tumors in SWXJ-9 mice with dehydroepiandrosterone. Cancer Res 1988;48:2788-2792.

39. Beamer WG, Shultz KL, Tennent BJ, Azumi N, Sundberg JP. Mouse model for malignant juvenile ovarian granulosa cell tumors. Toxicol Pathol 1998;26:704-710.

40. Beamer WG, Shultz KL, Tennent BJ, Nadeau JH, Churchill GA, Eicher EM. Multigenic and imprinting control of ovarian granulosa cell tumorigenesis in mice. Cancer Res 1998;58:3694-3699.

41. Tennent BJ, Beamer WG, Shultz LD, Adamson ED. Epidermal growth factor receptors in spontaneous ovarian granulosa cell tumors of SWR-derived mice. Int J Cancer 1989;44:477-482.

42. Tennent BJ, Shultz KL, Sundberg JP, Beamer WG. Ovarian granulosa cell tumorigenesis in SWR-derived F1 hybrid mice: preneoplastic follicular abnormality and malignant disease progression. Am J Obstet Gynecol 1990;163:625-634.

43. Beamer WG, Shultz KL, Tennent BJ, Shultz LD. Granulosa cell tumorigenesis in genetically hypogonadal-immunodeficient mice grafted with ovaries from tumor-susceptible donors. Cancer Res 1993;53:3741-3746.

44. Gell JS, Stannard MW, Ramnani DM, Bradshaw KD. Juvenile granulosa cell tumor in a 13-year-old girl with enchondromatosis (Ollier's disease): a case report. J Pediatr Adolesc Gynecol 1998; 11(3):147-150.

45. Lancaster EJ, Muthuphei MN. Bilateral juvenile granulosa--cell tumour with multiple metastases. A case report. Cent Afr J Med 1998;44:158-160.

46. Guthrie MJ. Tumorigenesis in intrasplenic ovaries in mice. Cancer 1957;10:190-203.

47. Armuth V, Berenblum I. Mechanism of ovarian carcinogenesis: effect of 7,12-dimethylbenz[a]anthracene administration on intrasplenic ovarian grafts in unilaterally ovariectomized C3HeB/Fe mice. J Natl Cancer Inst 1979;63:1047-1050.

48. Capen CC, Beamer WG, Tennent BJ, Stitzel KA. Mechanisms of hormone-mediated carcinogenesis of the ovary in mice. Mutat Res 1995;333:143-151.

49. Hilfrich J. Comparative morphological studies on the carcinogenic effect of 7,12-dimethylbenz(A)anthracene (DMBA) in normal or intrasplenic ovarian tissue of C3H mice. Br J Cancer 1975;32:588-595.

50. Rao AR. Effects of carcinogen and/or mutagen on normal and gonatotropin-primed ovaries of mice. Int J Cancer 1981;28:105-110.

51. Taguchi O, Michael SD, Nishizuka Y. Rapid induction of ovarian granulosa cell tumors by 7,12- dimethylbenz(a)anthracene in neonatally estrogenized mice. Cancer Res 1988;48:425-429.

52. Tennent BJ, Beamer WG. Ovarian tumors not induced by irradiation and gonadotropins in hypogonadal (hpg) mice. Biol Reprod 1986;34:751-760.

53. Nishizuka Y, Sakakura T, Taguchi O. Mechanism of ovarian tumorigenesis in mice after neonatal thymectomy. Natl Cancer Inst Monogr 1979;51:89-96.
54. Michael SD, Taguchi O, Nishizuka Y. Hormonal characterization of female SL/Ni mice: a small thymus gland strain exhibiting ovarian dysgenesis. J Reprod Immunol 1988;12:277-286.
55. Michael SD, Taguchi O, Nishizuka Y. Changes in hypophyseal hormones associated with accelerated aging and tumorigenesis of the ovaries in neonatally thymectomized mice. Endocrinology 1981;108:2375-2380.
56. Michael SD, Taguchi O, Nishizuka Y. Effect of neonatal thymectomy on ovarian development and plasma LH, FSH, GH and PRL in the mouse. Biol Reprod 1980;22:343-350.
57. Fujii K, Watanabe M. Comparative study of tumorigenicity in mice administered transplacentally or neonatally with metabolites of tryptophan and its related compounds. J Cancer Res Clin Oncol 1980;96:163-168.
58. Michael SD, De Angelo L, Kaikis-Astaras A. Plasma protein and hormone profiles associated with autoimmune oophoritis and ovarian tumorigenesis in neonatally thymectomized mice. Autoimmunity 1990;6:1-12.
59. Couse JF, Hewitt SC, Bunch DO, Sar M, Walker VR, Davis BJ, Korach KS. Postnatal sex reversal of the ovaries in mice lacking estrogen receptors alpha and beta. Science 1999;286:2328-2331.
60. Couse JF, Bunch DO, Lindzey J, Schomberg DW, Korach KS. Prevention of the polycystic ovarian phenotype and characterization of ovulatory capacity in the estrogen receptor-alpha knockout mouse. Endocrinology 1999;140:5855-5865.
61. Fisher CR, Graves KH, Parlow AF, Simpson ER. Characterization of mice deficient in aromatase (ArKO) because of targeted disruption of the cyp19 gene. Proc Natl Acad Sci USA 1998;95:6965-6970.
62. Matzuk MM, Finegold MJ, Su JG, Hsueh AJ , Bradley A. Alpha-inhibin is a tumour-suppressor gene with gonadal specificity in mice. Nature 1992;260:313-319.
63. Kumar TR, Wang Y, Matzuk MM. Gonadotropins are essential modifier factors for gonadal tumor development in inhibin-deficient mice. Endocrinology 1996;137:4210-4216.
64. Kumar TR, Wang Y, Lu N, Matzuk MM. Follicle stimulating hormone is required for ovarian follicle maturation but not male fertility. Nat Gen 1997;15:201-204.
65. Kumar TR, Palapattu G, Wang P, Woodruff TK, Boime I, Byrne MC, Matzuk MM. Transgenic models to study gonadotropin function: the role of follicle- stimulating hormone in gonadal growth and tumorigenesis. Mol Endocrinol 1999;13:851-865.
66. Krege JH, Hodgin JB, Couse JF, Enmark E , Warner M, Mahler JF, Sar M, Korach KS, Gustafsson JA, Smithies O. Generation and reproductive phenotypes of mice lacking estrogen receptor beta. Proc Natl Acad Sci USA 1998;95:15677-15682.
67. Keri RA, Lozada KL, Abdul-Karim FW, Nadeau JH, Nilson JH. Luteinizing hormone induction of ovarian tumors: oligogenic differences between mouse strains dictates tumor disposition. Proc Natl Acad Sci USA 2000;97:383-387.
68. Lee WL, Yuan CC, Lai CR, Wang PH. Hemoperitoneum is an initial presentation of recurrent granulosa cell tumors of the ovary. Jpn J Clin Oncol 1999;29:509-512.
69. Fontanelli R, Stefanon B, Raspagliesi F, Kenda R, Tomasic G, Spatti G, Riboldi G, Di Donato P, Pilotti S, De Palo G. Adult granulosa cell tumor of the ovary: a clinico pathologic study of 35 cases. Tumori 1998;84:60-64.
70. Zheng W, Lu JJ, Luo F, Zheng Y, Feng Y, Felix JC, Lauchlan SC, Pike MC. Ovarian epithelial tumor growth promotion by follicle-stimulating hormone and inhibition of the effect by luteinizing hormone. Gynecol Oncol 2000;76:80-88.
71. Zheng W, Magid MS, Kramer EE, Chen YT. Follicle-stimulating hormone receptor is expressed in human ovarian surface epithelium and fallopian tube. Am J Pathol 1996;148:47-53.

72. Stouffer RL, Grodin MS, Davis JR, Surwit EA. Investigation of binding sites for follicle-stimulating hormone and chorionic gonadotropin in human ovarian cancers. J Clin Endocrinol Metab 1984;59:441-446.

73. Burger HG, Baillie A, Drummond AE, Healy DL, Jobling T, Mamers P, Robertson DM, Susil B, Cahir N, Shen Y, Verity K, Fuller PJ, Groome NP, Findlay JK. Inhibin and ovarian cancer. J Reprod Immunol 1998;39:77-87.

74. Robertson DM, Cahir N, Burger HG, Mamers P, Groome N. Inhibin forms in serum from postmenopausal women with ovarian cancers. Clin Endocrinol 1999;50:381-386.

75. Robertson DM, Cahir N, Burger HG, Mamers P, McCloud PI, Pettersson K, McGuckin M. Combined inhibin and CA125 assays in the detection of ovarian cancer. Clin Chem 1999;45:651-658.

76. Lane AH, Lee MM, Fuller AF, Jr., Kehas DJ, Donahoe PK, MacLaughlin DT. Diagnostic utility of Mullerian inhibiting substance determination in patients with primary and recurrent granulosa cell tumors. Gynecol Oncol 1999;73:51-55.

77. Meyer JS, Rao BR, Valdes R, Jr., Burstein R, Wasserman HC. Progesterone receptor in granulosa cell tumor. Gynecol Oncol 1982;13:252-257.

78. Avizienyte E, Loukola A, Roth S, Hemminki A, Tarkkanen M, Salovaara R, Arola J, Butzow R, Husgafvel-Pursiainen K, Kokkola A, Jarvinen H, Aaltonen LA. LKB1 somatic mutations in sporadic tumors. Am J Pathol 1999;154:677-681.

79. Wang ZJ, Churchman M, Campbell IG, Xu WH, Yan ZY, McCluggage WG, Foulkes WD, Tomlinson IP. Allele loss and mutation screen at the Peutz-Jeghers (LKB1) locus (19p13.3) in sporadic ovarian tumours. Br J Cancer 1999;80:70-72.

80. Werness BA, Afify AM, Eltabbakh GH, Huelsman K, Piver MS, Paterson JM . p53, c-erbB, and Ki-67 expression in ovaries removed prophylactically from women with a family history of ovarian cancer. Int J Gynecol Pathol 1999;18:338-343.

81. Van dB, I, Dal Cin P, De Groef K, Michielssen P, Van den BH. Monosomy 22 and trisomy 14 may be early events in the tumorigenesis of adult granulosa cell tumor. Cancer Genet Cytogenet 1999;112:46-48.

82. Tanyi J, Rigo J, Jr., Csapo Z, Szentirmay Z. Trisomy 12 in juvenile granulosa cell tumor of the ovary during pregnancy. A report of two cases. J Reprod Med 1999;44:826-832.

83. Wise PM, Kashon ML, Krajnak KM, Rosewell KL, Cai A, Scarbrough K, Harney JP, McShane T, Lloyd JM, Weiland NG. Aging of the female reproductive system: a window into brain aging. Recent Prog Horm Res 1997;52:279-303.

84. Wise PM, Krajnak KM, Kashon ML. Menopause: the aging of multiple pacemakers. Science 1996;273:67-70.

85. Mosgaard BJ, Lidegaard O, Kjaer SK, Schou G, Andersen AN. Infertility, fertility drugs, and invasive ovarian cancer: a case- control study. Fertil Steril 1997;67:1005-1012.

86. Bristow RE, Karlan BY. Ovulation induction, infertility, and ovarian cancer risk. Fertil Steril 1996;66:499-507.

87. Willemsen W, Kruitwagen R, Bastiaans B, Hanselaar T, Rolland R. Ovarian stimulation and granulosa-cell tumour. Lancet 1993;341:986-988.

88. Rossing MA, Daling JR, Weiss NS, Moore DE, Self SG. Ovarian tumors in a cohort of infertile women. N Engl J Med 1994;331:771-776.

89. Venn A, Watson L, Lumley J, Giles G, King C, Healy D. Breast and ovarian cancer incidence after infertility and in vitro fertilisation. Lancet 1995;346:995-1000.

90. Unkila-Kallio L, Leminen A, Tiitinen A, Ylikorkala O. Nationwide data on falling incidence of ovarian granulosa cell tumours concomitant with increasing use of ovulation inducers. Hum Reprod 1998;13:2828-2830.

5

F₀ TRANSGENICS FOR STUDIES OF TRANSCRIPTIONAL CONTROL IN VIVO TISSUE AND DEVELOPMENTAL-SPECIFIC REGULATION OF THE HUMAN AND RAT GROWTH HORMONE/PROLACTIN/ PLACENTAL LACTOGEN GENE FAMILY

Peter A. Cattini and Mary Lynn Duckworth

University of Manitoba, Winnipeg, MB R3E 3J7 CANADA

INTRODUCTION

The highly specific and developmentally regulated expression of growth hormone (GH) and placental lactogen (PL) genes in the pituitary or placenta/decidua has naturally led to an interest in the DNA elements, transcription factors or mechanisms responsible for these patterns of synthesis. In the case of pituitary GH (GH-N) this has resulted in the identification of DNA sequences that dictate pituitary-specific expression. For PLs, the complexity of their pattern of expression, particularly the rodent genes, offers the opportunity not only to identify DNA sequences and associated factors that target the foetal/maternal placenta but also those that might restrict activity to a particular stage of development. The ability to target expression in a tissue- and developmental-specific manner could prove extremely useful for a variety of *in vivo* experiments.

The use of gene transfer *in vitro* has played an important role in the study of transcriptional regulation in general and the identification of tissue-specific DNA elements in particular. Specific gene fragments containing putative regulatory elements are tested for their ability to direct reporter gene expression in cell lines derived from tissues of interest. For the human and rat GH and PL genes, which are largely limited in expression to the pituitary somatotrophs and placental trophoblasts, this has usually involved the use of rat anterior pituitary tumor cells and human or rat choriocarcinoma cells, respectively. These experiments have met with varied success. In the case of the pituitary GH-N gene, DNA elements for pituitary cell-specific expression *in vitro* were identified in both human and rat promoter sequences. These sequences were not sufficient, however, to ensure efficient expression *in vivo*, at least for the human gene, hinting at further complexity. In studies of PL gene expression, available cell lines have allowed the identification of placental enhancer DNA-protein complexes in examples of human, rat and mouse genes. This list, however, is limited and does not reflect all the PL genes or their diverse patterns of gene expression observed during pregnancy.

Even when useful cell lines exist for *in vitro* studies, the nature of cell lines is such that they tend to provide little insight into the regulation of developmental expression of a specific gene. At best, they represent a snap shot of a particular stage of normal development and are of little use in studying the regulation of temporal aspects of gene expression. The generation of transgenic mice to study tissue-specific/developmental gene regulation is now a realistic and powerful complement to these *in vitro* approaches. Here we focus on the use of transgenics in combination with hybrid reporter genes directed by homologous and heterologous promoters to investigate basic regulatory mechanisms underlying spatial and temporal expression of the human GH-N and rodent prolactin-like PL gene families. In particular we discuss the usefulness of F_0 transgenics (so-called "transient transgenics") for the study of these genes. Emphasis is given to the approaches taken, including choice of reporter gene, promoter and mode of gene transfer and a comparison of their advantages and disadvantages are discussed.

THE GROWTH HORMONE AND PLACENTAL LACTOGEN GENE FAMILY

The human growth hormone (GH) and placental lactogen (PL, also known as chorionic somatomammotropin, CS) gene family consists of five genes located at a single locus on chromosome 17q22-24 (1). This locus is flanked upstream by the muscle SCN4A and lymphocyte-specific CD79b genes and downstream by the TRIP-1 and BAF60b genes (2). The five genes in the human GH-PL locus are arranged in the same transcriptional orientation in head to tail fashion and span 46,728 base pairs (bp). They are, from 5′ to 3′, pituitary GH-N (also known as GH1), PL-1 (also known as CS-L, CSHL1 and CS5), PL-4 (also known as CS-A, CSH1 and CS1), placental GH variant (GH-V or GH2) and PL-3 (also known as CS-B, CSH2 and CS2). The GH-

PL genes share greater than 90% nucleotide sequence similarity in their immediate flanking, intervening, and coding sequences and have clearly evolved by gene duplication from an ancestral GH-N gene (1). While the GH-N gene is present and expressed in the pituitary somatotrophs in all vertebrates, the GH-V and GH-like PL genes are found only in primates and expressed in the syncytiotrophoblast in the placenta.

Table 1. Rat Placental Prolactin Family

Rat placental protein	*Expression	Expressing cell types
Placental lactogen I (rPLI) (3)	Day 7 to 12	Primary and basal zone trophoblast giant cells.
Placental lactogen II (rPLII) (4,5)	Day 11 to term	Primary, basal zone and labyrinth trophoblast giant cells.
Prolactin-like protein B (rPLP-B) (5,6,7)	Day 7 to 13 Day 13 to term	Antimesometrial decidua. Spongiotrophoblasts.
Decidual/trophoblast prolactin-related protein (d/tPRP) (8,9)	Day 7 to 13 Day 13 to 16	Antimesometrial decidua Spongiotrophoblasts, basal zone giant cells.
Prolactin-like proteins A, C, Cv, Iv, D, H (rPLP's) (10-16)	Day 14 to term	Spongiotrophoblasts, basal zone trophoblast giant cells
Prolactin-like protein J (17)	Day 7 to 9	Decidual cells at implantation site.

* Day of mating is day 1.

Unlike the human PLs and their relationship with pituitary GH-N, the rodent PLs are more structurally and functionally related to pituitary prolactin (PRL). There is evidence accumulating, however, that like the human locus, rat and mouse PRL and PLs are linked on the same chromosome; chromosome 17 in the rat and chromosome 13 in the mouse (18,19). Analysis of a mouse YAC library has demonstrated that all known pituitary and placental members of the mouse PRL family except proliferin are located on a single 700 kilobase (kb) clone (19). All new PL members thus far identified are expressed in the developing placenta and, more rarely, the maternal antimesometrial decidua. PRL, itself, was recently shown to be expressed in this latter location in the rat (20). As outlined in Table 1, the placental members of the rat PRL family show very specific cellular and temporal expression patterns. The most common pattern is that shown by the rat PLP-A, -C, -Cv, -D, -E, H and PL-Iv genes which are expressed during the latter half of pregnancy in spongiotrophoblast cells and, to a much lesser extent, in giant cells of the basal zone. Other members, however, have unique expression patterns.

Cell Lines for *In Vitro* Analysis
Human GH/PL genes: The search for the mechanism underlying pituitary-specific GH-N promoter activity has been well served by the availability of rat anterior pituitary cell lines, such as GC and GH3, that are capable of expressing endogenous GH-N. Efforts to identify the minimal sequences required for pituitary-specific expression made use of hybrid reporter genes directed by the GH-N promoter together with increasing lengths of upstream flanking DNA (21,22). Promoter activity was then compared in pituitary tumor versus non-pituitary tumor cells after transient or stable gene transfer. The pituitary cell line of choice has been GC cells, which was first used in gene transfer studies with the human GH-N gene to investigate a mechanism of thyroid hormone regulation (23). With regard to pituitary-specific expression, most attention focused on fragments containing 500 bp or less of either human or rat GH-N 5′-flanking DNA. This was quickly justified by the observation that a nuclear homeodomain protein variously called Pit-1 or GHF-1 binds at two sites in this region, and directs efficient and pituitary-specific GH-N promoter activity in transfected GC cells (24,25). Subsequent cloning of Pit-1 (26,27), including from GC cells, and characterization revealed that it was not only required for expression of GH–N and PRL in the pituitary but also, more importantly, for normal pituitary development itself, affecting somatotrophs, mammotrophs and thyrotrophs (28,29).

Using monolayer cultures of human choriocarcinoma JEG-3 and BeWo cells with hybrid reporter genes and gene transfer, an efficient placenta-specific enhancer DNA element has been identified in the 3′-flanking of the PL-3 gene (30-34). This activity appears to require complex protein-DNA interactions involving a member or members of the transcriptional enhancer factor (TEF) family (35) but as yet not a placental equivalent to Pit-1 that is employed by GH-N and PRL in the pituitary. Regardless, these results suggest that all the necessary transcriptional components for efficient placental expression of the human PLs are present in these cells. It is unclear, however, why the same level of enhancer activity is not observed with equivalent PL-4 sequences (32,36). Human PL-4 is expressed at the same levels as PL-3 during early pregnancy and its levels have been reported to increase above those of PL-3 at term (1,37). In this regard the available human choriocarcinoma cell lines may offer poor models for the developing villous syncytiotrophoblast where these genes are normally expressed. It is worth noting that endogenous levels of PL RNA are extremely low in these human choriocarcinoma cells relative to placenta (38). Indeed, PL RNA can only be readily detected in JEG-3 cells by reverse transcriptase polymerase chain reaction (39). There is evidence to suggest that the PL-1, PL-3 and PL-4 enhancer regions become more active in primary human term placental cytotrophoblasts as they differentiate in culture (40); however, even here, we found endogenous levels of PL RNA expression to be very low (41).

Rodent PL genes: Initially the lack of good cell lines was also a major deterrent to regulation studies of the rat and mouse PRL-like PL genes. These genes are expressed variously in at least two cell types in the placenta - the giant cell and the spongiotrophoblast - as well as in cells in the early antimesometrial decidua. There was a reluctance to use the human choriocarcinoma lines for the rodent PRL-like genes, given the highly cell-specific nature of their expression. The development of the Rcho (42) and the related Rcho-1 (43) rat trophoblast cell lines in the early 1990s was an important advance which made *in vitro* transfection studies feasible for a large number of the PRL-like genes.

The Rcho cell lines grow as small rapidly proliferating cells that differentiate into the placental giant cell type. All the known rat genes that are expressed in this cell type are expressed to varying extents in the Rcho cell lines, suggesting that the appropriate transcription factors required for expression are present. These lines have been used to identify regulatory sequences of a number of the rat and mouse PRL-like genes as outlined in Table 2. The GG-AD antimesometrial decidual cell line (44), which expresses at least one member of the family, has not yet been extensively used for transfection studies. No cell lines currently exist to study spongiotrophoblast-specific expression.

Limitations of Cell Lines

In spite of the major advance that these cell lines represent for studying regulation of the pituitary and placental members of the GH-PL family, they have limitations when it comes to assessing spatial and, more serious, temporal control of gene expression. The problems associated with a limited number or lack of appropriate cell lines as well as use of cell lines in general, are perhaps best illustrated by attempts to study regulation of the placental members of the rodent PRL gene family. Rcho lines only express rat PRL family members normally seen in the placental giant cell. As shown in Table 1, several of the genes are highly expressed in spongiotrophoblasts and to a much lesser extent in giant cells. Only this lesser activation in giant cells can currently be studied in Rcho cell lines.

Several of the family members are transcribed exclusively in the spongiotrophoblast and/or the maternal antimesometrial decidua. The GG-AD rat decidual cell line expresses the rat PLP-B gene, but not the d/tPRP gene (44); whether it expresses the rat PLP-J gene has not yet been reported. This cell line could potentially be useful for defining regulatory elements responsible for expression of rat PLP-B in decidua, but would not be useful to study expression of this gene in the late term placental spongiotrophoblast cells. Identification of regulatory sequences in the d/tPRP gene, which is also expressed in these cell types, has been carried out using primary decidual and spongiotrophoblast cultures (48,49). These, however, are time-consuming to isolate from pregnant rat or mouse tissue and primary cell cultures are often more difficult to transfect than cell lines.

The most serious limitation of these cell lines is that they cannot be readily used to study the factors involved in regulating developmental changes that occur in the expression of these genes during pregnancy. These include changes in both time of expression and cell type. None of the Rcho cell lines have been reported to turn off transcription of the rat PLI gene, as occurs *in vivo*. Our original Rcho cultures expressed rat PLI as soon as giant cells differentiated in the population while rat PLII was only highly expressed about 12-14 days later, suggesting that these cells were continuing to differentiate in culture (53). This characteristic has proved to be difficult to maintain with continued passaging of the cells. Our interest in understanding the factors involved in the complete developmental expression of the rat PRL-like PL gene family, as well as pituitary-specific regulation of the human GH-N gene promoter *in vivo*, has led us to investigate the potential of transgenic mice in these studies as an alternative and/or complement to *in vitro* transfection studies.

Transgenic Mice for Regulation Studies
Many of the difficulties encountered by using cell lines for the study of cell-specific and temporal regulation of genes can be addressed by the use of transgenic mice. Transgenic mice will allow regulatory elements to be tested in all cells at all times during embryonic and postnatal development making them a valuable model for cell/tissue specific regulation studies. We and

Table 2. Summary of Regulation Studies on Rodent PRL-Related Placental Genes

Gene	Reporter	5′ DNA	*In Vitro*	*In Vivo*	Result/Expression
Mouse PLII (45)	SV40 large T antigen	569 bp 2.7 kb		2 TG lines 3 TG lines	nd* Placenta; possibly some in brain
Rat PLP-A (46)	CAT	975 bp 4.6 kb	Rcho GC Rcho GC		5X control nd 27X control nd
**Rat PLP-A	CAT Luc	975 bp 4.6 kb		6 TG lines 9 TG lines	nd Placenta; ectopic in brain
Mouse PLI (47)	CAT CAT CAT CAT	274 bp 2.7 kb 64 bp All fragments	Rcho-1 Rcho-1 Rcho-1 CHO, Cos-7 Mouse L cells		Active Active <274 nd nd
Rat d/tPRP (48,49)	Luc Luc Luc Luc Luc	3960 bp 3960 bp 3960 bp 3960 bp 93 bp	Rcho-1 Primary spongio-trophoblasts. Primary decidual cells. U1, CUSV2, V4 (uterine stroma). HRP-1, GH3, L929		Active Active Active Active<decidua nd Rcho-1 and decidual cells only
Rat PLI (50,**)	Luc β-gal	294bp, 1.4 kb 1.4 kb	Rcho	8 TG lines	Similarly active nd.
Rat PLII (50)	Luc Luc Luc	765bp 4.5 kb, 3.0 kb 4.5 kb, 3.0 kb	Rcho Rcho GC		nd Similarly active nd
Rat PLII (51)	Luc	3.0 kb		9 F$_0$ TG	Placenta; ectopic in some fetuses.
Mouse PLII (52)	CAT	2.0 kb		9 F$_0$ TG	Placenta; ectopic in some fetuses.

CAT, chloramphenicol acetyl transferase; Luc, Luciferase; β-gal, β-galactosidase; TG, transgenic; *nd, no detectable activity; ** unpublished data

others have noted that expression in a cell line does not guarantee appropriate expression in transgenic mice, suggesting that these animals are the most stringent test of the role of regulatory elements in authentic gene expression. For example, in 1986 we showed that 500 bp of human GH-N promoter sequences was able to confer pituitary-specific expression on a reporter gene in GC cells after gene transfer (22). It had been reported as early as 1984, however, that 500 bp or 5000 bp of GH-N 5′-flanking DNA was not sufficient to target or see efficient human GH-N gene expression in the pituitary of transgenic mice (Table 3). Interestingly, only 181 bp of rat GH-N upstream flanking DNA was sufficient to target reporter gene expression to the pituitary of transgenic mice (54). This raises the as yet unanswered question of whether the difference between the results obtained with the human versus rat promoter relates not so much to promoter strength but the sensitivity of the detection assay provided by the reporter gene employed. Regardless, the use of less than 7000 bp of human GH-N 5′-flanking DNA was not sufficient to yield a level of pituitary GH-N expression comparable to that observed with the endogenous mouse GH-N gene (55).

Table 3. Human GH-N Promoter Length and Pituitary Expression in Transgenic (TG) Mice

Promoter/Flanking DNA (bp)	Number TG Mice (expressing/total)	Expression of the GH-N Reporter Gene	Ref.
496	0/8	not detected	(56)
5,000	0/16	not detected	(56)
<7,000	4/9	low/inconsistent	(55)
22,500	4/4	high	(57)
41,500	5/5	high	(57)

The lack of efficient human GH-N promoter activity in the pituitary of transgenic mice comparable to that seen with equivalent rat GH-N gene sequences suggested that additional genetic information is required. Two likely, but not mutually exclusive, possibilities were requirements for a distal enhancer region similar to that seen in the human PRL gene (58) or a locus control region (LCR) as described for a limited number of systems including the globin gene family (55,59). The presence of either of these regulatory domains would be expected to affect chromatin organization and, experimentally, be reflected in the presence of deoxyribonuclease hypersensitive sites (HS) in the chromatin of pituitary nuclei. In a comprehensive series of experiments, Jones et al. (57) demonstrated the presence of two pituitary-specific HS (HS I and HS II) centred about 15 kb upstream of the human GH-N transcription initiation site in the 5′-flanking DNA of the adjacent B-lymphocyte-specific CD79b gene (60). Two additional hypersensitive sites HS III and HS V were located 27.5 and 32.0 kilobases upstream, respectively, in the skeletal muscle SCN4A gene, but were not pituitary-specific (57).

Assessment of human GH-N transgenes containing 5′-flanking DNA including HS I and II with or without HS III and HS V, resulted in efficient expression in the transgenic mouse pituitary (Table 3). When a fragment containing HS III and HS V but not HS I and HS II was tested, expression was predominantly in the kidney (57). It was concluded, however, that inclusion of HS III and HS V with HS I and HS II was necessary to obtain the functional GH-N LCR with both the ability to open chromatin and enhance transcription (56,57). It is also important to note that HS I and HS II were reconstructed in the transgenic mouse pituitary chromatin indicating the ability of the mouse transcriptional machinery to recognise and remodel the human genetic information.

This study in which a human GH-N transgene of 40-50 kb was employed, also illustrates another advantage of the transgenic approach. Unlike in transfection studies, where DNA uptake can be a problem, the size of the transgene is not an issue when making transgenic mice. Transgenes greater than 100 kb have been injected successfully into the pronucleus of one cell mouse embryos (61-63). This is a particularly important advantage when studying regulation of the human and rat GH-PL gene families where large gene loci are involved. It was essential for the characterization of the GH-N LCR in which it was necessary to use a human GH-N transgene with 41.5 kb of 5′-flanking DNA (57).

In contrast, to human GH-N, appropriate expression of human PL in the mouse placenta has not been observed (56,57). This may, as originally appears to be the case with GH-N, simply reflect the failure to include sufficient flanking DNA in the transgene. The situation with rat PLs is more promising as most of the PRL family members have now been shown to be conserved between the rat and the mouse, so that lingering doubts that rat genes might not express correctly in the mouse have now largely been dispelled.

Transgenic Mice - Other Considerations
The major drawback of using transgenic mice routinely for regulatory studies has been the requirement for specialised technical skills and equipment to carry out the injections, and the need for a well-regulated animal facility. Such facilities and expertise are more widely available now,

however, and this is becoming much less of an impediment. Nonetheless, although the production of transgenic mice is now feasible for many laboratories, a further disadvantage has been the length of time and high costs associated with developing and maintaining lines of mice for testing. This is an important issue when one wishes to carry out deletion analyses to map regulatory regions that are essential for tissue-specific expression - precisely how we wished to use transgenic mice for our studies of the human GH-N and placental members of the rat PRL gene family. To alleviate the high costs of these experiments, we decided to analyse founder transgenic embryos or " F_0 transgenics".

F_0 Transgenics

Founder transgenic embryos have been called "F_0" transgenics (64),"transient" transgenics (51,65) and "founder" transgenics (65,66). They represent the fetal/placental units that have developed directly from injected embryos and which are usually harvested prior to birth. We favor the term " F_0 transgenics". "Transient" could be misunderstood to mean transient expression, a term that may be more appropriately used in future for just such transgenics. The term "founder" can then be reserved to describe the original animal used to establish transgenic lines.

F_0 transgenics offer an important new approach to examine gene regulation *in vivo*, particularly for the study of embryonic and extraembryonic genes. A large number of constructs can be rapidly analyzed without the need for expensive and time consuming breeding programs. As with all transgenics, however, each animal is unique with regards to integration site and copy number, both of which can effect expression levels. It is essential that at least 10 to 20% of the surviving embryos be transgenic to ensure efficient use of animals for statistically meaningful data.

Although F_0 transgenic animals, similar to cell lines, represent only one developmental time, once important regulatory regions have been identified, these same constructs can be used to establish new founder lines for developmental studies.

Choosing a Reporter Gene

A number of reporter genes are now available for analysis of regulatory regions and elements, each with its advantages and disadvantages. The enzyme-assay based chloramphenicol acetyl transferase (*cat*) and luciferase reporter genes have been widely used for cell culture experiments. Latterly luciferase has been favored because of its ease of assay without radioactivity and its superior sensitivity to CAT (67). We and others have successfully used luciferase in F_0 transgenic studies as outlined in the next sections. Luciferase is easily assayed in tissue extracts, and its sensitivity makes it particularly useful for potentially weak promoters in the context of the whole animal. Its main disadvantages are that it is a relatively unstable protein, and that, unlike CAT, there are currently few if any good antibodies for immunohistochemical studies to determine cell-specific expression.

One of the most useful reporter genes for studying gene expression at the level of the cell, particularly during embryonic but also postnatal development, has been *E.coli* β-galactosidase (68,69). β-Galactosidase is assayed by a simple colourimetric enzyme assay using the chromophore substrate, 5-bromo-4-chloro-3-indolyl-β-D-galactopyranoside (X-gal) (65). Sensitivity, however, is not as great as luciferase and some tissues, notably placenta, express high levels of the mammalian form of this enzyme, making it less useful for the study of weak promoters. The recently discovered and extensively modified "green" fluorescent proteins show promise as cell-specific reporter genes, although they are only just beginning to be used for this purpose in transgenic animals (70,71). These proteins are particularly useful for visualizing gene expression in living cells. They have long half-lives and do not require the addition of substrates for visualization. A gene product combining the activities of both luciferase and green fluorescent protein (72) would provide an ideal reporter in many instances.

HOMOLOGOUS PROMOTER STUDIES OF THE PROLACTIN FAMILY PROTEINS EXPRESSED IN RODENT PLACENTA

Summary of Regulation Studies on PRL Family Genes
Although one of the most compelling features of the placental members of the rodent PRL gene family is their very specific developmental expression patterns, most regulation studies to date have been carried out only in cell lines. Table 2 summarizes gene regulation analyses on members of the rat and mouse gene families. The development of the Rcho cell lines has led to most studies focusing on the regulation of genes that have been shown to be expressed in these cells (PLP-A, PLI, PLII, rPLP-Cv and d/tPRP).

Regulation of the Rat PLP-A Gene
In vitro: We were among the first to use Rcho cells when we undertook to study elements important for regulating expression of the rPLP-A gene in the giant cell type (46). We showed that both a 975 bp and a 4.6 kb 5′ proximal fragment would direct expression of the *cat* reporter gene in Rcho

cells, but not in the rat pituitary GC cells. The larger fragment produced approximately five times higher level of CAT activity than the 975 bp fragment, suggesting that although both fragments were sufficient for directing placental giant cell expression there were further enhancing elements within the larger fragment.

In vivo: As a more stringent test of the placental specificity of these rPLP-A fragments we produced and bred lines of transgenic mice (Duckworth and Fresnoza, unpublished data). The same 975 bp 5′ rPLP-A/*cat* reporter gene construct used *in vitro* was also used to produce the transgenic mouse lines. Since it was not possible to isolate the 4.5 kb 5′/*cat* construct free of vector sequences, which could interfere with tissue-specific expression, we created a new luciferase construct for generating transgenics. Only the 4.5 kb fragment showed consistent reporter expression in placenta. In addition, we detected ectopic luciferase expression, particularly in the brain. A more detailed analysis showed that individual lines expressed in different regions of the brain probably as a result of position effects of the transgene insertion site. This fact alone suggests that further regulatory elements are required for authentic rPLP-A expression.

There could be at least two reasons for the apparent inability of the shorter rPLP-A 5′ fragment to direct expression in the whole placenta. The obvious one is lack of essential regulatory elements within the 975 bp. Another reason, however, could be the relative insensitivity of the *cat* reporter gene as compared to luciferase within the context of the whole placenta. Given the approximately five fold lower activity of the 975 bp construct compared to the 4.5 kb in the Rcho cells, the level of CAT expression may be below the level of detection of the assay in whole placental extract. An rPLP-A 975 bp 5′/luciferase construct remains to be tested in transgenic mice.

In situ hybridization analysis of placentas from the most highly expressing rPLP-A 5′/luciferase lines showed little if any specific hybridization to an antisense luciferase probe, in either the giant cell or the spongiotrophoblast. From the point of view of choice of a reporter gene for transgenic experiments, these experiments serve to illustrate both an advantage and disadvantage of using a reporter as sensitive as luciferase.

Regulation of the Placental Lactogen I Gene
In vitro: Others and we have demonstrated by *in situ* hybridization that there is a tightly controlled switch in expression between PLI and PLII that occurs at mid-pregnancy in the same giant cells of the basal zone (53). Transfection

studies in Rcho cells have been carried out for both the rat and mouse PLI genes (50,52). Results from both species indicate that less than 300 bp of proximal 5′ flanking DNA is sufficient to direct reporter gene expression (luciferase and *cat*, respectively). Sequences within this region are highly related between the rat and mouse, suggesting that similar regulatory elements are likely to be important in this expression (50).

In vivo: Rcho cell lines express rPLI as soon as differentiated giant cells appear in the culture (53). Unlike in the placenta, however, expression of rPLI is not turned off, even if cultures are maintained for long periods. To assess these regulatory elements for ability to direct expression in early versus late placenta, and to identify the cell type(s) in which placental expression occurred, we developed transgenic mouse lines using an rPLI 5′ 1.4 kb/β-galactosidase construct (Duckworth, Sun and Fresnoza, unpublished data). This fragment expressed similarly to the shorter fragment in Rcho cells (50). We were unable to detect reporter expression in any cell type in the placentas of several lines of transgenic mice. This result may again be due to the lower sensitivity of the β-galactosidase reporter, but it serves to illustrate an important point. The testing of regulatory elements in transgenic animals provides a more stringent examination of the importance of those elements not only for location and time, but also level of expression, than do *in vitro* experiments.

Regulation of the PLII Gene

In vitro: Our most recent work has focused on understanding the regulatory elements required for the placental-specific expression of the rPLII gene. Based on experience with the rPLP-A and rPLI genes, we chose the sensitive luciferase gene as our reporter. Preliminary studies carried out in Rcho cells indicated that both approximately 4.5 and 3.0 kb of proximal 5′ flanking DNA would direct expression of the luciferase reporter gene in Rcho cells, but that 765 bp of proximal 5′ flanking DNA was insufficient (51). None of the constructs were active in the rat pituitary GC cell line, suggesting that elements on these fragments were able to regulate placental-specific expression.

In Vivo Studies Using F_0 Transgenics

To test the ability of the 3.0 kb fragment rPLII 5′ flanking fragment to direct placental-specific expression during pregnancy we used the luciferase construct to develop F_0 transgenic mice. The decision to use F_0 transgenics in this instance was based on previous experience with the time-consuming and expensive development of transgenic lines for studying rPLP-A and rPLI, and our ability to produce at least 20% transgenic animals from injected embryos. Table 4 shows the results of these experiments. The information in this table was first presented in Shah et al. (51) and is reprinted here with the permission of The Endocrine Society. Out of a total of 45 fetal/placental units analysed at day 14 to 16, nine were transgenic. All transgenic placentas expressed luciferase. Expression levels in placenta varied widely and in the case of four of the embryos, varying amounts of luciferase activity were detected in at least one of the three general regions of the body. These results indicated more rigorously than the transfection experiment results that elements within the 3.0 kb rPLII 5′ flanking fragment were important in placental expression but also suggested that further elements, presumably elsewhere in the gene, were also required for placental-specific expression. We have gone on to identify an enhancing element within this region at -1793 to -1729 and are currently testing the role of this sequence in placental expression (73).

Table 4. Luciferase Activity in rPLII 5′ Flanking F_0 Transgenic Placentas and Fetuses

Transgenic Placenta	Day of Pregnancy[a]	Luciferase Activity (units/mg protein X 10^3)[b]			
		Placenta	Head	Thorax	Abdomen
P1	Day 14	33.6	nd	nd	nd
P2		0.77	0.3	0.2	0.76
P3	Day 15	3.6	nd	nd	nd
P4		70.6	0.07	0.04	nd
P5*		0.13	nd	nd	nd
P6		2.8	nd	nd	nd
P7		0.9	3.3	nd	nd
P8	Day 16	5.2	nd	nd	nd
P9		68.5	447	759	1186

[a] Day 0 is the day of mating
[b] nd - no detectable luciferase activity.
* Indicates a fetus which appeared to carry <1 copy of the transgene, potentially a mosaic.

F_0 transgenics have also been used to assess a 2.0 kb proximal 5′ flanking region of the mouse PLII gene (52). For these studies *cat* was used as the reporter gene. All transgenic placentas expressed CAT, and in all cases the placental expression was greater than total fetal expression. Rat and mouse PLII 5′ flanking sequences are highly related and enhancing activity has been localized to essentially the same region in the two species. Interestingly, this region contains a consensus Ets binding site in the rPLII sequence, but not in the mouse; nonetheless, this same region has been shown by mutagenesis experiments to be essential for enhancing activity. These differences remain to be resolved.

How Else Might We Use F_0 Transgenics?

It still remains to be determined whether the rPLII 5′ enhancing element identified in our studies is sufficient to direct placental-specific expression of a reporter gene. This type of question has most frequently been addressed using *in vitro* transfection studies. As outlined in the following section, F_0 transgenic mice have now proven to be powerful model systems for dissecting the tissue-specific activity of small, defined regulatory elements of the human GH-N gene when used in conjunction with heterologous promoters.

CHARACTERIZATION OF A REGULATORY ELEMENT USING A HETEROLOGOUS PROMOTER AND F_0 TRANSGENIC MICE

What is a Heterologous Promoter?

A heterologous promoter is a sequence capable of driving transcription with the goal of (i) promoting expression of a foreign (not of the same origin as the promoter) sense or antisense DNA fragment, or (ii) testing a foreign DNA fragment for activity independent of its own (homologous) promoter. Although viral promoters such as Herpes-simplex thymidine kinase (TK), Simian virus 40 (SV40), Rous sarcoma virus (RSV) and cytomegalovirus (CMV) are often used as heterologous promoters, eukaryotic gene promoters also serve as heterologous promoters.

Why Use a Heterologous Promoter?

Using a viral promoter as a heterologous promoter to drive expression of a DNA fragment is most often done to allow expression in a wider range of cell types and/or to generate higher levels of expression than could be provided by the homologous promoter. The CMV and RSV promoters are very potent and commonly used for this purpose. At the other end of the spectrum, a tissue-specific promoter can be used as a heterologous promoter to target/limit expression of a foreign DNA. Some advantages and disadvantages of using a heterologous versus homologous promoter to test a fragment for an effect on promoter activity in cell transfection studies are presented in Table 5.

Table 5. Using Homologous Versus Heterologous Promoters to Test Putative Regulatory DNA Elements for Activity

Homologous		Heterologous	
Advantages	Disadvantages	Advantages	Disadvantages
Activity can be limited to cells of specific tissue origin (eg, allowing a tissue-specific element to be localised).	Activity can be limited to cells of specific tissue origin (appropriate cell line might not be available).	Promoters that are active in multiple cell types/lines can be used to test putative elements.	The promoter activity might be too strong to be affected by the putative regulatory sequences
Assessment is done on most biologically relevant promoter.	Efficient expression may require flanking DNA in excess of 1000 bp.	Promoters can be relatively short, less than 100 bp.	The activity might be promoter specific
	The presence and interaction of multiple regulatory sites might make it difficult to dissect one activity from another in the promoter region	Putative regulatory elements can be tested individually or in combinations in an attempt to reconstitute an activity.	

There are three main reasons to use a heterologous promoter:

(i)   The first is that promoter activity can be ensured in almost any cell with a relatively short DNA fragment/promoter. A commonly used TK promoter includes only 81 bp of upstream flanking DNA (74).

(ii)   Secondly, an attempt can be made to test a DNA fragment for possible enhancer or repressor function even when no appropriate cell line is available to allow expression of the homologous promoter.

(iii)   Thirdly, if it is suspected that a regulatory region contains multiple elements, they can be tested individually or in combination for independent function.

It was for this latter reason that we used a heterologous promoter to dissect multiple regulatory regions in the human GH-N upstream flanking DNA and assess a remotely located sequence for independent enhancer function.

Will Heterologous Promoters Behave the Same in Transgenic Mice as in Cell Cultures After Gene Transfer?

The answer to this is clearly no. As already reviewed, a 500 bp human GH-N promoter will preferentially direct expression of a gene in pituitary tumor cells in culture (4) but would appear to be a poor choice to target expression of the same gene by pituitary somatotrophs in transgenic mice (56,57). Even viral promoters that are commonly held to be ubiquitously and equally expressed in cells of different tissue origins in culture, display more restrictive patterns of expression in transgenic mouse tissues. For example, the RSV promoter that is active in a wide range of cells in culture (4) is largely limited to striated muscle cells in transgenic mice (75-77). Similarly, the CMV promoter and its enhancer appear to be a poor choice for overexpression of a gene product in various tissues, including lung, liver and pancreas, of transgenic mice (78).

Identification of Remotely as well as Proximally Located Pit-1 Elements in the Human GH-N Locus

As indicated above (in "Transgenic Mice for Regulation Studies"), inclusion of sequences located about 15 kb upstream of the human GH-N transcription start site and containing HS I and HS II in the transgene was necessary for efficient GH-N promoter activity in transgenic mouse pituitary. A 1.6 kb fragment containing HS I and HS II can be isolated conveniently by *Bgl*II digestion (1.6G) and was shown to enhance the activity of a minimal human GH-N promoter (-496/+1) *in vitro* (66) as well as be a requirement for efficient pituitary-specific expression *in vivo* (57). We isolated and sequenced the 1.6G fragment (nucleotide designation 1-1605, Genbank accession number AF010280) and identified multiple A/T-rich sequences resembling Pit-1-like DNA elements, that we showed were capable of binding Pit-1 (66). This result has subsequently been confirmed by others (79). We went on to isolate the Pit-1 elements in a 203 bp subfragment of the 1.6G fragment (nucleotides 1298-1500) and tested it for enhancer activity in transiently transfected pituitary tumor GC cells. The 203 bp subfragment displayed about 90% of the enhancer activity observed with the full-length 1.6G fragment (66). The question we then asked was whether the 203 subfragment as well as the 1.6G fragment, containing HS I and HS II and Pit-1 elements, could function as an enhancer in the pituitary *in vivo* in the absence of the GH-N promoter. More specifically, can the Pit-1 elements and the region containing HS I and HS II act as a pituitary-specific enhancer independent of the Pit-1 elements located in the proximal (200 bp) promoter region?

Selection of the TK Promoter to Assess an Enhancer Element in Transgenic Mice

To answer this question, the 1.6G fragment had to be tested upstream of a promoter capable of expression in the transgenic mouse pituitary but without a requirement for Pit-1. As such, candidate promoters would include those expected to be ubiquitously expressed and, based on sequence analysis at least, which did not possess a putative Pit-1 element. It was also important to select a promoter that if expressed at all, was expressed at a sufficiently low level to be positively influenced by the presence of a stimulatory DNA element. This was important since our goal was to assess the 1.6G fragment for independent enhancer activity. The reverse applies if a possible repressor element is assessed. In this case care would be taken to select a sufficiently active promoter in the appropriate tissue/cell of interest. We selected the minimal TK promoter (-81/+52) because (i) it is relatively weak compared to other promoters suggesting that any stimulation would likely be observed and (ii) its small size was expected to reduce complications resulting from the presence and possible contribution of additional regulatory elements. The firefly luciferase gene was selected as the reporter gene for reasons of sensitivity and because the amount of fetal tissue available for sampling, for example, the pituitary itself, would be limiting (see "Choosing a Reporter Gene" for a discussion of reporter selection).

F_0 transgenic mice were generated initially using the minimal TK promoter and luciferase (TK-luciferase) gene as a transgene to confirm low levels of "background" promoter activity in the pituitary. Fetal tissues were harvested at embryonic day 19 or term. Little or no luciferase activity was detected in the pituitary and other tissues screened from TK-luciferase mice when compared with levels in non-transgenic mice (Table 6). Luciferase activity was standardised to protein. Much of the information presented in Table 6 was first presented in reference (66) and is reprinted with permission from The Endocrine Society.

Assessment of the 1.6G Fragment Sequences for Enhancer Activity in F₀Transgenic Mice

F_0 transgenic mice were generated using the 1.6G fragment (nucleotides 1-1605) ligated upstream of the TK-luciferase gene as a transgene. The 1.6G fragment, containing HS I and HS II as well as Pit-1 binding sites, was able to stimulate TK promoter activity efficiently in the pituitary (Table 6), and to a lesser extent in brain and testis, of F_0 transgenic mice. From a methodological point of view the results of this analysis show that:

(i)   the TK promoter in combination with the luciferase reporter gene can be used to assess a fragment of DNA for possible enhancer activity (when inserted upstream); and

(ii) F_0 transgenic mice offer a viable method of assessing a DNA fragment for an effect on promoter activity *in vivo*.

Once identified, it is possible to use F_0 transgenics in combination with more sophisticated methods such as deletion and site-directed mutagenesis to further characterise a functional element. For example, we ligated 203 bp (nucleotides 1298-1500) or 82 bp (nucleotides 1344-1425) subfragments of the 1.6G fragment upstream of the TK-luciferase gene and used these hybrid genes to generate F_0 transgenic mice. The 203 bp subfragment contains all the characterised Pit-1 binding sites, and was able to stimulate TK promoter activity efficiently in the mouse pituitary (Table 6). In contrast, the 82 bp subfragment is deleted of the Pit-1-like element at nucleotide position 1431-1438, and no pituitary enhancer activity was observed (Table 6).

Table 6. Luciferase (Luc) Reporter Gene Activity in F_0 Transgenic (TG) Mouse Pituitary.

Length and Region of the 1.6G (GH-N LCR) Fragment (nucleotides 1-1605) Tested	Number of TG Mice	Pituitary Specific Luc Gene Expression
none (minimal TK promoter alone)	2	not detected
1605 bp (nucleotides 1-1605)	3	high
203 bp (nucleotides 1298-1500)	5	high (4)/low (1)
82 bp (nucleotides 1344-1425)	2	not detected
M1/M3-203 bp (nucleotides 1298-1500 but 1371-1378 and 1436-1441 mutated)	3	not detected
M3-203 bp (nucleotides 1298-1500 but 1436-1441 mutated)	3	not detected

Site-directed mutagenesis was done to assess the contribution of Pit-1 sequences to the pituitary enhancer activity *in vivo*. The Pit-1 consensus element has been described as 5-WWTATNCAT-3 (80). The M1 mutation included modification of the Pit-1-like element 5'-ATGTTTATA-3' at nucleotides 1368-1376 to 5'-ATGgcggcc-3', and the M3 mutation included modification of the Pit-1-like element 5'-TTTTTCAT-3' at nucleotides 1431-1438 to 5'-TTTTTTgtc-3' (66). F_0 transgenic mice were generated using the 203 bp fragment containing the M1 and M3 mutations or the M3 mutation alone. Six of the seven F_0 transgenic mice generated using the 203 bp subfragment with double (M1 and M3) or single (M3) mutations displayed only background levels of activity, comparable to that seen using the minimal TK promoter alone, in the pituitary as well as other tissues (Table 6).

Taken together with the results of our other studies (66), these observations show that:

(i)   the 1.6G fragment and, more specifically, sequences in the 203 bp subfragment, constitute a distal enhancer located in the human GH-N LCR with preferential pituitary activity in F_0 transgenic mice; and

(ii)   this enhancer activity is related to multiple A/T-rich sequences capable of binding the pituitary-specific factor Pit-1.

FINAL COMMENTS

The rPLII F_0 transgenic experiments were completed in less than three months, a fraction of the time that would have been spent on identifying founders and developing transgenic lines. They tested, in the most stringent manner, the ability of the 3.0 kb 5' flanking region to reproduce at least some of the aspects of rPLII expression - placental specificity and level of expression. The F_0 transgenics represent a powerful tool to identify and dissect tissue specific regulatory regions. They will be particularly useful for analyzing the regulation of the rodent PRL family members that are expressed in placental spongiotrophoblasts and antimesometrial decidual cells during pregnancy and for which there are no cell lines or limited cell lines available.

The identification of the pituitary-specific factor Pit-1 as a key component of the LCR for GH-N has significantly increased our understanding of the expression and regulation of this important hormone. F_0 transgenics were key to the success of this venture, providing the means to observe preferential pituitary-enhancer activity and as a result, to localize the distal enhancer region. Further characterization of this enhancer and additional upstream regions, marked by hypersensitive sites that make up the GH-N LCR, can now be attempted in an economically viable and timely manner. We can look forward to the use of F_0 transgenics in combination with homologous and heterologous promoters in the identification and characterization of minimal sequences required for integration-independent as well as pituitary-enhanced GH-N synthesis. With this approach, there is also the hope of assessing sequences *in vivo* that have been implicated in human PL gene expression but until now have been limited to experiments in culture or by economics.

Clearly the approaches described are not limited to the expression and regulation of the GH-N and PL genes but are widely applicable providing the opportunity for an explosion of data to be generated in both endocrine and non-endocrine systems alike.

REFERENCES

1. Chen EY, Liao YC, Smith DH, Barrera-Saldãna HA, Gelinas RE, Seeburg PH. The human growth hormone locus: nucleotide sequence, biology, and evolution. Genomics. 1989;4:479-497.
2. Surabhi RM, Daly LD, Cattini PA. Evidence for evolutionary conservation of a physical linkage between the human BAF60b, a subunit of SWI/SNF complex, and thyroid hormone receptor interacting protein-1 genes on chromosome 17. Genome. 1999;42:545-549.
3. Robertson MC, Croze F, Schroedter IC, Friesen HG. Molecular cloning and expression of rat placental lactogen I complementary deoxyribonucleic acid. Endocrinology 1990;127:702-710.
4. Duckworth ML, Kirk KL, Friesen HG. Isolation and identification of a cDNA clone of rat placental lactogen II. J Biol Chem 1986;261:10871-10878.
5. Duckworth ML, Schroedter IC, Friesen HG. Cellular localization of rat placental lactogen II and rat prolactin-like proteins A and B by in situ hybridization. Placenta 1990;11:143-155.
6. Duckworth ML, Peden LM, Friesen HG. A third prolactin-like protein expressed by the developing rat placenta: complementary deoxyribonucleic acid sequence and partial sequence of the gene. Mol Endcrinol 1988;2:912-920.
7. Croze F, Kennedy TG, Schroedter IC, Friesen HG. Expression of rat prolactin-like protein B in deciduoma of pseudopregnant rat and in decidua during early pregnancy. Endocrinology 1990;127:2665-2672.
8. Roby KF, Deb S, Gibori G, Szpirer C, Levan G, Kwok SCM, Soares MJ. Decidual prolactin-related protein. J Biol Chem 1993;268:3136-3142.
9. Orwig K, Dai G, Rasmussen CA, Soares MJ. Decidual/trophoblast prolactin-related protein: characterization of gene structure and cell-specific expression. Endocrinology 1997;138:2491-2500.
10. Duckworth ML, Peden LM, Friesen HG. Isolation of a novel prolactin-like cDNA clone from developing rat placenta. J Biol Chem 1986;261:10879-10884.

11. Deb S, Roby KF, Faria TN, Szpirer C, Levan G, Kwok SCM, Soares MJ. Molecular cloning and characterization of prolactin-like protein C complementary deoxyribonucleic acid. J Biol Chem 1991;266:23027-23032.

12. Dai G, Liu B, Szpirer C, Levan G, Kwok SCM, Soares MJ. Prolactin-like protein-C variant: complementary deoxyribonucleic acid, unique six exon structure, and trophoblast cell-specific expression. Endocrinology 1996;137:5009-5019.

13. Robertson MC, Schroedter IC, Friesen MC. Molecular cloning and expression of rat placental lactogen-Iv, a variant of rPL-I present in late pregnant rat placenta. Endocrinology 1991;129:2746-2756.

14. Deb S, Faria TN, Roby KF, Larsen D, Kwok SCM, Talamantes F, Soares MJ. Identification and characterization of a new member of the prolactin family, placental lactogen-I variant. J Biol Chem 1991;266:1605-1610.

15. Iwatsuki K, Shinozaki M, Hattori N, Hirasawa K, Itagaki S, Shiota K, Ogawa T. Molecular cloning and characterization of a new member of the rat placental prolactin (PRL) family, PRL-like protein D (PLP-D). Endocrinology 1996;137:3849-3855.

16. Iwatsuki K, Oda M, Sun W, Tanaka S, Ogawa T, Shiota K. Molecular cloning and characterization of a new member of the rat placental prolactin (PRL) family, PRL-like protein H. Endocrinology 1998;139:4976-4983.

17. Toft DJ, Linzer DI. Prolactin (PRL)-like protein J, a novel member of the PRL/growth hormone family is exclusively expressed in maternal decidua. Endocrinology 1999;140:5095-5101.

18. Soares MJ, Müller H, Orwig KE, Peters TJ, Dai G. The Uteroplacental Prolactin Family and Pregnancy. Biol Reprod 1998;58:273-284.

19. Lin J, Poole J, Linzer DI. Two novel members of the prolactin/growth hormone family are expressed in the mouse placenta. Endocrinology 1997;138:5535-5540.

20. Prigent-Tessier A, Tessier C, Hirosawa-Takamori M, Boyer C, Ferguson-Gottschall S, Gibori G. Rat decidual prolactin: identification, molecular cloning, and characterization. J Biol Chem 1999;53:37982-37989.

21. Nelson C, Crenshaw EB 3rd, Franco R, Lira SA, Albert VR, Evans RM, Rosenfeld MG. Discrete cis-active genomic sequences dictate the pituitary cell type-specific expression of rat prolactin and growth hormone genes. Nature 1986;322:557-562.

22. Cattini PA, Peritz L, Anderson TR, Baxter JD, Eberhardt NL. The 5′-flanking sequences of the human growth hormone gene contain an element responsible for cell-specific expression. DNA 1986;5:503-509.

23. Cattini PA, Anderson TR, Mellon PL, Baxter JD, Eberhardt NL. The human growth hormone gene is negatively regulated by triiodothyronine when transfected into rat pituitary tumor cells. J Biol Chem 1986;261:13367-13372.

24. Lefevre C, Imagawa M, Dana S, Grindlay J, Bodner M, Karin M. Tissue-specific expression of the human growth hormone gene is conferred in part by the binding of a specific trans-acting factor. EMBO J 1987;6:971-981.

25. West BL, Catanzaro DF, Mellon SH, Cattini PA, Baxter JD, Reudelhuber TL. Interaction of a tissue-specific factor with an essential rat growth hormone gene promoter element. Mol Cell Biol 1987;7:1193-1197.

26. Bodner M, Castrillo JL, Theill LE, Derinck T, Ellisman M, Karin M. The pituitary-specific transcription factor GHF-1 is a homeobox-containing protein. Cell 1988;55:505-518.

27. Ingraham HA, Chen R, Mangalam HJ, Elsholtz HP, Flynn SE, Lin CR, Simmons DM, Swanson L, Rosenfeld MG. A tissue-specific transcription factor containing a homeodomain specifies a pituitary phenotype. Cell 1988;55:519-529.

28. Li S, Crenshaw EB 3rd, Rawson EJ, Simmons DM, Swanson LW, Rosenfeld MG. Dwarf locus mutants lacking three pituitary cell types result from mutations in the POU-domain gene pit-1. Nature 1990;347:528-533.

29. Simmons DM, Voss JW, Ingraham HA, Holloway JM, Broide RS, Rosenfeld MG, Swanson LW. Pituitary cell phenotypes involve cell-specific Pit-1 mRNA translation and synergistic interactions with other classes of transcription factors. Genes Dev 1990;4:695-711.

30. Rogers BL, Sobnosky MG, Saunders GF. Transcriptional enhancer within the human placental lactogen and growth hormone multigene cluster. Nucl Acids Res 1986;14:7647-7659.

31. Walker WH, Fitzpatrick SL, Saunders GF. Human placental lactogen transcription enhancer. J Biol Chem 1990;265:12940-12948.

32. Jacquemin P, Oury C, Peers B, Morin A, Belayew A, Martial JA. Characterization of a strong tissue-specific enhancer downstream from the three human genes encoding placental lactogen. Mol Cell Biol 1994;14:93-103.

33. Lytras A, Cattini PA. Human chorionic somatomammotropin gene enhancer activity is dependent on the blockade of a repressor mechanism. Mol Endocrinol 1994;8:478-489.

34. Jiang S-W, Eberhardt NL. The human chorionic somatomammotropin gene enhancer is composed of multiple DNA elements that are homologous to several SV40 enhansons. J Biol Chem 1994;269:10384-10392.

35. Jacquemin P, Martial JA, Davidson I. Human TEF-5 is preferentially expressed in placenta and binds to multiple functional elements of the human chorionic somatomammotropin-B gene enhancer. J Biol Chem 1997;272:12928-12937.

36. Lytras A, Surabhi RM, Zhang JF, Jin Y, Cattini PA. "Repair" of the chorionic somatomammotropin-A "enhancer" region reveals a novel functional element in the chorionic somatomammotropin-B enhancer. Mol Cell Endocrinol 1996;119:1-10.

37. MacLeod JN, Lee AK, Liebhaber SA, Cooke NE. Developmental regulation and alternative splicing of the placentally expressed transcripts from the human growth hormone gene cluster. J Biol Chem 1992;267:14219-14226.

38. Nickel BE, Cattini PA. Tissue-specific expression and thyroid hormone regulation of the endogenous placental growth hormone variant and chorionic somatomammotropin genes in a human choriocarcinoma cell line. Endocrinology 1991;128:2353-2359.

39. Lytras A, Bock ME, Dodd, JG, Cattini PA. Detection of placental growth hormone variant and chorionic somatomammotropin RNA expression in human trophoblastic neoplasms by reverse transcriptase-polymerase chain reaction. Endocrinology.. 1994;134:2461-2467.

40. Jacquemin P, Alsat E, Oury C, Belayew A, Muller M, Evain-Brion D, Martial JA. The enhancers of the human placental lactogen B, A, and L genes: progressive activation during in vitro trophoblast differentiation and importance of the DF-3 element in determining their respective activities. DNA Cell Biol 1996;15:845-854.

41. Klassen ME, Nachtigal MW, Cattini PA. Human chorionic somatomammotropin gene expression in primary placental cell cultures. Placenta 1989;10:321-329.

42. Verstuyf A, Sobis H, Goebels J, Fonteyn E, Cassiman JJ, Vandeputte M. Establishment and characterization of a continuous in vitro line from a rat choriocarcinoma. Int J Cancer 1990;45:752-756.

43. Faria TN, Soares MJ. Trophoblast cell differentiation:establishement, characterization, and modulation of a rat trophoblast cell line expressing members of the placental prolactin family. Endocrinology 1991;129:2895-2906.

44. Srivastava RK, Gu Y, Zilberstein M, Ou JS, Mayo KE, Chou JY, Gibori G. Development and characterization of a simian virus 40-transformed, temperature-sensitive rat antimesometrial decidual cell line. Endocrinology 1995;136:1913-9.

45. Shida MM, Jackson-Grusby LL, Ross, SR, Linzer DIH. Placental-specific expression from the mouse placental lactogen II gene promoter. Proc Natl Acad Sci USA 1992;89:3864-3868.

46. Vuille J-C, Cattini PA, Bock, ME, Verstuyf A, Schroedter IC, Duckworth ML, Friesen HG. Rat prolactin-like protein A partial gene and promoter structure: promoter activity in placental and pituitary cells. Mol Cell Endocrinol 1993;96:91-98.

47. Shida MM, Ng Y-K, Soares MJ, Linzer DIH. Trophoblast-specific transcription from the mouse placental lactogen-I gene promoter. Mol Endocrinol 1993;7:181-188.

48. Orwig KE, Dai G, Rasmussen CA, Soares MJ. Decidual/trophoblast prolactin-related protein: characterization of gene structure and cell-specific expression. Endocrinology 1997;138:2491-2500.

49. Orwig K, Soares MJ. Transcriptional activation of the decidual/trophoblast prolactin-related protein gene. Endocrinology 1999;1404032-4039.

50. Sun Y, Robertson MC, Duckworth ML. The effects of epidermal growth factor /transforming growth factor α on the expression of placental lactogen I and II in a rat choriocarcinoma cell line. Endocr J 1998;45:297-306.

51. Shah P, Sun Y, Szpirer C, Duckworth ML. Rat placental lactogen II gene: characterization of gene structure and placental-specific expression. Endocrinology 1998;139:967-973.

52. Lin J, Linzer DIH. Identification of trophoblast-specific regulatory elements in the mouse placental lactogen II gene. Mol Endocrinol 1998;12:418-427.

53. Duckworth ML, Robertson MC, Schroedter IC, Szpirer C, Friesen HG. Molecular genetics and biology of the rat placental prolactin family. In: Trophoblast cells: pathways for maternal-embryonic communication. (Eds: Soares MJ, Handwerger S, Talamantes F), pp169-190. New York: Springer-Verlag, 1993.

54. Lira SA, Crenshaw EB 3rd, Glass CK, Swanson LW, Rosenfeld MG. Identification of rat growth hormone genomic sequences targeting pituitary expression in transgenic mice. Proc Natl Acad Sci U S A 1988;85:4755-4759.

55. Cooke NE, Liebhaber SA. Molecular biology of the growth hormone-prolactin gene system. In: Vitamins and Hormones. (Ed: Litwack G) 50:385-459. San Diego: Academic Press, 1995.

56. Hammer RE, Palmiter RE, Brinster RL. Partial correction of murine hereditary growth disorder by germ-line incorporation of a new gene. Nature 1984;370:65-67.

57. Jones BK, Monks BR, Liebhaber SA, Cooke NE. The human growth hormone gene is regulated by a multicomponent locus control region. Mol Cell Biol 1995;15:7010-7021

58. Peers B, Voz ML, Monget P, Mathy-Hartert M, Berwaer M, Belayew A, Martial JA. Regulatory elements controlling pituitary-specific expression of the human prolactin gene. Mol Cell Biol 1990;10:4690-4700.

59. Grosveld F, van Assendelft B, Greaves DR, Kollias G. Position-independent, high level expression of the human β-globin gene in transgenic mice. Cell 1987;51:975-985.

60. Bennani-Baiti IM, Asa SL, Song D, Iratni R, Liebhaber SA, Cooke NE. DNase I-hypersensitive sites I and II of the human growth hormone locus control region are a major developmental activator of somatotroph gene expression. Proc Natl Acad Sci USA 1998;95:10655-10660.

61. Lakshmanan G, Lieuw KH, Grosveld F, Engel JD. Partial rescue of GATA-3 by yeast artificial chromosome transgenes. Dev Biol 1998;204:451-463.

62. Huxley C. Exploring gene functions: use of yeast artificial chromosome transgenesis. Methods 1998;14:199-210.

63. Schedl A, Larin Z, Montoliu L, Thies E, Lelsey G, Lehrach H, Schütz G. A method for the generation of YAC transgenic mice by pronuclear microinjection. Nucl Acids Res 1993;21:4783-4787.

64. Calzonetti T, Stevenson L, Rossant J. A novel regulatory region is required for trophoblast-specific transcription in transgenic mice. Dev Biol 1995;171:615-626.

65. Gilthorpe JD, Rigby PWJ. Reporter genes for the study of trancriptional regulation in transgenic mouse embryos. In: Methods in Molecular Biology. (Eds: Sharpe PT, Mason I) Vol 37: Molecular Embryology. 1999;159-182.

66. Jin Y, Surabhi RM, Fresnoza A, Lytras A, Cattini PA. A role for A/T-rich sequences and Pit-1/GHF-1 in a distal enhancer located in the human growth hormone locus control region with preferential pituitary activity in culture and transgenic mice. Mol Endocrinology 1999;13:1249-1266.

67. Alam J, Cook J. Reporter genes: application to the study of mammalian gene transcription. Anal Biochem 1990;188:245-254.

68. Goring DR, Rossant J, Clapoff S, Breitman ML, Tsui L-C. In situ detection of β-galactosidase in lenses of transgenic mice with a γ-crystallin/lacZ gene. Science 1987;235:456-458.

69. Whiting J, Marshall H, Cook M, Krumlauf R, Rigby PWJ, Stott D, Allemann RK. Multiple spatially specific enhancers are required to reconstruct the pattern of Hox-2.6 gene expression. Genes Dev 1991;5:2048-2059.

70. Zhuo L, Sun B, Zhang C-L, Fine A, Chiu S-Y, Messing A. Live astrocytes visualized by green fluorescent protein in transgenic mice. Dev Biol 1997;187:36-42.

71. Spergel DJ, Krüth U, Hanley DF, Sprengel R, Seeburg PH. GABA- and glutamate-activated channels in green fluorescent protein-tagged gonadotropin-releasing hormone neurons in transgenic mice. J Neurosci 1999;19:2037-2050.

72. Day RN, Kawecki M, Berry D. Dual-function reporter protein for analysis of gene expression in living cells. BioTechniques 1998;25:848-856.

73. Sun Y, Duckworth ML. Identification of a placental-specific enhancer in the rat placental lactogen II gene that contains binding sites for members of the Ets and AP-1 (activator protein-1) families of transcription factors. Mol Endocrinol 1999;13:385-399.

74. Nordeen S. Luciferase reporter gene vectors for analysis of promoters and enhancers BioTechniques 1988;6:454-458.

75. Jackson T, Allard MF, Sreenan CM, Doss LK, Bishop SP, Swain JL. The c-myc proto-oncogene regulates cardiac development in transgenic mice. Mol Cell Biol 1990;10:3709-3716.

76. Conti FG, Powell R, Pozzi L, Zezze G, Faraggiana T, Gannon F, Fabbrini A. A novel line of transgenic mice (RSV/LTR-bGH) expressing growth hormone in cardiac and striated muscle. Growth Regul 1995;5:101-108.

77. Sheikh F, Kardami E, Cattini PA. Increased number of cardiomyocytes is detected in postnatal primary cultures from FGF-2 transgenic mice. J Mol Cell Cardiol 1999;31:A20.

78. Schmidt EV, Christoph G, Zeller R, Leder P. The cytomegalovirus enhancer: a pan-active control element in transgenic mice. Mol Cell Biol 1990;10:4406-4411.

79. Shewchuk BM, Asa SL, Cooke NE, Liebhaber SA. Pit-1 binding sites at the somatotrope-specific DNase I hypersensitive sites I, II of the human growth hormone locus control region are essential for in vivo hGH-N gene activation. J Biol Chem 1999;274:35725-35733.

6

NEUROENDOCRINE AND REPRODUCTIVE FUNCTIONS IN TRANSGENIC MICE WITH ALTERED GROWTH HORMONE SECRETION AND IN GROWTH HORMONE RECEPTOR GENE DISRUPTED MICE

Varadaraj Chandrashekar, Karen T. Coschigano,
Andrzej Bartke and John J. Kopchick
[1]Southern Illinois University School of Medicine (VC, AB), Carbondale, Illinois
62901-6512, USA, and [2]Edison Biotechnology Institute, & College of Osteopathic
Medicine (KTC, JJK), Ohio University, Athens, Ohio 45701, USA

INTRODUCTION

It has been established that growth hormone (GH) exerts a profound anabolic effect in mammals. GH secretion by the pituitary gland is regulated by a complex control system. Two hypothalamic peptide hormones, GH-releasing hormone (GHRH) and GH-release inhibiting hormone (somatostatin), exert stimulatory and inhibitory influences, respectively, on the somatotropes of the pituitary gland. These two hypothalamic hormones are subjected to modulation by a host of neurotransmitters, other hypothalamic neuropeptides, endocrine glands, GH and insulin-like growth factor-I (IGF-I), and their binding proteins. GH plays an important role in induction of linear growth, fat metabolism, increase in lactation, enhancement of immune response, and in bone mineralization. Recombinant technology has made it possible to produce large amounts of GH that is used in the agriculture industry and for treating some human ailments. Thus, administration of bovine GH (bGH) has been shown to increase milk production in dairy cattle (1-3). Also treatment of growing pigs with porcine GH markedly stimulated muscle growth and reduced fat deposition (3). In humans, administration of hGH to GH-deficient children improves the growth rate (4, 5), in aged individuals it improves immune response (6), and in patients with AIDS, treatment with GH increases body weight and lean body mass (7). With all the above activities it is paramount to continue to evaluate the effects of GH, in addition to its anabolic action. In our laboratory, we have attempted to evaluate the effects of GH/IGF-I on systems that control reproduction in mice with altered GH secretion/action.

The role of GH in the control of neuroendocrine and gonadal functions in mammals is poorly understood. However, a number of studies have suggested the existence of a relationship between GH and the reproductive system. Suppression of GH secretion in female rats has been shown to delay puberty and reduce ovarian luteinizing hormone (LH) receptors (8, 9), and GH augments follicle-stimulating hormone (FSH)-induced differentiation of ovarian granulosa cells (10). Recently, it has been shown in FSH-deficient mice that, despite the absence of FSHβ subunit, ovarian follicles develop up to preantral stage (11). Furthermore GH, in the absence of FSH, can induce folliculogenesis in isolated ovarian preantral follicles obtained from immature mice, strongly suggesting a role of GH in the control of ovarian function (12). Childs et al. in their extensive studies have identified GH antigens in pituitary cells containing FSH or LH mRNAs and in cells containing GnRH receptors indicating that either GH cells are transitory gonadotrophs, or GH is present in these pituitary cells possibly helping to control their function (13, 14). Furthermore, GH-binding protein antigens were identified in pituitary cells that contained LH and FSH, indicating a possible paracrine effect of GH on the function of the gonadotropes (15). Thus, GH may function as a "co-gonadotropin" (13, 14). Most recently, it has been shown that somatotropes may be converted to transitional gonadotropes just before proestrus secretory activity. The LH and FSH antigen content of the GHRH target cells from proestrus rats demonstrate that the LHβ and FSHβ mRNAs are indeed translated (16). Additionally, the increased expression of prolactin (PRL) antigens by these cells signifies that these convertible somatotropes may also be somatomammotropes (16). Furthermore, female GH-deficient Ames and Snell dwarf mice are infertile and their neuroendocrine functions are altered (17-19). These studies suggest that there is a functional interrelationship among somatotropes, gonadotropes and, lactotropes in female rodents.

A role of GH in reproduction in women has also been suggested. In infertile women with hyposensitive ovaries, GH treatment increases the ovarian sensitivity to gonadotropin (20, 21). GH has been shown to increase the human chorionic gonadotropin (hCG)-induced progesterone secretion by isolated human luteal cells (22) and in some cases, anovulation is associated with reduced GH secretion (23, 24). Excess GH secretion can also have an effect on reproduction. In women, acromegaly can affect menstrual cycles, possibly due to alteration in the pituitary function (25).

In the male, a limited number of studies so far conducted have shown that GH can influence neuroendocrine and testicular functions. Treatment in hypophysectomized rats with GH increases the LH receptor content of the testis (26) and enhances the testicular responsiveness to gonadotropin treatment (27). In adult rats, a lack of GH secretion results in a delay in testicular growth and differentiation of germinal cells (28). Although GH-deficient dwarf rats are fertile, their testes are small, suggesting a reduced daily sperm output in these animals (29). Subsequent studies have shown that the sperm motility is also impaired (30) and GH or IGF-I treatment increases motility of immature spermatozoa in the GH-deficient dwarf rats (31, 32), indicating the importance of GH and IGF-I in male reproduction. Biological neutralization of the endogenously secreted GH or administration of GH results in alterations in neuroendocrine function in rats (33). Additionally, in GH-deficient boys, the development of genital organs is poor and treatment of GH increases the growth of the genitalia (34). In men, congenital GH resistance due to mutated GH receptors (Laron syndrome) is associated with a delay in sexual maturation (35, 36). Also co-treatment of GH with gonadotropin in infertile men increased serum testosterone levels and normalized sperm parameters (37). In these patients, treatment with gonadotropins alone had no effect on these parameters. These studies suggest that GH plays a role in the control of neuroendocrine and reproductive functions in the male. It is known that GH acts on the liver and possibly other tissues to stimulate the production of IGF-I and several of the major GH effects are mediated by IGF-I. In this chapter we present data related to neuroendocrine and gonadal functions in GH-deficient mice and in various lines of transgenic mice either producing excess GH/IGF-I or mice resistant to GH due to the absence of GH receptors.

There are some controversies with regard to the role of GH in pituitary and testicular functions. Administration of GH to GH-deficient young adult males resulted in significant increases in total and free IGF-I levels in all patients, but had no significant effect on the pituitary FSH and LH response to gonadotropin-releasing hormone (GnRH), as well as basal and hCG-stimulated levels of androgens (38). Therefore, this study concluded that the effects of GH administration do not appear to involve major alterations in the pituitary-gonadal axis. It was also shown in baboons that treatment of GH or IGF-I did not alter gonadotropin stimulation of testicular function (39). In cynomolgus monkeys, long-term treatment of GH did not alter spermatogenesis (40), and suppression of GH by active immunization against GHRH failed to affect ongoing spermatogenesis in rats (41).

Therefore, it is imperative to keep in mind that not all effects of GH are stimulatory and the duration of exposure of the hypothalamo-hypophyseal-gonadal system to GH might influence the secretions of gonadotropins and gonadal steroids. Also, the concentrations of IGF-I binding proteins and the ratio of these binding proteins and IGF-I within the body might play an important role in evoking the effects of IGF-I as it relates to the neuroenodcrine-gonadal system.

EFFECTS OF GH-DEFICIENCY

It has been demonstrated that testicular weight, seminiferous tubular diameter and germ cell number are reduced in GH-deficient Snell dwarf mice (42). However, administration of GH during the postnatal development resulted in normalization of these parameters. Male Ames dwarf mice are also deficient in GH/IGF-I secretion (43). It has been shown that administration of bGH to these mice induces IGF-I secretion (43). In these mice, GH treatment also enhanced plasma LH levels, and the effect of GnRH on LH secretion was significantly increased but this LH response was lower than in normal siblings that previously received vehicle. In addition, in these GH/IGF-I deficient animals, the castration-induced increase in plasma LH levels was attenuated. Pretreatment of dwarf mice with GH resulted in increased production of androstenedione and testosterone by the isolated testis treated with hCG. These results indicated that the alterations in hypothalamic-pituitary-testicular function in GH-deficient mice is due to the lack of IGF-I secretion.

In female dwarf mice, the ovariectomy-induced increase in LH secretion is decreased and GH treatment increased circulating LH levels (18). In these mice, the plasma LH response to GnRH treatment was reduced and GH administration normalized this response. The negative feedback effect of estrogen on LH secretion was decreased in dwarf mice. These studies suggest that GH plays an important role in the control of neuroendocrine function. However, male and female Ames dwarf mice are also PRL and TSH-deficient. Therefore, some of the effects observed in Ames dwarf mice might have been also due to absence of PRL and thyroid hormones. It would be important to study the effects of GH/IGF-I in animals with only IGF-I deficiency. In this regard, GH receptor gene knockout mice were produced recently. These mice are resistant to GH action and are IGF-I-deficient. Therefore, these mice are a good experimental model to assess the influence of IGF-I in neuroendocrine and gonadal functions. Some of our studies in these animals will be presented later in this chapter.

CONSEQUENCES OF EXCESS GH SECRETION

Pioneering investigations reporting the introduction of heterologous GH genes into mouse embryos and observation of expression of the foreign GH gene in resulting transgenic offspring (44-49) is the 'hallmark' of transgenic technology, opening doors to study the impact of GH secretion on the various functions of the body. GH-producing transgenic mice created by using mouse metallothionein-I (MT) promoter express the foreign GH in multiple organs including liver, kidney, skin, gonads, and adenohypophyses (44-46, 50) with expression starting during fetal development and continuing during the entire lifespan. In mice with phosphoenopyruvate carboxykinase (PEPCK) promoter GH fusion gene, the expression of the GH gene begins around the time of birth and usually is limited to the liver, kidney, and adipose tissue (48). In this section we describe the consequences of hGH, bGH, and hGHRH gene expression on neuroendocrine and reproductive functions in mice.

Although transgenic male mice expressing the MT-human GH (MT-hGH) gene are fertile, their seminal vesicles are significantly enlarged (51). The enlargement of the seminal vesicles might be due to the PRL-like activity of the secreted hGH (52). In young adult transgenic mice, the plasma testosterone levels and sperm production per gram of testicular mass are similar to those observed in normal mice (53). In these transgenic mice, plasma LH levels were significantly increased (54-56). However, the LH response to castration, and testosterone treatment were attenuated (54). These LH responses are possibly due to the increased noradrenergic activity resulting in an increase in stimulation of GnRH neurons and increased secretion of GnRH. In male mice expressing the MT-hGH gene there is increased expression of LHβ mRNA in the pituitary gland (57). Again these alterations are due to the consequence of PRL-like activity of hGH (52), similar to the results observed in experimental induction of hyperprolactinemia in mice (58). In contrast to the PRL-like activity of hGH, bGH has no lactogenic effect. Our studies in the male transgenic mice expressing the bGH gene with the mouse MT promoter showed no significant effect on the pituitary-hypothalamic-testicular function (59), confirming the PRL-like effect of hGH in mice bearing the hGH gene. Thus, expression of the MT-hGH gene can alter neuroendocrine function in male mice.

In intact female transgenic mice expressing the MT-hGH gene secreting moderate amounts of hGH, plasma LH levels measured at estrus were elevated, while FSH levels were decreased relative to their normal siblings (60). The ovariectomy-induced rise in circulating gonadotropin levels was attenuated in these transgenic mice. However, plasma FSH levels in ovariectomized, estradiol-treated transgenic mice were significantly decreased. Furthermore, FSH and LH responses to a single dose of GnRH were significantly reduced in ovariectomized-estrogen primed mice bearing the hGH gene (60). Subsequent studies have indicated that these female mice are sterile, possibly due to the stimulation of the hypothalamic tuberoinfundibular dopaminergic neurons resulting in suppression of PRL release associated with luteal failure (61). Treatment of pregnant mice secreting hGH with PRL by PRL-secreting ectopic pituitary transplants lead to normal pregnancies (61). Female mice expressing hGH gene with PEPCK promoter secrete large amounts of hGH and reproduce normally (62). The inconsequential effect of very high levels of hGH secretion on reproduction in these mice might have been due to the effective replacement of the suppressed PRL secretion and the PRL-like effect of hGH in rodents (52). These observations indicate that moderate secretion of the foreign hGH alters hypothalamic-pituitary function and severely affects reproduction in female mice.

Unlike lactogenic action of hGH, bGH is purely somatotropic in function. In order to assess the differential effects of hGH and bGH, we have examined the effects of excess bGH secretion on reproduction in transgenic mice expressing the bGH fusion genes driven by either the MT or PEPCK promoter. Mice bearing the PEPCK-bGH gene secrete high levels of bGH relative to animals expressing the MT-bGH gene (63). Most of the female mice from MT-bGH line are fertile and their reproductive parameters are nearly normal except for the absence of gestations from postpartum estrus and the consequent significant increase in the average interval between pregnancies (64). Expression of the MT-bGH gene resulted in attenuated FSH response to GnRH treatment, while LH response was similar to that in controls (18). The ovariectomy-induced increases in plasma FSH and LH levels were decreased in transgenic mice. The absolute circulating FSH levels were also decreased in estrogen-treated transgenic mice, while plasma LH levels were similar in normal mice and in mice bearing the MT-bGH gene. These ovariectomized transgenic mice were hyperprolactinemic (18). High levels of bGH secretion in PEPCK-bGH females resulted in high incidence of pregnancy failure (65), due to luteal deficiency (66). The luteal failure in these mice is attributed to the failure of stimuli associated with mating to induce the normal pattern of twice daily PRL surges (66). It has been shown that pregnancy could be maintained in mated PEPCK-bGH mice by treating with progesterone or PRL, or a dopaminergic antagonist that releases PRL (67). When normal pregnancies are carried to term in small percent of PEPCK-bGH mice, litter size was significantly increased, relating to the increased ovulation rate in these animals. A recent study has demonstrated that the percentage of ovarian follicles containing apoptotic cells was lower in transgenic mice bearing the PEPCK-bGH gene than in normal mice (68). The percentage of follicles undergoing apoptosis was lower in these transgenic mice versus control animals in preovulatory and early antral follicles. The percentage of healthy preovulatory follicles was also higher in transgenic versus normal mice. These results indicate that GH overexpression in PEPCK-bGH mice significantly decreases follicle apoptosis, and atresia in the mouse ovary, therefore leading to increased propensity for ovulation in these animals (68).

Transgenic male mice expressing MT-bGH are usually fertile. Their plasma gonadotropin, PRL, and testosterone levels were similar to those observed in their normal siblings (59). Despite reductions in pituitary FSHβ mRNA and LHβ mRNA levels (69), PEPCK-bGH male mice produced normal spermatozoa and were fertile (70). These studies suggest that excess bGH secretion has little or no consequence on male reproduction.

In transgenic mice bearing the ovine (o) MT 1a-oGH fusion gene (oMT1a-oGH), oGH is not expressed in animals maintained on a standard laboratory mouse diet and tap water, but can be stimulated by providing $ZnSO_4$ solution for drinking (71). Therefore, this transgenic model is valuable experimental animal to stimulate or inhibit the production of the heterologous GH by simply providing or withholding the consumption of $ZnSO_4$. Transgenic female mice expressing oMT1a-oGH fusion gene can cycle, mate, and support early embryonic development; but they fail to maintain pregnancy due to luteal insufficiency (72). Treatment of these mice with progesterone resulted in maintenance of pregnancy and normal lactation (72).

Availability of transgenic mice expressing the hGHRH gene (73) has made it possible to examine the effects of endogenously produced, homologous mouse GH. These animals produce large amounts of GH and the pituitary is enlarged (74). The source of GH is from the in situ pituitary gland and unlike other transgenic mice indicated above. As expected, we found that these animals secrete high levels of IGF-I (74). Interestingly, their testicular weights were significantly increased. Although the basal LH levels were similar in transgenic mice bearing the hGHRH gene and in their normal siblings, the LH response to GnRH treatment was attenuated in these transgenic mice. Excess secretion of GH resulted in significant increases in testosterone levels. A possible explanation of these results may be that excess GH/IGF-I secretion apparently increased the sensitivity of the Leydig cells of the testis to LH action and thus influenced the testosterone secretion in transgenic mice (74). Recently, it was shown that the number of lactotropes in the male hGHRH animals was increased two fold, yet their plasma PRL levels were not changed (75). This is due to the increased hypothalamic dopamine synthesis and release coupled with an increase in D_2 dopamine receptor gene expression and functional sensitivity of the pituitary gland (75). Expression of the hGHRH gene also resulted in decreased LHβ mRNA levels in the pituitary glands (75). However, the total LHβ mRNA levels per pituitary gland was significantly higher in transgenic mice than in normal mice.

NEUROENDOCRINE AND GONADAL FUNCTIONS IN GH RECEPTOR GENE DISRUPTED MICE

Recently, GH receptor gene knockout (GHR-KO) mice were produced (76). They represent a good experimental model to assess the role of IGF-I in neuroendocrine and reproductive functions. Although these mice secrete large amounts of GH, due to the absence of GHRs, IGF-I is not produced. These mice have been designated as an 'accurate' model for Laron syndrome (77).

Although most female GHR-KO mice are fertile, the age at first conception is greatly delayed in homozygous x homozygous matings (76). The age of vaginal opening is significantly delayed in GHR-KO vs. normal mice, but it can be advanced by treatment with recombinant human IGF-I (78). In pregnant GHR-KO mice, fetal size is reduced and pregnancy is prolonged while placental weight is, unexpectedly, increased. Alterations in fetal and placental weight are related to maternal rather than fetal genotype. Moreover, litter size and body weight of newborn pups are significantly reduced in GHR-KO vs. normal females. Reduction in litter size reflects both dam and sire effects. Therefore, GH resistance and consequent reduction in peripheral IGF-I levels is associated with the delay of female puberty, alterations in fetal and placental growth, delay of parturition, and reduced litter size (78).

Our studies in male mice indicate that the absence of GHRs was associated with decreased weights of testes and male sex accessory structures (79). There was a significant increase in plasma PRL levels (79). Although the basal plasma LH levels were similar in GHR-KO mice relative to those in their normal siblings, the circulating LH response to GnRH treatment was significantly attenuated. Plasma testosterone levels were unaffected by disruption of the GHR gene. However, basal and LH-stimulated testosterone releases from the isolated testes of GHR-KO mice were decreased (79). The rate of fertility in GHR-KO male mice was also reduced (79). These results indicate that the lack of GHRs (with GH resistance and lack of IGF-I secretion) induces hyperprolactinemia and alters the effect of GnRH on LH secretion as well as testicular function. Our recent study has shown that the plasma testosterone response to LH treatment was attenuated in GHR-KO mice, while circulating androstenedione levels were not different than in their normal siblings (80). This suggests that within the testes of GHR-KO mice the key enzyme 17β-hydroxysteroid dehydrogenase, that converts androstenedione to testosterone, is defective/or less responsive to exogenous LH. This indicates that IGF-I may play an important role in testicular steroidogenesis.

There are some similarities in reproductive characteristics of male GHR-KO mice with IGF-I gene disrupted mice. Similar to GHR-KO mice, the weights of male sex accessory structures were reduced in targeted IGF-I gene disrupted mice (81). In addition, their circulating testosterone levels were reduced. As in GHR-KO mice (80), the *in vitro* testosterone response to LH treatment was also suppressed in the absence of IGF-I gene (81). However, the IGF-I gene disrupted mice were infertile. In contrast, fertility is reduced, but not totally suppressed in GHR-KO mice. The mechanism responsible for the maintenance of fertility in GHR-KO mice is unknown. Growth hormone treatment increased IGF-I secretion and consequently elevated the total number of viable spermatozoa in GH-deficient dwarf rats (31), and the absence of mating behavior in mice with IGF-I gene null mutation (81) strongly suggest that IGF-I is important for normal male reproduction.

COMMENTS AND CONCLUSIONS

Studies referred to in this chapter indicate that GH/IGF-I influence the neuroendocrine and gonadal functions in both sexes. Thus, GH deficiency and GH excess can alter gonadotropin, PRL, and gonadal steroid secretions. It is important to note that some of the specific effects of GH are different in female and male mice. For example, expression of MT-bGH gene has no effect on PRL secretion in males, but female mice of this line are hyperprolactinemic. The effects of excess hGH in female transgenic mice are more severe than in male mice bearing the same gene. In addition, the species of origin of GH is important to achieve alterations in the neuroendocrine function. Unlike MT-bGH mice, MT-hGH mice, despite subnormal secretion of PRL, are physiologically hyperprolactinemic, possibly due to the PRL-like activity of hGH in rodents. The hypothalamic-pituitary system was significantly activated in MT-hGH mice as evidenced by increases in the secretion of LH and in norepinephrine turnover in the median eminence of the hypothalamus (54, 60, 82).

In MT-bGH female mice secreting increased amounts of PRL, the expression of dopamine type 2 receptors in the anterior pituitary gland was significantly reduced, whereas expression of the estrogen receptors was significantly increased (83). It is interesting that opposite alterations were observed in the expression of dopamine type 2 and estrogen receptors in the pituitaries of MT-hGH females (84) in which plasma PRL levels are reduced (60). These data indicate that bGH and hGH differentially act on pituitary dopamine and estrogen receptors to alter PRL secretion.

IGF-I is present in the hypothalamo-hypophyseal system (85) and IGF-I receptors are present in highest concentration in the median eminence of the brain (86). It has been suggested that IGF-I of peripheral origin acts centrally to accelerate the initiation of female puberty (87). In both rat and mouse ovaries, IGF-I mRNA is selectively expressed by granulosa cells in growing, "healthy" follicles. Type I IGF receptor mRNA was also present in granulosa cells (88).

There is a growing body of evidence that IGF-I is also involved in testicular function. Production of IGF-I by testicular tissues has been confirmed by the presence of IGF-I mRNA in the testis (89). It has been shown that this peptide is secreted by Sertoli and Leydig cells (90-93). Furthermore, IGF-I receptors have been localized in the Sertoli (94) and Leydig cells (95-97), as well as secondary spermatocytes and spermatids (98). Although IGF-I production is typically considered to be under the control of GH, its secretion can be stimulated by FSH (Sertoli cells, 90, 91) and by hCG (Leydig cells, 91). The observation that IGF-I potentates LH action in androgen secretion by the testis (99) and its important role in initiation of puberty in females (78) suggests that IGF-I is a critical factor in gonadal function. Evidence presented in this chapter strongly suggests that GH/IGF-I influence hypothalamic-pituitary function. They may act as 'co-gonadotropins' in the regulation of gonadal function.

Acknowledgments

NIH Grants HD20001, HD20033, and RR15012 supported studies conducted in our laboratory and described in this chapter. JJK is supported by Sensus Corporation and the State of Ohio's Eminent Scholar Program which includes a grant from Milton and Lawrence Goll. We are grateful to Dr. T. E. Wagner and Ms. J.S. Jun, Edison Biotechnology Center, Ohio University, Athens, Ohio, for generously supplying founder male transgenic mice, which were used to develop colonies of transgenic mice. Dr. J. F. Hyde, University of Kentucky College of Medicine, Lexington, Kentucky, kindly provided a male transgenic founder mouse expressing the hGHRH gene. We thank Dr. G. D. Niswender, Colorado State University, Fort Collins, CO, Dr. A. F. Parlow, Pituitary Hormone and Antisera Center, Harbor-UCLA Medical Center, Torrance, CA, and the National Hormone and Pituitary Program, Rockville, MD for generously providing reagents used in testosterone, IGF-I, and pituitary hormone RIAs. Eli Lilly Company, Indianapolis, IN, generously supplied the recombinant human IGF-I.

REFERENCES:

1. Bauman DE, Everett RW, Weiland WH, Collier RJ. Production responses to bovine somatotropin in northeast dairy herds. J Dairy Sci 1999;82:2564-2573.
2. Bauman DE. Bovine somatotropin and lactation: from basic science to commercial application. Domest Anim Endocrinol 1999;17:101-116.
3. Etherton TD, Bauman DE. Biology of somatotropin in growth and lactation of domestic animals. Physiol Rev 1998;78:745-761.
4. Ohlsson C, Bengtsson BA, Isaksson OG, Andreassen TT, Slootweg MC. Growth hormone and bone. Endocr Rev 1998;19:55-79.
5. Ohlsson C, Jennische E. Effects in skeletal muscle of supraphysiological growth hormone stimulation. Eur J Endocrinol 1995;133:678-679.
6. Gelato MC. Ageing and immune function: a possible role for growth hormone. Horm Res 1996;45:46-49.
7. Schambelan M, Mulligan K, Grunfeld C, Daar ES, LaMarca A, Kotler DP, Wang J, Bozzette SA, Breitmeyer JB. Recombinant human growth hormone in patients with HIV-associated wasting. A randomized, placebo-controlled trial. Serostim Study Group. Ann Intern Med 1996;125:873-882.
8. Ramaley JA, Phares CK. Delay of puberty onset in females due to suppression of growth hormone. Endocrinology 1980;106:1989-1993.
9. Advis JP, White SS, Ojeda SR. Activation of growth hormone short loop negative feedback delays puberty in the female rat. Endocrinology 1981;108:1343-1352.
10. Jia XC, Kalmijn J, Hsueh AJ. Growth hormone enhances follicle-stimulating hormone-induced differentiation of cultured rat granulosa cells. Endocrinology 1986;118:1401-1409.

11. Kumar TR, Wang Y, Lu N, Matzuk MM. Follicle stimulating hormone is required for ovarian follicle maturation but not male fertility. Nat Gen 1997;15:201-204.

12. Liu X, Andoh K, Yokota H, Kobayashi J, Abe Y, Yamada K, Mizunuma H, Ibuki Y. Effects of growth hormone, activin, and follistatin on the development of preantral follicle from immature female mice. Endocrinology 1998;139:2342-2347.

13. Childs GV, Unabia G, Rougeau D. Cells that express luteinizing hormone (LH) and follicle-stimulating hormone (FSH) beta-subunit messenger ribonucleic acids during the estrous cycle: the major contributors contain LH beta, FSH beta, and/or growth hormone. Endocrinology 1994;134:990-997.

14. Childs GV, Unabia G, Miller BT. Cytochemical detection of gonadotropin-releasing hormone-binding sites on rat pituitary cells with luteinizing hormone, follicle-stimulating hormone, and growth hormone antigens during diestrous up-regulation. Endocrinology 1994;134:1943-1951.

15. Harvey S, Baumbach WR, Sadeghi H, Sanders EJ. Ultrastructural colocalization of growth hormone binding protein and pituitary hormones in adenohypophyseal cells of the rat. Endocrinology 1993; 133:1125-30.

16. Childs GV, Unabia G, Miller BT, Collins TJ. Differential expression of gonadotropin and prolactin antigens by GHRH target cells from male and female rats. J Endocrinol 1999;162:177-187.

17. Bartke A Reproduction of female dwarf mice treated with prolactin. J Reprod Fertil 1966;11:203-206.

18. Chandrashekar V, Bartke A. Influence of hypothalamus and ovary on pituitary function in transgenic mice expressing the bovine growth hormone gene and in growth hormone-deficient Ames dwarf mice. Biol Reprod 1996;54:1002-1008.

19. de Reviers MM, Viguier-Martinez MC, Mariana JC. FSH, LH and prolactin levels, ovarian follicular development and ovarian responsiveness to FSH in the Snell dwarf mouse. Acta Endocrinol (Copenh) 1984;106:121-126.

20. Homburg R, West C, Torresani T, Jacobs HS. Cotreatment with human growth hormone and gonadotropins for induction of ovulation: a controlled clinical trial. Fertil Steril 1990;53:254-260.

21. Homburg R, Farhi J. Growth hormone and reproduction. Curr Opin Obstet Gynecol 1995;7:220-223.

22. Lanzone A, Di Simone N, Castellani R, Fulghesu AM, Caruso A, Mancuso S. Human growth hormone enhances progesterone production by human luteal cells in vitro: evidence of a synergistic effect with human chorionic gonadotropin. Fertil Steril 1992;57:92-96.

23. Ovesen P, Moller J, Moller N, Christiansen JS, Orskov H, Jorgensen JO. Growth hormone secretory capacity and serum insulin-like growth factor I levels in primary infertile, anovulatory women with regular menses. Fertil Steril 1992;57:97-101.

24. Strobl JS, Thomas MJ. Human growth hormone. Pharmacol Rev 1994;46:1-34.

25. Kaltsas GA, Mukherjee JJ, Jenkins PJ, Satta MA, Islam N, Monson JP, Besser GM, Grossman AB. Menstrual irregularity in women with acromegaly. J Clin Endocrinol Metab 1999;84:2731-2735.

26. Zipf WB, Payne AH, Kelch RP. Prolactin, growth hormone, and luteinizing hormone in the maintenance of testicular luteinizing hormone receptors. Endocrinology 1978;103:595-600.

27. Swerdloff RS, Odell WD. Modulating influences of FSH, GH, and prolactin on LH-stimulated testosterone secretion. In: Troen P, Nankin HR (eds) The testis in normal and infertile men 1977, Raven Press, New York, pp. 395-401.

28. Arsenijevic Y, Wehrenberg WB, Conz A, Eshkol A, Sizonenko PC, Aubert ML. Growth hormone (GH) deprivation induced by passive immunization against rat GH-

releasing factor delays sexual maturation in the male rat. Endocrinology 1989;124:3050-3059.

29. Spiteri-Grech J, Bartlett JM, Nieschlag E. Regulation of testicular insulin-like growth factor-I in pubertal growth hormone-deficient male rats. J Endocrinol 1991;131:279-285.

30. Gravance CG, Breier BH, Vickers MH, Casey PJ. Impaired sperm characteristics in postpubertal growth-hormone-deficient dwarf (dw/dw) rats. Anim Reprod Sci 1997;49:71-76.

31. Breier BH, Vickers MH, Gravance CG, Casey PJ. Growth hormone (GH) therapy markedly increases the motility of spermatozoa and the concentration of insulin-like growth factor-I in seminal vesicle fluid in the male GH-deficient dwarf rat. Endocrinology 1996;137:4061-4064.

32. Vickers MH, Casey PJ, Champion ZJ, Gravance CG, Breier BH. IGF-I treatment increases motility and improves morphology of immature spermatozoa in the GH-deficient dwarf (dw/dw) rat. Growth Horm IGF Res 1999;9:236-240.

33. Chandrashekar V, Bartke A. The role of growth hormone in the control of gonadotropin secretion in adult male rats. Endocrinology 1998;139:1067-1074.

34. Sheikholislam BM, Stempfel RS Jr. Hereditary isolated somatotropin deficiency: effects of human growth hormone administration. Pediatrics 1972;49:362-374.

35. Laron Z, Sarel R. Penis and testicular size in patients with growth hormone insufficency. Acta Endocrinol (Copenh) 1970;63:625-633.

36. Laron Z, Klinger B. Effect of insulin-like growth factor-I treatment on serum androgens and testicular and penile size in males with Laron syndrome (primary growth hormone resistance). Eur J Endocrinol 1998;138:176-180.

37. Shoham Z, Conway GS, Ostergaard H, Lahlou N, Bouchard P, Jacobs HS. Cotreatment with growth hormone for induction of spermatogenesis in patients with hypogonadotropic hypogonadism. Fertil Steril 1992;57:1044-1051.

38. Juul A, Andersson AM, Pedersen SA, Jorgensen JO, Christiansen JS, Groome NP, Skakkebaek NE Effects of growth hormone replacement therapy on IGF-related parameters and on the pituitary-gonadal axis in GH-deficient males. A double-blind, placebo-controlled crossover study. Horm Res 1998;6:269-278.

39. Crawford BA, Handelsman DJ. Recombinant growth hormone and insulin-like growth factor I do not alter gonadotrophin stimulation of the baboon testis in vivo. Eur J Endocrinol 1994;131:405-412.

40. Sjogren I, Ekvarn S, Zuhlke U, Vogel F, Bee W, Weinbauer GF, Nieschlag E. Lack of effects of recombinant human GH on spermatogenesis in the adult cynomolgus monkey (*Macaca fascicularis*). Eur J Endocrinol 1999;140:350-357.

41. Awoniyi CA, Veeramachaneni DN, Roberts D, Tucker KE, Chandrashekar V, Schlaff WD. Suppression of growth hormone does not affect ongoing spermatogenesis in rats. J Androl 1999;20:102-108.

42. Matsushima M, Kuroda K, Shirai M, Ando K, Sugisaki T, Noguchi T. Spermatogenesis in Snell dwarf, little and congenitally hypothyroid mice. Int J Androl 1986;9:132-140.

43. Chandrashekar V, Bartke A. Induction of endogenous insulin-like growth factor-I secretion alters the hypothalamic-pituitary-testicular function in growth hormone-deficient adult dwarf mice. Biol Reprod 1993;48:544-551.

44. Palmiter RD, Brinster RL, Hammer RE, Trumbauer ME, Rosenfeld MG, Birnberg NC, Evans RM. Dramatic growth of mice that develop from eggs microinjected with metallothionein-growth hormone fusion genes. Nature 1982;300:611-615.

45. Palmiter RD, Norstedt G, Gelinas RE, Hammer RE, Brinster RL. Metallothionein-human GH fusion genes stimulate growth of mice. Science 1983;222:809-814.

46. Hammer RE, Pursel VG, Rexroad CE Jr, Wall RJ, Bolt DJ, Ebert KM, Palmiter RD, Brinster RL Production of transgenic rabbits, sheep and pigs by microinjection. Nature 1985;315:680-683.

47. Hammer RE, Pursel VG, Rexroad CE Jr, Wall RJ, Bolt DJ, Palmiter RD, Brinster RL. Genetic engineering of mammalian embryos. J Anim Sci 1986;63:269-278.

48. McGrane MM, de Vente J, Yun J, Bloom J, Park E, Wynshaw-Boris A, Wagner T, Rottman FM, Hanson RW. Tissue-specific expression and dietary regulation of a chimeric phosphoenolpyruvate carboxykinase/bovine growth hormone gene in transgenic mice. J Biol Chem 1988;263:11443-11451.

49. Selden RF, Wagner TE, Blethen S, Yun JS, Rowe ME, Goodman HM. Expression of the human growth hormone variant gene in cultured fibroblasts and transgenic mice. Proc Natl Acad Sci U S A 1988;85:8241-8245.

50. Stefaneanu L, Kovacs K, Horvath E, Losinski NE, Mayerhofer A, Wagner TE, Bartke A. An immunocytochemical and ultrastructural study of adenohypophyses of mice transgenic for human growth hormone. Endocrinology 1990;126:608-615.

51. Prins GS, Cecim M, Birch L, Wagner TE, Bartke A. Growth response and androgen receptor expression in seminal vesicles from aging transgenic mice expressing human or bovine growth hormone genes. Endocrinology 1992;131:2016-2023.

52. Rivera EM, Forsyth IA, Folley SJ. Lactogenic activity of mammalian growth hormones in vitro. Proc Soc Exp Biol Med 1967;124:859-865.

53. Bartke A, Naar EM, Johnson L, May MR, Cecim M, Yun JS, Wagner TE. Effects of expression of human or bovine growth hormone genes on sperm production and male reproductive performance in four lines of transgenic mice. J Reprod Fertil 1992;95:109-118.

54. Chandrashekar V, Bartke A, Wagner TE. Endogenous human growth hormone (GH) modulates the effect of gonadotropin-releasing hormone on pituitary function and the gonadotropin response to the negative feedback effect of testosterone in adult male transgenic mice bearing human GH gene. Endocrinology 1988;123:2717-2722.

55. Chandrashekar V, Bartke A. Effects of age and endogenously secreted human growth hormone on the regulation of gonadotropin secretion in female and male transgenic mice expressing the human growth hormone gene. Endocrinology 1993;132:1482-1488.

56. Chandrashekar V, Bartke A, Wagner TE. Interactions of human growth hormone and prolactin on pituitary and Leydig cell function in adult transgenic mice expressing the human growth hormone gene. Biol Reprod 1991;44:135-140.

57. Tang K, Bartke A, Gardiner CS, Wagner TE, Yun JS. Gonadotropin secretion, synthesis, and gene expression in human growth hormone transgenic mice and in Ames dwarf mice. Endocrinology 1993;132:2518-2524.

58. Klemcke HG, Bartke A. Effects of chronic hyperprolactinemia in mice on plasma gonadotropin concentrations and testicular human chorionic gonadotropin binding sites. Endocrinology 1981;108:1763-1768.

59. Chandrashekar V, Bartke A, Wagner TE. Effects of gonadotropin releasing hormone on luteinizing hormone and testosterone secretion in male transgenic mice with bovine growth hormone gene expression. Annual Meeting of The Endocrine Society, Washington DC, 1989, p81, Abstract.

60. Chandrashekar V, Bartke A, Wagner TE. Neuroendocrine function in adult female transgenic mice expressing the human growth hormone gene. Endocrinology 1992;130:1802-1808.

61. Bartke A, Steger RW, Hodges SL, Parkening TA, Collins TJ, Yun JS, Wagner TE. Infertility in transgenic female mice with human growth hormone expression: evidence for luteal failure. J Exp Zool 1988;248:121-124.

62. Milton S, Cecim M, Li YS, Yun JS, Wagner TE, Bartke A. Transgenic female mice with high human growth hormone levels are fertile and capable of normal lactation without having been pregnant. Endocrinology 1992;131:536-538.

63. Bartke A, Chandrashekar V, Turyn D, Steger RW, Debeljuk L, Winters TA, Mattison JA, Danilovich NA, Croson W, Wernsing DR, Kopchick JJ. Effects of growth hormone overexpression and growth hormone resistance on neuroendocrine and reproductive functions in transgenic and knock-out mice. Proc Soc Exp Biol Med 1999;222:113-123.

64. Naar EM, Bartke A, Majumdar SS, Buonomo FC, Yun JS, Wagner TE. Fertility of transgenic female mice expressing bovine growth hormone or human growth hormone variant genes. Biol Reprod 1991;45:178-187.

65. Cecim M, Kerr J, Bartke A. Effects of bovine growth hormone (bGH) transgene expression or bGH treatment on reproductive functions in female mice. Biol Reprod 1995;52:1144-1148.

66. Cecim M, Kerr J, Bartke A. Infertility in transgenic mice overexpressing the bovine growth hormone gene: luteal failure secondary to prolactin deficiency. Biol Reprod 1995;52:1162-1166.

67. Cecim M, Fadden C, Kerr J, Steger RW, Bartke A. Infertility in transgenic mice overexpressing the bovine growth hormone gene: disruption of the neuroendocrine control of prolactin secretion during pregnancy. Biol Reprod 1995;52:1187-1192.

68. Danilovich NA, Bartke A, Winters TA. Ovarian follicle apoptosis in bovine growth hormone transgenic mice. Biol Reprod 2000;62:103-107.

69. Tang K, Bartke A, Gardiner CS, Wagner TE, Yun JS. Gonadotropin secretion, synthesis, and gene expression in two types of bovine growth hormone transgenic mice. Biol Reprod 1993;49:346-353.

70. Bartke A, Cecim M, Tang K, Steger RW, Chandrashekar V, Turyn D. Neuroendocrine and reproductive consequences of overexpression of growth hormone in transgenic mice. Proc Soc Exp Biol Med 1994;206:345-359.

71. Shanahan CM, Rigby NW, Murray JD, Marshall JT, Townrow CA, Nancarrow CD, Ward KA. Regulation of expression of a sheep metallothionein 1a-sheep growth hormone fusion gene in transgenic mice. Mol Cell Biol 1989;9:5473-5479.

72. Pomp D, Geisert RD, Durham CM, Murray JD. Rescue of pregnancy and maintenance of corpora lutea in infertile transgenic mice expressing an ovine metallothionein 1a-ovine growth hormone fusion gene. Biol Reprod 1995;52:170-178.

73. Mayo KE, Hammer RE, Swanson LW, Brinster RL, Rosenfeld MG, Evans RM. Dramatic pituitary hyperplasia in transgenic mice expressing a human growth hormone-releasing factor gene. Mol Endocrinol 1988;2:606-612.

74. Chandrashekar V, A. Bartke. Pituitary and testicular function in adult transgenic mice expressing the human growth hormone-releasing hormone gene. Biol Reprod 56; Supplement 1, p198, 1997, Abstract.

75. Moore JP Jr, Cai A, Hostettler ME, Arbogast LA, Voogt JL, Hyde JF. Pituitary hormone gene expression and secretion in human growth hormone-releasing hormone transgenic mice: focus on lactotroph function. Endocrinology 2000;141:81-90.

76. Zhou Y, Xu BC, Maheshwari HG, He L, Reed M, Lozykowski M, Okada S, Cataldo L, Coschigamo K, Wagner TE, Baumann G, Kopchick JJ. A mammalian model for Laron syndrome produced by targeted disruption of the mouse growth hormone receptor/binding protein gene (the Laron mouse). Proc Natl Acad Sci USA 1997;94:13215-13220.

77. Kopchick JJ, Laron Z. Is the Laron Mouse an Accurate Model of Laron Syndrome? Mol Genet Metab 1999;68:232-236.

78. Danilovich N, Wernsing D, Coschigano KT, Kopchick JJ, Bartke A. Deficits in female reproductive function in GH-R-KO mice; role of IGF-I. Endocrinology 1999;140:2637-2640.

79. Chandrashekar V, Bartke A, Coschigano KT, Kopchick JJ. Pituitary and testicular function in growth hormone receptor gene knockout mice. Endocrinology 1999;140:1082-1088
80. Chandrashekar V, Bartke A, Kopchick JJ. In vivo effects of luteinizing hormone on testicular endocrine function in growth hormone receptor gene disrupted mice. Biol Repro 60: Supplement 1, p113, 1999, Abstract no63.
81. Baker J, Hardy MP, Zhou J, Bondy C, Lupu F, Bellve AR, Efstratiadis A. Effects of an Igf1 gene null mutation on mouse reproduction. Mol Endocrinol 1996;10:903-918.
82. Steger RW, Bartke A, Parkening TA, Collins T, Yun JS, Wagner TE. Neuroendocrine function in transgenic male mice with human growth hormone expression. Neuroendocrinology 1990;52:106-111.
83. Vidal S, Stefaneanu L, Thapar K, Aminyar R, Kovacs K, Bartke A. Lactotroph hyperplasia in the pituitaries of female mice expressing high levels of bovine growth hormone. Transgenic Res 1999;8:191-202.
84. Vidal S, Stefaneanu L, Kovacs K, Yamada S, Bartke A. Pituitary estrogen receptor alpha and dopamine subtype 2 receptor gene expression in transgenic mice with overproduction of heterologous growth hormones. Histochem Cell Biol 1999;111:235-241.
85. Aguado F, Fernandez T, Martinez-Murillo R, Rodrigo J, Cacicedo L, Sanchez-Franco F. Immunocytochemical localization of insulin-like growth factor I in the hypothalamo-hypophyseal system of the adult rat. Neuroendocrinology 1992;56:856-863.
86. Bohannon NJ, Corp ES, Wilcox BJ, Figlewicz DP, Dorsa DM, Baskin DG. Characterization of insulin-like growth factor I receptors in the median eminence of the brain and their modulation by food restriction. Endocrinology 1988;122:1940-1947.
87. Hiney JK, Srivastava V, Nyberg CL, Ojeda SR, Dees WL. Insulin-like growth factor I of peripheral origin acts centrally to accelerate the initiation of female puberty. Endocrinology 1996;137:3717-3728.
88. Adashi EY, Resnick CE, Payne DW, Rosenfeld RG, Matsumoto T, Hunter MK, Gargosky SE, Zhou J, Bondy CA. The mouse intraovarian insulin-like growth factor I system: departures from the rat paradigm. Endocrinology 1997;138:3881-3890.
89. Casella SJ, Smith EP, van Wyk JJ, Joseph DR, Hynes MA, Hoyt EC, Lund PK. Isolation of rat testis cDNAs encoding an insulin-like growth factor I precursor. DNA 1987;6:325-330.
90. Cailleau J, Vermeire S, Verhoeven G. Independent control of the production of insulin-like growth factor I and its binding protein by cultured testicular cells. Mol Cell Endocrinology 1990;69:79-89.
91. Naville D, Chatelain PG, Avallet O, Saez JM. Control of production of insulin-like growth factor I by pig Leydig and Sertoli cells cultured alone or together. Cell-cell interactions. Mol Cell Endocrinol 1990;70:217-224.
92. Vannelli BG, Barni T, Orlando C, Natali A, Serio M, Balboni GC. Insulin-like growth factor-I (IGF-I) and IGF-I receptor in human testis: an immunohistochemical study. Fertil Steril 1988;49:666-669.
93. Hansson HA, Nilsson A, Isgaard J, Billig H, Isaksson O, Skottner A, Andersson IK, Rozell B. Immunohistochemical localization of insulin-like growth factor I in the adult rat. Histochemistry 1988;89:403-410.
94. Borland K, Mita M, Oppenheimer CL, Blinderman LA, Massague J, Hall PF, Czech MP. The actions of insulin-like growth factors I and II on cultured Sertoli cells. Endocrinology 1984;114:240-246.
95. Handelsman DJ, Spaliviero JA, Scott CD, Baxter RC. Identification of insulin-like growth factor-I and its receptors in the rat testis. Acta Endocrinol (Copenh) 1985;109:543-549.

96. Lin T, Haskell J, Vinson N, Terracio L. Characterization of insulin and insulin-like growth factor I receptors of purified Leydig cells and their role in steroidogenesis in primary culture: a comparative study. Endocrinology 1986;119:1641-1647.

97. Lin T, Wang DL, Calkins JH, Guo H, Chi R, Housley PR. Regulation of insulin-like growth factor-I messenger ribonucleic acid expression in Leydig cells. Mol Cell Endocrinol 1990;73:147-152.

98. Tres LL, Smith EP, Van Wyk JJ, Kierszenbaum AL. Immunoreactive sites and accumulation of somatomedin-C in rat Sertoli-spermatogenic cell co-cultures. Exp Cell Res 1986;162:33-50.

99. Lin T, Haskell J, Vinson N, Terracio L. Direct stimulatory effects of insulin-like growth factor-I on Leydig cell steroidogenesis in primary culture. Biochem Biophys Res Commun 1986;137:950-956.

7

TRANSGENIC AND KNOCKOUT MODELS OF PROLACTIN ACTION IN FEMALE REPRODUCTION

Nadine Binart and Paul A. Kelly

*INSERM Unité 344 - Endocrinologie Moléculaire Faculté de Médecine Necker,
75730 Paris Cedex 15, France*

INTRODUCTION

Generation of transgenic mice that lack functional copies of one or several endogenous genes, or that overexpress an exogenous gene, as well as the phenotypic characterization of spontaneous mutations, has recently allowed physiologists to investigate the functional role of these genes *in vivo*. In this review, we describe phenotypes that specifically affect the fertility of female mice. It is important to have available physiological models that mimic events that occur during human development, in order to better understand, treat, and prevent ovarian and general reproductive failure in women.

Female fertility depends on a precise series of steps in the ovary leading to regulated allocation and maturation of oocytes, and proliferation and differentiation of the surrounding somatic cells during folliculogenesis. The proliferation of female oogonia occurs only during prenatal life in the mouse. At birth, the female has a finite population of oocytes, some begin to grow, while others remain quiescent. Gametogenic and endocrine functions of the gonads are regulated primarily by pituitary gonadotropins. Folliculogenesis is controlled by intragonadal factors initiating follicular growth and oocyte development, granulosa and thecal cell components of the follicle at early stages (1) and by extragonadal factors synchronizing granulosa cell and thecal function later in folliculogenesis. The development of mouse models provided by molecular technology has allowed us to understand the effects of the suppression or overexpression of a selected gene in a tissue or the entire organism (2).

Prolactin (PRL) along with growth hormone (GH) and placental lactogen (PL) form a family of hormones which probably evolved by the duplication of an ancestral gene (3). PRL is mainly synthesized by the pituitary in all vertebrates, whereas placental lactogen is produced by the placenta in mammals. PRL has more actions than all other pituitary hormones combined (4). The initial step in the action of PRL is the binding to a specific membrane receptor, the PRL receptor (PRLR). Similar to the ligand, the PRLR has also been shown to be a member of the same family as the GH receptor, and as well part of the larger category of receptors, known as class-1 cytokine receptor superfamily (5).

In this chapter we will describe transgenic and knockout mice models in which PRL has been deleted or overexpressed or where the PRLR has been removed. This chapter briefly describes the sites of PRL synthesis, the interaction of PRL with its receptor and the main signaling cascades activated in target cells. The main female reproductive phenotypes, including mammary development of PRLR and PRL knockout (KO) mice will be discussed, along with those described for other transgenic models.

PROLACTIN

The amino acid sequence of PRL identified a protein of 199 amino acids (6). Genetic, structural, binding and functional studies of PRL and PL as well as of the more recently identified somatolactin and PRL-related proteins have clearly confirmed that they all belong to a unique family of proteins (7). Post-translational modifications of mature PRL, including glycosylation, phosphorylation or proteolytic cleavage, have been reviewed (8). PRL is mainly synthesized by lactotrophic cells of the anterior pituitary and its expression is essentially controlled by the negative regulation of dopamine, although numerous other factors are able to stimulate or inhibit PRL synthesis/secretion. In addition to being secreted by the pituitary gland, PRL is also produced by numerous other cells and tissues (9). PRL gene expression has been confirmed in various regions of the brain, decidua, myometrium, lacrimal gland, thymus, spleen, circulating lymphocytes and lymphoid cells of bone marrow, mammary epithelial cells and tumors, skin fibroblasts, and sweat glands. Interestingly, hypophysectomized rats retain ~20% of biologically active PRL in the circulation, which increases to ~50% of normal levels with time. Neutralization of circulating PRL with anti-PRL antibodies results in immune dysfunction and death (10), suggesting that PRL from extrapituitary origin is important, and under some circumstances, can compensate for pituitary PRL.

PROLACTIN RECEPTOR

The PRLR was identified as a specific, high affinity, membrane-anchored protein after which, the cDNA encoding the rat PRLR was isolated in our laboratory (11) and shown to have a single-pass transmembrane chain. Sequence comparisons with newly identified membrane receptors led to the identification of a new family of receptors including both PRLR and GHR (12) termed Class-1 cytokine receptors. The gene encoding the human PRLR contains at least 10 exons (13 in the mouse) for an overall length > 100 kb (13,14). Multiple isoforms of membrane-bound PRLR resulting from alternative splicing of the primary transcript have been identified in several species (4). These different PRLR isoforms vary in the length and composition of their cytoplasmic tail and are referred to as short, intermediate (rat Nb2 form, not shown) or long PRLR with respect to their size (Figure 1). In mice, one long and three short isoforms have been identified, the short forms only differing by a limited number of amino acids in the C-terminal part of their cytoplasmic tail (15).

Typically, the extracellular domain (ECD) of a cytokine receptor is composed of a module of ~200 amino acids (aa), referred to as the cytokine receptor homology region (5). This region can be divided into two subdomains, each showing analogies with the fibronectin type III module (17). Two highly conserved features are found in cytokine receptor ECDs: the first is two pairs of disulfide-linked cysteines in the N-terminal subdomain, and the second a "WS motif" (Trp-Ser-any amino acid-Trp-Ser) found in the membrane proximal region of the C-terminal subdomain (Figure 1). The cytoplasmic domain of cytokine receptors displays more restricted sequence similarity than the extracellular domain. Box 1 is a membrane-proximal region composed of amino acids highly enriched in prolines and hydrophobic residues (aa 243-250 in PRLR; Figure 1).

Tissue Distribution of the PRLR

Prolactin receptors have been identified in a number of cells and tissues of embryonic and adult mammals. The expression of short and long isoforms of receptor have been shown to vary as a function of the stage of the estrous cycle, pregnancy and lactation (4). PRL binding sites or receptors are widely distributed throughout vertebrate tissues. Recent studies showed that the mRNA encoding the short and long isoforms was widely expressed in tissues from all three germ layers and that, in addition to the classical target organs of PRL, tissues not known previously to contain PRLR, such as olfactory neuronal epithelium and bulb, fetal adrenal cortex, gastrointestinal and bronchial mucosae, renal tubular epithelia, choroid plexus, thymus, liver pancreas and epidermis, also express the PRLR mRNA. The wide distribution of the PRLR is obviously correlated with the extremely large spectrum of activities of its ligand (4).

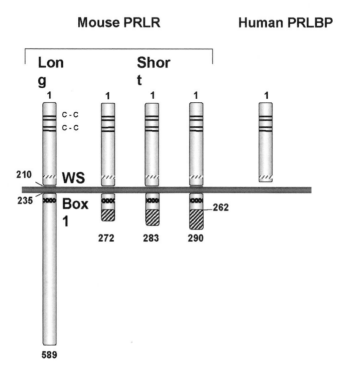

Fig 1. Schematic representation of soluble and membrane isoforms of the PRLR. Although the mechanism of PRLBP generation remains unclear (alternative splicing or proteolysis or both), an mRNA encoding a soluble PRLBP of 206 aa has been isolated in the human breast cancer cell line BT-474 (16). It should be noted that there is no circulating PRLRBP. All four forms have identical extracellular, ligand binding domains. The main consensus features (disulfide C-C, WS motif and Box 1) are represented. The C-terminal region from amino acid 262 differs among the three short isoforms (hatched box), compared to that found in the long form.

INTERACTION BETWEEN PRL AND ITS RECEPTOR

Dimerization of the PRLR upon ligand binding has now been clearly established and at least two regions of hPRL are involved in binding of the hormone to the PRLR. Activation of the PRLR involves ligand-induced sequential receptor dimerization. In a first step, interaction of PRL binding site 1 with one PRLR occurs and leads to the formation of an inactive $H_1:R_1$ (1 hormone, 1 receptor) complex. Formation of this complex appears to be a prerequisite for PRL binding site 2 to interact with another PRLR, then allowing formation of an active trimeric complex ($H_1:R_2$), composed of one molecule of hormone and one receptor homodimer.

Signal Transduction Pathways

All of the actions of PRL result from the interaction of PRL with its receptor on the numerous target cells, which leads to the activation of a cascade of intracellular events. Hormonal stimulation of the PRLR leads to tyrosine phosphorylation of several cellular proteins, including the receptor itself (18). Of the four members of Janus tyrosine kinase family (19), JAK2 is the kinase associated with the PRLR. This association is constitutive (20) i.e., activation, but not recruitment of the kinase is induced by ligand binding. The PRLR-JAK2 interaction involves the membrane-proximal region of the PRLR cytoplasmic domain, in agreement with the ability of the short PRLR isoform to associate with the kinase (4). It is usually assumed that activation of Janus kinases occurs by trans-phosphorylation of tyrosines upon ligand-induced oligomerization of cytokine receptors, which brings two JAK molecules close to each other (19). JAK2 phosphorylates tyrosine residues of different target proteins, the receptor itself and a family of transducing proteins termed Stats (Signal Transducers and Activators of Transcription) (21). Phosphotyrosines are potential binding sites for proteins containing consensus domains involved in protein-protein interactions, such as SH2 or PTB domains. Stat proteins are such signaling molecules recruited by phosphorylated tyrosines of the PRLR-JAK2 complex. Three members of the Stat family have been identified thus far as transducer molecules of the PRLR: Stat1, Stat3 and, Stat5, the latter being the major Stat activated by the PRLR. Once activated, Stats dimerize and migrate to the nucleus where they specifically recognize DNA sequences within the promoters of target genes.

Although the JAK-Stat cascade is presumably the most important signaling pathway used by cytokine receptors, other transducing pathways are also involved in signal transduction by these receptors. Signaling through MAP kinases (MAPK) involves the Shc/SOS/Grb2/Ras/Raf/MAPK cascade. Several recent data suggest that the JAK-Stat and the MAPK cascades are interconnected. PRL also induces a rapid tyrosine phosphorylation of insulin-receptor substrate-1 (IRS-1) and of the 85 kDa subunit of the phosphatidyl-inositol (PI)-3' kinase (22). Both PI-3' kinase and IRS-1 appear to associate with the PRLR in a PRL-dependent manner.

In agreement with the fact that most transducer molecules are activated by tyrosine phosphorylation, involvement of tyrosine phosphatases to modulate or down-regulate the signaling cascades is expected. Moreover, a family of cytokine inducible inhibitors of signaling (CIS/SOCS proteins) has been identified These proteins act not only as negative feedback regulators but also inhibit response to cytokines different from those used to induce their expression. (23-26).

Experimental Models
Classically, two experimental models have been primarily used for the examination of the role of PRL and its receptor *in vivo*. The first involves lowering circulating PRL by either pituitary ablation, achieved by hypophysectomy, or the administration of dopamine D2 receptor agonists, and the second is based on the use of mice strains lacking pituitary somatotrophic and lactotrophic cells. The possibility of incomplete PRL suppression and/or the possible suppression of other pituitary and non-pituitary hormones leave each of these approaches compromised. To circumvent these technical limitations, we have generated PRLR deficient mice by gene targeting (27). PRL KO mice have been developed by others (28). The use of the PRLR knockout model offers the possibility of directly examining the effects of PRLR-mediated signaling *in vivo*. In the following section, we will first briefly describe the generation of these KO mice and then summarize the phenotypes that have been characterized to date in the areas of reproduction/lactation and behavior.

PRL was originally isolated by its ability to stimulate mammary development and lactation in rabbits and soon thereafter to stimulate the production of crop milk in pigeons. PRL was also shown to be luteotrophic, that is to promote the formation and action of the corpus luteum. Subsequently, a number of additional activities have been associated with this hormone in various vertebrate species. We have recently updated the list of PRL activities in vertebrates and found up to 300 different functions involving this hormone (4).

GENERATION OF MICE DEFICIENT IN THE PROLACTIN RECEPTOR GENE AND IN THE PROLACTIN GENE

The technique of gene targeting in mice (2), has been used to develop the experimental model in which the effects of the complete absence of any lactogen or PRL-mediated effects can be studied (27). No PRLR mRNA containing exon 5 was transcribed and no PRLR protein could be detected in the liver of PRLR$^{-/-}$ animals. All data indicate that the exon 5 deletion caused the complete absence of functional receptor in PRLR$^{-/-}$ animals.

Concomitantly, the PRL gene was mutated by others using a targeted insertion of a neomycin (neo) resistance gene into the region encoding the second α-helix of the PRL protein (28). The neo cassette insertion truncated the pre-PRL polypeptide at serine 117 in the second α -helix and added a 12 amino acid extension before the first stop codon in neo. A resulting strain of mice was then generated to better understand the actions of PRL. It should be pointed out that mice lacking the PRLR would be unable to respond to both PRL and placental lactogens (complete lactogen KO), whereas PRL KO mice would still be able to respond to the effects of PL during pregnancy.

Reproductive Phenotypes of PRL KO and PRLR KO Mice

Homozygous PRLR$^{-/-}$ female mice were completely infertile. After mating with male mice of established fertility, no litters were produced following more than 15 matings. Each female mated repeatedly at irregular intervals, without entering a state of pseudopregnancy. Estrous cycles were irregular and individual females failed to establish any consistent pattern of cycling. These observations led the authors to conclude that PRL is essential for female reproduction (28).

PRLR$^{-/-}$ females never became pregnant and showed a number of reproductive deficiencies: an alteration of estrous cyclicity, and an absence of pseudopregnancy. Fewer eggs were fertilized, oocytes at the germinal vesicle stage were released from the ovary and fragmented embryos were found. Single cell fertilized eggs were recovered, suggesting that, for most oocytes, an arrest of development occurred immediately after fertilization. Only a few blastocysts were recovered in the uterus of PRLR$^{-/-}$ mice. The lack of blastocyst implantation leading to the complete sterility in PRLR$^{-/-}$ females may be due to a deficient environment in both the oviduct and uterus. The failure of a significant proportion of eggs to undergo germinal vesicle breakdown within maturing PRLR$^{-/-}$ follicles directly demonstrates the important influence of PRL and its receptor on oocyte maturation or atresia.

Divergent effects of PRL on the rate of implantation and development of mouse embryos have been reported (see references in Ref. 4). A number of factors in the oviduct, which influence preimplantation development, may be affected by the PRLR mutation. For example, estrogen and progesterone can influence the rate of ovum transport and preimplantation development (29).

Uterine preparation for embryo implantation is dependent upon continued estrogen and progesterone secretion by the corpus luteum of the ovary, which is supported by a functional pituitary in rodents during the first half of pregnancy (30). PRL has been shown to stimulate progesterone synthesis by dispersed ovarian cells from mid-pregnant mice (31), demonstrating that lactogenic hormones can directly stimulate ovarian progesterone secretion. Furthermore a nidatory ovarian estrogen surge is required to allow embryo implantation (32). Thus PRLR$^{-/-}$ females probably cannot support the implantation of blastocysts because the corpus luteum does not receive functionally active pituitary PRL support, and thus progesterone- and estrogen-dependent signals for implantation cannot occur. We recently reported that the defect of the preimplantation egg development in these mice could be completely rescued by the addition of progesterone. However, although implantation occurs, the maintenance of full term pregnancy is not complete (33).

The PRL gene and its receptor are expressed in uterus (34) suggesting this might also represent a potential source of ligand for a paracrine or autocrine effect. Overall, our observations indicate that preventing PRL action by disruption of the PRLR gene alters the maternal-decidual transformation in response to the implanting blastocyst, demonstrating an essential role of PRL in reproduction. PRLR expression has also been reported in human endometrial tissue. As shown in Figure 2, we recently studied the spatio-temporal expression of the receptor gene in the mouse embryo and uterus (35). On days 6-8 of pregnancy, uterine PRLR expression is restricted to a subpopulation of undecidualized cells adjacent to the uterine crypt and in the antimesometrial stroma. Although the function of PRLR in these cells is unknown, we cannot exclude their contribution to normal decidual function. PRL is known to be expressed in the decidualized human endometrium and secreted into amniotic fluid. By using *in situ* hybridization histochemistry techniques, PRL specific hybridization signals were distributed over the decidual cells in early and term pregnancy. It would be interesting to determine whether production of PRL or the expression of PRLR is altered in pathological conditions associated with female sterility.

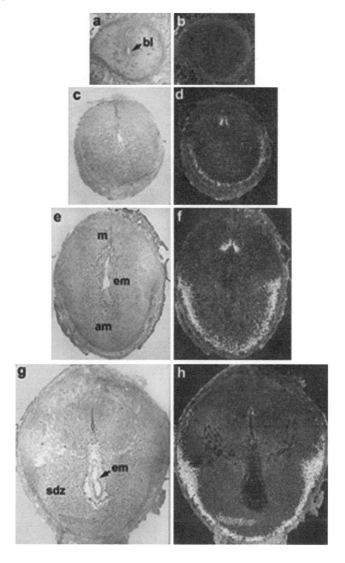

Fig 2. *In situ* hybridization of PRLR mRNA in the periimplantation mouse embryo and uterus. Bright-field (left panel) and corresponding dark-field

(right panel) photomicrographs of representative sections of implantation sites on day 5 (a, b), day 6 (c, d), day 7 (e, f), and day 8 (g, h) are shown at 20X. bl, blastocyst, em, embryo; sdz, secondary decidual zone; m, mesometrial, am, antimesometrial (35).

Furthermore, the expression pattern of progesterone-dependent genes such as amphiregulin, COX-1, and Hoxa-10 was similar in wild-type and steroid-supplemented PRLR$^{-/-}$ mice. These results suggest that the correction of reproductive deficits by progesterone in PRLR$^{-/-}$ mice is accomplished by proper expression of progesterone-dependent genes that are essential in early pregnancy. We did not detect alterations in expression patterns of specific uterine genes that regulate growth and differentiation, positional identity, or vascular tone and permeability. Thus, the rescue of pregnancy failure by progesterone and the cause of pregnancy loss at a later stage in PRLR$^{-/-}$ mice cannot be ascribed to an aberrant spatial expression of genes that normally contribute to the establishment of pregnancy. In summary, PRLR-mediated progesterone production in the ovary appears to be critical for implantation and decidualization.

In order to more fully understand the molecular mechanisms involved in PRL-mediated events in the ovary, oviduct and uterus, we are using approaches such as Differential Display and Serial Analysis of Gene Expression or differential screening to identify genes that are either induced or repressed in the PRLR deficient mice and characterize new gene targets of PRL regulation. In-depth molecular characterization should help to elucidate the actual control mechanisms as well as PRL-dependent signaling pathways which are involved in the early stages of pregnancy.

Mammary Gland Development
At the onset of puberty, the mammary rudiments of ductal architecture transform and ductal elongation and bifurcations begin. PRLR$^{+/-}$ females were unable to lactate their first litter. Histological examination of their mammary glands showed that lactational performance was correlated with the degree of mammary gland development. These results indicate that two functional alleles of the PRLR are required for efficient lactation and that this phenotype in heterozygotes is primarily due to a deficit in the degree of mammary gland development. We suggest that epithelial cell proliferation during pregnancy and the post partum period depends on a threshold of PRLR expression which is not achieved with just one functional allele, given that the level of PRLR is closely controlled in mammary gland (36). Although progesterone is essential

for lobular proliferation, direct actions of PRL are essential for mature mammary gland organogenesis. Heterozygous mice became pregnant and carried their offspring to term, suggesting that progesterone secretion is normal. In these mice where the level of the receptor is reduced, mammary gland proliferation is insufficient to insure lactation at the first pregnancy but further estrous cycles or a single pregnancy lead to the development of a mammary gland capable of producing milk. This is the phenotype seen in a mixed genetic background and in inbred 129/SvJ mice. In contrast, the failure to lactate is much more severe in inbred C57/Bl6 mice, which in general fail to lactate even after multiple pregnancies (37).

The glands of PRLR$^{-/-}$ females showed reduced ductal sidebranching, whereas the terminal end buds of the major mammary ducts and side branches in wild-type animals had differentiated to yield alveolar buds. During pregnancy, alveolar buds previously formed on the ductal tree give rise to lobulo-alveolar structures capable of milk production. Analysis of mammary gland from virgin -/- females compared to those of progesterone-treated animals at 18.5 day of pregnancy by whole mount analysis showed that ductal branching is able to develop. The increase of ductal branching is extensive after administration of progesterone, however the treatment by progesterone and estradiol does not improve alveolar development. Histological analysis of PRLR$^{-/-}$mammary glands was compared to wild type animals at the same stage of pregnancy (Figure 3). The formation of alveolar buds is present but lobuloalveolar development is markedly reduced. Other elusive lactogenic stimuli must signal in parallel and activate the epithelium in particular, both EGF and GH have been suggested to contribute to the developmental program of the mammary gland (37).

In nulliparous PRL$^{-/-}$ mice, the mammary glands are composed of a primary and secondary branched ductal system with numerous terminal end buds that persisted in mice that were even several months old (38). In contrast, phenotypically normal nulliparous littermates (PRL$^{+/-}$) had mammary glands with extensive sprouting of lobular buds along the length of the ducts. Based on these observations, the authors infer that complete PRL deficiency results in arrest of mammary organogenesis at an immature pubertal state where the epithelial component of the gland consists of a basic ductal system and terminal end buds.

In conclusion, the phenotypes of animals lacking functional genes encoding PRLR or PRL confirm the essential role of these molecules in the signaling pathway(s) leading to mammary gland development and/or reproductive function, whereas estradiol and progesterone receptors are probably involved to a lesser extent in lobuloalveolar mammary development.

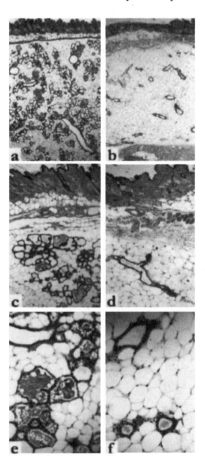

Fig 3. Histological analysis of mammary glands from pregnant PRLR$^{+/+}$ and PRLR$^{-/-}$ progesterone treated females at day 18.5. Hematoxylin/eosin-stained sections through the skin, subcutaneous fat, mammary fat pad and epithelium. PRLR$^{+/+}$ female (a, c, e) and PRLR$^{-/-}$ female (b, d, f) treated with progesterone at the same stage of pregnancy. Magnification: 40 (a and b), 100 (c and d) and 250 (e and f) (33).

Maternal Behaviour Phenotype of PRLR KO Mice

With the evolution of live birth synchronized with lactation, maternal behavior toward young is essential for species survival in mammals. In rodents, such behaviors include nest building, pup retrieval and grouping, anogenital licking and crouching over the pups in the nest to provide warmth and nourishment. It has also been shown that these behaviors can be induced in virgin females with repeated exposure to pups (39,40).

There has been increasing correlative evidence of hormonal control of the onset and maintenance of nurturing behavior (41,42). PRL, levels of which increase markedly in the blood just before parturition, is often implicated as playing a key role in the regulation of maternal behavior.

The use of the PRLR knockout model offers the possibility of directly examining the effects of PRLR-mediated signaling on maternal behavior in the mouse. We studied pup-directed maternal behavior in PRLR KO mice (43). Each female tested was exposed to three foster pups (1 to 3 days old), which were placed in the opposite corner of the cage from the nest. For the following 30 minutes, the test female was continually observed and the following endpoints were recorded: latency to retrieve each pup and place it into the nest, total number of pups retrieved, and latency to crouch over all three pups for more than 5 minutes at a time.

Since PRLR knockout mice are infertile, it was not possible to study behavior in PRLR$^{-/-}$ pregnant or lactating animals. However, primiparous heterozygous females were studied. A profound deficit was observed in maternal care when challenged with foster pups. The typical behavior of wild-type females is to immediately begin retrieving the pups to the nest and crouching over them, remaining with the pups in the nest for the remainder of the test. PRLR heterozygous mutant mice took notice, made immediate contact with and sniffed the pups, as did the wild-type animals. PRLR$^{+/-}$ females then typically returned to the nest, returning once or twice to sniff the pups before retrieving one or at most two pups over the next several minutes before settling into the nest and ignoring the unretrieved pups for the remainder of the test. Wild-type females typically displayed full maternal behavior within several minutes, while none of the PRLR$^{+/-}$ mice exhibited such behavior within the testing period.

PRLR$^{-/-}$ and PRLR$^{+/-}$ nulliparous females showed a clear deficiency in pup-induced maternal behavior. Full maternal behavior, defined as retrieving all three pups and crouching over them in the nest, was displayed by wild-type virgin females with a latency of one or two days, while PRLR$^{+/-}$ virgin females showed an average latency of 4 days before displaying this behavior, and PRLR$^{-/-}$ virgins did not exhibit such behavior at all during the 6 day testing period.

Morris maze studies revealed normal configural learning in the heterozygous and homozygous animals. In addition, eating, locomotor activity, sexual behavior and exploration, all processes regulated by the hypothalamus, were normal in PRLR mutant mice, suggesting specificity in the response. Although the direct target of PRL and its receptor in the brain involved in the mediation of maternal behavior is not known, the use of the PRLR knockout model clearly establishes the PRL and its receptor as regulators of maternal behavior in the mouse, and provides a new tool to understand this process.

Interestingly, PRL$^{-/-}$ nulliparous female mice were tested for maternal behavior toward foster pups and the majority of females retrieved 2-day-old foster pups and crouched over them in a nursing position 30 min after placement of the pups. Thus, the loss of PRL in contrast to a deficit of the receptor does not cause any profound defect in spontaneous maternal behavior.

OTHER TRANSGENIC AND KO MOUSE MODELS

Mice transgenic for hGH (a hormone capable of binding to the PRLR and showing full lactogenic activity) have higher levels of IGF-I (44). Thus, mammary tumor formation observed in these mice could be a consequence of the activation of GHR, PRLR, IGF-IR, or an activation of a combination of them.

Currently, two lines of transgenic mice overexpressing the rat PRL gene are available which should help clarify the receptor specificity (45). These animals have elevated levels of PRL (one line secretes 150 ng/ml and the other, 13 ng/ml) that can only bind to the PRLR but show normal IGF-I levels. All of these animals develop mammary carcinomas at 11-15 months of age. To date, no effect on reproductive function has been reported. Recently, to determine the *in vivo* effects of human PRL in lactating mammary glands, we placed the hPRL cDNA under the control of the Whey Acidic Protein (WAP) TRE (46). Founder lines, in which hPRL was detected in the milk by Western analysis, have been created. Analysis indicated hPRL concentrations of 0.5 and 2.3 g/l in two separate lines. These animals provide both a source for producing large amounts of hPRL and a model to further study the effects of hPRL on mammary gland development and nursing.

Finally, the group of F. Talamantes is in the process of disrupting the placental lactogen –I and –II genes. The bioactivities of these hormones are essentially identical, therefore the functional significance of the sequential expression of very high levels of mPL-I at mid-pregnancy and lower levels of mPL-II during the latter half of pregnancy remains unknown (47). ES cell clones carrying the targeted mutations have been used to generate chimeric mice for both mPL-I and –II, which are currently being evaluated for germline transmission of the disrupted allele. This approach will provide interesting models to examine the functional contribution of placental lactogens.

PATHOLOGIES ASSOCIATED WITH PRL

To date, no pathology linked to mutations of genes encoding PRL or the PRLR have been reported. Either these genes are not important, or they are essential to the survival of the species. As described above, knockout of the PRLR gene in mice is not lethal, but does produce major reproductive defects in females, which would of course affect reproductive function and survival. Despite the lack of known genetic diseases related to PRL or its receptor, PRL is linked with some pathological states, such as tumor formation and reproductive or immunological defects. In humans, hyperprolactinemia is known to be associated with amenorrhea, galactorrhea, and impotence (48). The inhibitory effects on the reproductive processes may be due to both central and peripheral actions of PRL.

The breast is the major target tissue of PRL and the hormone is believed to participate to the proliferation of breast tumor cells. Although this is well accepted for *in vitro* mammary cell cultures, the status of PRL in human breast cancer remains much more controversial *in vivo* (49,50). It is currently assumed that an involvement of PRL could occur through a proliferative autocrine/paracrine loop of mammary produced PRL (49,51,52). Moreover, we detected a higher expression of the PRLR in human breast tumors compared to normal adjacent mammary tissue, which also argues for an increased sensitivity of cancer cells to this hormone (53).

One of our current projects is to analyze the ability of hPRL antagonists (54) to inhibit wt hPRL-induced proliferation of breast cancer cell, both *in vitro* (cell cultures) and *in vivo* using transgenic mice models.

CONCLUSIONS

PRL receptors or binding sites are widely distributed throughout the body. In fact, it is difficult to find a tissue that does not express any PRL receptor mRNA or protein. In agreement with this wide distribution of receptors is the fact that now over 300 separate actions of PRL have been reported in various vertebrates, including effects on water and salt balance, growth and development, endocrinology and metabolism, brain and behavior, reproduction, and immunology and protection. Clearly, a large proportion of these actions are directly or indirectly associated with the process of reproduction, including many behavioral effects. It will be important to correlate known effects with local production of PRL in order to differentiate classical endocrine from autocrine/paracrine effects. The fact that extrapituitary PRL can, under some circumstances, compensate for pituitary PRL raises the interesting possibility that there may be other effects of PRL than those originally observed in hypophysectomized rats. Knockout mouse models should be an interesting system to look for effects activated by PRL or other lactogenic hormones. Future research will center on further expanding the already long list of PRL actions and attempt to better understand the mechanisms of action of this intriguing hormone.

REFERENCES

1. Adashi EY. Intraovarian peptides. Stimulators and inhibitors of follicular growth and differentiation. Endocrinol Metab Clin North Am 1992;21:1-17.
2. Thomas K, Capecchi M. Site-directed mutagenesis by gene targeting in mouse embryo-derived stem cells. Cell 1987;51:503-512.
3. Miller WL, Eberhardt NL. Structure and evolution of the growth hormone gene family. Endocr Rev 1983;4:97-130.
4. Bole-Feysot C, Goffin V, Edery M, Binart N , Kelly PA. Prolactin and its receptor: actions, signal transduction pathways and phenotypes observed in prolactin receptor knockout mice. Endocr Rev 1998;19:225-268.
5. Wells JA, De Vos AM. Hematopoietic receptor complexes. Annu Rev Biochem 1996;65:609-634.
6. Li CH, Dixon JS, Lo TB, Pankov YM, Schmidt KD. Amino acid sequence of ovine lactogenic hormone. Nature 1969;224:695-696.
7. Goffin V, Shiverick KT, Kelly PA, Martial JA. Sequence-function relationships within the expanding family of prolactin, growth hormone, placental lactogen and related proteins in mammals. Endocr Rev 1996;17:385-410.
8. Sinha YN. Structural variants of prolactin: occurence and physiological significance. Endocr Rev 1995;16:354-369.

9. Ben-Jonathan N, Mershon JL, Allen DL, Steinmetz RW. Extrapituitary prolactin: distribution, regulation, functions, and clinical aspects. Endocr Rev 1996;17:639-669.

10. Nagy E, Berczi I. Hypophysectomized rats depend on residual prolactin for survival. Endocrinology 1991;128:2776-2784.

11. Boutin JM, Jolicoeur C, Okamura H, Gagnon J, Edery M, Shirota M, Banville D, Dusanter-Fourt I, Djiane J, Kelly PA. Cloning and expression of the rat prolactin receptor, a member of the growth hormone/prolactin receptor gene family. Cell 1988;53:69-77.

12. Bazan F. A novel family of growth factor receptors: a common binding domain in the growth hormone, prolactin, erythropoietin and IL-6 receptors, and p75 IL-2 receptor β-chain. Biochem Biophys Res Commun 1989;164:788-795.

13. Arden KC, Boutin JM, Djiane J, Kelly PA, Cavenee WK. The receptors for prolactin and growth hormone are localized in the same region of human chromosome 5. Cytogenet Cell Genet 1990;53:161-165.

14. Ormandy CJ, Binart N, Helloco C, Kelly PA . Mouse prolactin receptor gene: genomic organization reveals alternative promoter usage and generation of isoforms via alternative 3'-exon splicing. DNA Cell Biol 1998;17:761-770.

15. Davis JA, Linzer DH. Expression of multiple forms of the prolactin receptor. Mol Endocrinol 1989;3:674-680.

16. Fuh G, Wells JA. Prolactin receptor antagonists that inhibit the growth of breast cancer cell lines. J Biol Chem 1995;270:13133-13137.

17. Kelly PA, Djiane J, Postel-Vinay MC, Edery M. The prolactin/growth hormone receptor family. Endocr Rev 1991;12:235-251.

18. Rui H, Djeu JY, Evans GA, Kelly PA, Farrar WL. Prolactin receptor triggering: evidence for rapid tyrosine kinase activation. J Biol Chem 1992;267:24076-24081.

19. Ihle JN, Kerr IM. Jaks and Stats in signaling by the cytokine receptor superfamily. Trends Genet 1995;11:69-74.

20. Lebrun JJ, Ali S, Sofer L, Ullrich A, Kelly PA. Prolactin-induced proliferation of Nb2 cells involves tyrosine phosphorylation of the prolactin receptor and its associated tyrosine kinase. J Biol Chem 1994;269:14021-14026.

21. Ihle JN. STATs: Signal Transducers and Activators of Transcription. Cell 1996;84:331-334.

22. Berlanga JJ, Gualillo O, Buteau H, Applanat M, Kelly PA, Edery M. Prolactin activates tyrosyl phosphorylation of insulin receptor substrate-1 and phosphatidylinositol-3-OH kinase. J Biol Chem 1997;272:2050-2052.

23. Yoshimura A, Ohkubo T, Kiguchi T, Jenkins NA, Gilbert DJ, Copeland NG, Hara T, Miyajima A. A novel cytokine-inducible gene CIS encodes an SH2-containing protein that binds to tyrosine-phosphorylated interleukin 3 and erythropoietin receptors. EMBO J 1995;14:2816-2826.

24. Starr R, Willson TA, Viney EM, Murray LJ, Rayner JR, Jenkins BJ, Gonda TJ, Alexander WS, Metcalf D, Nicola NA, Hilton DJ. A family of cytokine-inducible inhibitors of signalling. Nature 1997;387:917-921.

25. Endo TA, Masuhara M, Yokouchi M, Suzuki R , Sakamoto H, Mitsui K, Matsumoto A, Tanimura S, Ohtsubo M, Misawa H, Miyazaki T, Leonor N, Taniguchi T , Fujita T, Kanakura Y, Komiya S, Yoshimura A. A new protein containing an SH2 domain that inhibits JAK kinases. Nature 1997;387:921-924.

26. Naka T, Narazaki M, Hirata M, Matsumoto T , Minamoto S, Aono A, Nishimoto N, Kajita T, Taga T, Yoshizaki K, Akira S, Kishimoto T. Structure and function of a new STAT-induced STAT inhibitor. Nature 1997;387:924-929.

27. Ormandy CJ, Camus A, Barra J, Damotte D, Lucas BK, Buteau H, Edery M, Brousse N, Babinet C, Binart N, Kelly PA. Null mutation of the prolactin receptor gene produces multiple reproductive defects in the mouse. Genes Dev 1997;11:167-178.

28. Horseman ND, Zhao W, Montecino-Rodriguez E, Tanaka M, Nakashima K, Engle SJ, Smith F, Markoff E, Dorshkind K. Defective mammopoiesis, but normal hematopoiesis, in mice with a targeted disruption of the prolactin gene. EMBO J 1997;16:6926-6935.

29. Roblero L, Garavagno A. Effect of oestradiol-17β and progesterone on oviductal transport and early development of mouse embryos. J Reprod Fertil 1979;57:91-95.

30. Astwood E, Greep R. A corpus luteum-stimulating substance in the rat placenta. Proc Soc Exp Biol Med 1938; 38:713-716.

31. Galosy S, Talamantes F. Luteotropic actions of placental lactogens at midpregnancy in the mouse. Endocrinology 1995;136:3993-4003.

32. O'Neill C, Quinn P. Interaction of uterine flushings with mouse blastocysts in vitro as assessed by the incorporation of (3H)uridine of the embryo-derived platelet-activating factor in mice. J Reprod Fertil 1981;62:257-262.

33. Binart N, Helloco C, Ormandy CJ, Barra J, Clement-Lacroix P, Baran N, Kelly PA. Rescue of preimplantatory egg development and embryo implantation in prolactin receptor-deficient mice after progesterone administration. Endocrinology 2000;141:2691-2697.

34. Tanaka S, Koibuchi N, Ohtake H, Ohkawa H, Kawatsu T, Tadokoro N, Kumasaka T, Inaba N, Yamaoka S. Regional comparison of prolactin gene expression in the human decidualized endometrium in early and term pregnancy. Eur J Endocrinol 1996;135:177-183.

35. Reese J, Binart N, Brown N, Ma WG, Paria BC, Das SK, Kelly PA, Dey SK. Implantation and decidualization defects in prolactin receptor (PRLR)- deficient mice are mediated by ovarian but not uterine PRLR. Endocrinology 2000;141:1872-1881.

36. Ormandy CJ, Sutherland RL. Mechanisms of prolactin receptor regulation in mammary gland. Mol Cell Endocrinol 1993;91:C1-C6.

37. Gallego MI, Binart N, Robinson GW, Okagaki R, Coschigano K, Perry J, Kopchick J, Oka T, Kelly PA, Hennighausen L. Prolactin, growth hormone and epidermal growth factor activate Stat5 in different compartments of mammary tissue and exert different and overlapping developmental effects. Dev Biol (in press).

38. Horseman ND. Prolactin and mammary gland development. J Mammary Gland Biol Neoplasia 1999;4:79-88.

39. Rosenblatt JS. Nonhormonal basis of maternal behavior in the rat. Science 1967;156:1512-1514.

40. Cosnier J, Couturier C. Maternal behavior induced in adult castrated rats. C R Seances Soc Biol Fil 1966;160:789-791.

41. Bridges RS, Numan M, Ronsheim PM, Mann PE, Lupini CE. Central prolactin infusions stimulate maternal behavior in steroid-treated, nulliparous female rats. Proc Natl Acad Sci USA 1990;87:8003-8007.

42. Bridges RS, Dibiase R, Loundes DD, Doherty PC. Prolactin stimulation of maternal behavior in female rats. Science 1985;227:782-784.

43. Lucas BK, Ormandy C, Binart N, Bridges RS , Kelly PA. Null mutation of prolactin receptor gene produces a defect in maternal behavior. Endocrinology 1998;139:4102-4107.

44. Mathews L, Hammer RE, Brinster RL, Palmiter RD. Expression of insulin-like growth factor 1 in transgenic mice with elevated levels of growth hormone is correlated with growth. Endocrinology 1988;123:433-437.

45. Wennbo H, Gebre-Medhin M, Gritli-Linde A, Ohlsson C, Isaksson OG, Tornell J. Activation of the prolactin receptor but not the growth hormone receptor is important for induction of mammary tumors in transgenic mice. J Clin Invest 1997;100:2744-2751.

46. Hovanec GM, Lewis CJ, Kelder B, Llovera M , Goffin V, Kelly PA, Kopchick JJ. Expression of human prolactin and a human prolactin antagonist (G129R) in cultured mammalian cells and transgenic animals. Proceedings of the Endocrine Society 2000;page 154.

47. Moffat J, McCarty MM, Bowen RA, Simms M, Anthony RV, Talamantes F. Cloning of mouse placental lactogen-I gene and targeted disruption of placental lactogen-I and -II genes. Proceedings of the Endocrine Society 2000; page 99.

48. Blackwell RE. Hyperprolactinemia: evaluation and management. Endocrinol Metab Clin North Am 1992;21:105-124.

49. Clevenger CV, Plank TL. Prolactin as an autocrine/paracrine factor in breast cancer. J Mammary Gland Biol Neopl 1997;2:59-68.

50. Vonderhaar BK. Prolactin: the forgotten hormone of human breast cancer. Pharmacol Ther 1998;79:169-178.

51. Ginsburg E, Vonderhaar BK. Prolactin synthesis and secretion by human breast cancer cells. Cancer Res 1995;55:2591-2595.

52. Clevenger CV, Chang WP, Ngo W, Pasha TM, Montone KT, Tomaszewski JE. Expression of prolactin and prolactin receptor in human breast carcinoma. Am J Pathol 1995;146:695-705.

53. Touraine P, Martini JF, Zafrani B, Durand JC, Labaille F, Malet C, Nicolas A, Trivin C, Postel-Vinay MC, Kuttenn F, Kelly PA. Increased expression of prolactin receptor gene assessed by quantitative polymerase chain reaction in human breast tumors versus normal breast tissues. J Clin Endocrinol Metab 1998;83:667-674.

54. Goffin V, Kinet S, Ferrag F, Binart N, Martial JA, Kelly PA. Antagonistic properties of human prolactin analogs that show paradoxical agonistic activity in the Nb2 bioassay. J Biol Chem 1996;271:16573-16579.

8

GENETIC MUTANTS WITH DYSREGULATION OF CORTICOTROPIN PATHWAYS

Susan E. Murray[1]*, Sarah C. Coste[1]*, Iris Lindberg[2] and
Mary P. Stenzel-Poore[1]

[1]*Oregon Health Sciences University, Portland, OR, [2]Louisiana State University
Medical Center, New Orleans, LA, USA*

*Both authors contributed equally to this work

INTRODUCTION

This review emphasizes emerging concepts in regulation of the hypothalamic-pituitary-adrenal (HPA) axis. We first discuss relevant physiologic pathways and molecular mechanisms that control the HPA axis. Imbalances at many levels can disrupt HPA axis homeostasis leading to conditions such as Cushing's syndrome, pituitary hyperplasia, and anxiety disorders. We then describe recently developed animal models with specific alterations in corticotropin pathways. The primary elements of these pathways have been engineered for upregulation, downregulation or deficiency and thus provide a powerful collection of models for studying HPA axis regulation. These models provide new views into features critical for maintaining control of the HPA axis. They have expanded our understanding of the degree of plasticity and compensation that exists within this system. In addition, they reveal novel mechanisms whereby inappropriate or absent regulatory components lead to abnormal endocrine manifestations and behavioral states.

THE HYPOTHALAMIC-PITUITARY-ADRENAL AXIS

CRH Pathways

Corticotropin-releasing hormone (CRH) is the primary hypothalamic hormone responsible for initiating stress-induced release of adrenocorticotropin hormone (ACTH) from the anterior pituitary and subsequent release of adrenal glucocorticoids (1). This cascade of HPA axis activation mediates the endocrine response to stress. Subsequent actions of glucocorticoids affect a wide range of physiologic and metabolic processes, most notably those influencing the mobilization of energy stores and glucose uptake. Such actions facilitate adaptations that improve the ability of an individual to adjust homeostasis and increase its survival. In addition, glucocorticoids dampen expression and release of CRH and ACTH, thereby providing an important source of negative feedback regulation in the stress response (2).

CRH is widely distributed throughout the brain, in addition to its well known location in the parvocellular region of the paraventricular nucleus (PVN) of the hypothalamus (3). The fact that CRH exists in numerous extra-hypothalamic sites (e.g. neocortex, amygdala, olfactory bulb, nucleus of the solitary tract) coupled with a growing appreciation of the multiple effects of CRH in the CNS as well as the periphery supports the idea that CRH plays a central role in stress-responsive systems that extend well beyond the `HPA axis neuroendocrine hormone response (4). In addition, urocortin (Ucn), a newly discovered peptide hormone that shares extensive structural similarity to CRH, exists in several sites in the CNS (e.g. Edinger-Westphal nucleus, lateral superior olivary nucleus and limited abundance in the PVN) as well as peripheral locations (e.g. intestinal and immune systems) (5, 6). Importantly, Ucn is a potent ligand of CRH receptors and while very little is known about the actions of Ucn at present, it is likely that Ucn, in addition to CRH, mediates specific actions within a stress-responsive system.

There are two major classes of high affinity CRH receptors that are G-protein regulated and positively coupled to adenylate cyclase (reviewed in (3)). Both receptors, referred to as CRH-R1 and CRH-R2, possess putative seven-transmembrane spanning domains and share ~70% homology (7-11). The two receptors are the product of distinct genes and while CRH-R1 generally exists as a single functional form (CRH-R1α), CRH-R2 is expressed in three distinct functional subtypes (CRH-R2α, CRH-R2β, and CRH-R2γ) that differ in their N-termini due to alternative splicing (3, 12). CRH-R1 and CRH-R2 differ in their binding profiles: CRH-R1 binds CRH and Ucn with similar affinity, while CRH-R2 binds Ucn with approximately 40-fold greater affinity than CRH (6). In addition, the distribution of the two receptors is distinct (3). CRH-R1 is expressed predominantly in the pituitary and specific regions of the brain including cerebral cortex, cerebellum, brain stem and septum with lower levels detectable in the hypothalamus. In contrast, CRH-R2 is expressed in the periphery, most notably in the heart and skeletal muscle, as well as the gastrointestinal tract, lung and epididymis. CRH-R2 is also expressed in limited regions of the brain including the lateral septum, hypothalamus, choroid plexus and within cerebral arterioles. The specific pharmacological properties and non-overlapping anatomical localization of CRH-R1 and CRH-R2 suggests they may subserve distinct functions in mediating the actions of CRH and Ucn.

CRH also binds a CRH-binding protein (CRH-BP) with an affinity equal to or higher than CRH receptors (6, 13-15). Binding of CRH by CRH-BP blocks the activity of CRH *in vitro* (14, 16). The precise role of CRH-BP *in vivo* is not entirely clear; however it is believed that CRH-BP modulates CRH levels available for binding to CRH receptors. In addition, CRH-BP may aid in clearance of CRH, particularly in humans, where CRH-BP is made in the liver and released into the circulation (15, 16). Unlike humans, rodents do not express CRH-BP in the liver or placenta. However, CRH-BP is found in rodents and humans in various regions of the brain (e.g. cerebral cortex, amygdala, hypothalamus) and importantly, within corticotropes in the pituitary (17). Due to its colocalization with either CRH or CRH receptors in the brain and pituitary, CRH-BP is thought to regulate CRH input to the HPA axis and central CRH pathways mediating behavior.

Regulation of CRH Pathways
Control of HPA axis activation exists at several levels to ensure rapid and specific hormone release that is followed by restoration of HPA axis activation to basal tone. Regulation of CRH and ACTH occurs through such diverse mechanisms as transcriptional modulation, peptide hormone processing and CRH-BP interactions.

Glucocorticoids impose a vital form of negative regulation on CRH and ACTH expression. Transcription of CRH in the hypothalamus and proopiomelanocortin (POMC) in pituitary corticotropes is suppressed by glucocorticoids, providing an essential means of re-establishing basal HPA axis hormone levels (18). In addition, glucocorticoids have been shown to regulate ACTH secretion (19). Glucocorticoid feedback actions are mediated by binding to intracellular receptors: Type 1, mineralocorticoid receptors (MR) found within the septal and hippocampal regions of the brain and Type 2, glucocorticoid receptors (GR) found in the hippocampus, hypothalamic PVN and on anterior pituitary corticotropes (reviewed in (18)). These receptors are ligand-regulated transcription factors within the nuclear hormone receptor super family. Similar to other members of its family, GR has a modular structure consisting of a DNA binding domain, a ligand binding domain and two transactivation motifs (AF1 and AF2). In the absence of ligand, receptor is retained in the cytosol in an inactive state associated with several heat shock proteins. Upon ligand binding, GR translocates to the nucleus where it controls the transcription rate of target genes. One mechanism involves activation or repression of gene transcription by homodimer binding of the receptor to glucocorticoid responsive elements in DNA promoter regions. In addition, GR also regulates transcription by interfering with functions of other transcription factors via protein-protein interactions such as activating protein 1 and nuclear factor-κB. Multiple interactions at the DNA and protein-protein level provide diversity and complexity of transcriptional control by glucocorticoids.

Regulation of the HPA axis is also accomplished through tissue-specific proteolytic processing of POMC-derived peptides. The POMC precursor is processed to hormones involved in the HPA axis, such as ACTH, through the action of specific processing enzymes - the prohormone convertases PC1 and PC2. Differential expression of PC1 and PC2 dictates the tissue profile of proteolytic products obtained from POMC. PC1 in the anterior and intermediate lobes initially processes POMC at multiple dibasic sites to yield ACTH, β-lipotropin, and several amino-terminal peptides. However, in intermediate lobe melanotrophs, which express both enzymes, PC2 cleaves PC1-generated peptide intermediates, such as ACTH, yielding smaller peptide products. These peptides include α-MSH and β-endorphin which possess very different biological activities from the PC1 generated products (20). While the transcriptional and enzymatic regulation of PC1 and PC2 is not fully understood, POMC expression itself is under complex transcriptional control (reviewed in (21)). In the anterior lobe, glucocorticoids inhibit POMC transcription, while in the intermediate lobe glucocorticoids upregulate POMC synthesis. Furthermore, dopaminergic input negatively controls both POMC synthesis and secretion (reviewed in (21)). Thus, the sum of corticotropic peptides secreted by the pituitary depends on neuronal and hormonal input and is affected by agents that promote pituitary tissue hyperplasia or affect the synthesis, processing and release of POMC.

The HPA axis may also be regulated by CRH-BP via its ability to modulate free CRH levels in the brain and circulation. Several sites of CRH-BP expression colocalize with CRH neurons (e.g. central nucleus of the amygdala, lateral septal nucleus) or CRH target cells, most notably pituitary corticotropes (17). It is important to note that CRH-BP binds 40-90% of the total CRH, thus the availability of "free CRH" may be determined, in large part, by CRH-BP (22). Although the precise function of CRH-BP remains speculative, the recent development of animal models that overexpress CRH-BP or lack the gene have given some clues regarding its role in maintaining HPA axis homeostasis.

Dysregulation of CRH Pathways in Human Disease States
Dysregulation of the HPA axis and central CRH pathways has been linked to several human pathological states, both physiologic and psychologic. Cushing's syndrome results from long-term exposure to glucocorticoids. It is commonly caused by altered ACTH or glucocorticoid secretion or by exogenous cortisol treatment. Cushing's disease, which occurs in a subgroup of patients with Cushing's syndrome, is defined as hyperadrenocorticalism resulting from excessive secretion of pituitary ACTH. Excess glucocorticoid levels are

associated with metabolic effects including muscle wasting, abnormal fat deposits, brittle bones, thin skin, hair loss and hyperglycemia (23). The most common source of ACTH in patients with Cushing's syndrome is pituitary tumors that are insensitive to glucocorticoid negative feedback; however, some patients display pituitary hyperplasia, ectopic ACTH-producing tumors, or CRH-secreting tumors (23). Pituitary hyperplasia may be the result of dysregulated hypothalamic input although this point is debated.

HPA axis and central CRH pathway dysregulation has also been implicated in several affective disorders such as anorexia (24) and major melancholic depression (25) which are characterized by hyperarousal and severe anxiety. These patients have high circulating cortisol levels (often as high as those seen in Cushing's patients) with concurrently high levels of CRH in cerebrospinal fluid (24). This suggests that negative regulation of CRH may be impaired since such elevations in cortisol ordinarily would be expected to suppress CRH levels. In contrast to conditions of presumed CRH hyperactivation, recent evidence suggests that lowered CRH levels may be associated with Alzheimer's disease (22, 26). These patients have low levels of CRH in cerebrospinal fluid (27) and post mortem examination showed that CRH immunoreactivity was decreased in the cortex (22, 26). In addition, brains from patients with Alzheimer's disease showed that a greater proportion of CRH was bound to CRH-BP potentially leading to a further drop in available CRH. Associations between apparent CRH dysregulation and affective or cognitive changes are intriguing; however, these observations need to be considered cautiously because causal links are missing. The recent development of appropriate animal models with well defined defects in the HPA axis and CRH pathways should provide much needed information on critical issues of causality.

Dysregulation of Corticotropin Pathways: Animal Models
Several animal models have been developed recently that represent a unique dissection of corticotropin pathways (Table 1). These transgenic and knockout mice demonstrate in some cases, known principles of HPA axis regulation and central CRH pathways. In other cases, discoveries of previously unrecognized functions for specific elements in corticotropin regulation are featured and as a consequence, our understanding of CRH pathways has been remodeled. Taken together, these mouse mutants underscore the leading importance of feedback regulation while sustaining HPA axis tone and the capacity to adapt to new stimuli.

Table 1. Animal models of corticotropin pathway dysregulation

Model	HPA axis		References
	basal	*stress*	
CRH Pathways			
CRH Tg	increased	delayed	(28)
CRH KO	decreased	hypo-responsive	(29)
CRH-R1 KO	decreased	hypo-responsive	(30, 31)
CRH-R2 KO	normal	altered kinetics	(32-34)
Altered CRH Regulation			
CRH-BP Tg	normal	normal	(35, 36)
CRH-BP KO	normal	normal	(37)
GR KO	normal/increased[1]	hyper-responsive	(38-40)
Pituitary Dysregulation			
D2R KO	increased	n.d.	(41, 42)
7B2 KO	increased	n.d.	(43)
LIF KO	normal	hypo-responsive	(44)
LIF Tg	increased	n.d.	(45)

[1]Three models are considered; the HPA axis phenotype depends on the model (see text). n.d. not determined.

DISRUPTION OF CRH PATHWAYS

CRH-Overexpressing Transgenic Mice (CRH-Tg)

A transgenic model of chronic CRH hypersecretion was created by Stenzel-Poore et al. using a chimeric CRH transgene comprised of the metallothionein promoter driving the rat CRH genomic gene (28). The hallmark of these mice is chronic HPA axis activation due to central overproduction of CRH leading to elevated ACTH release and increased circulating glucocorticoid levels. The physical phenotype of these mice mimics Cushing's syndrome: thin skin, hair loss, brittle bones, truncal obesity and a characteristic buffalo hump (28). These features presumably arise from high circulating corticosterone levels similar to those seen in patients treated with exogenous glucocorticoids. Adrenalectomized CRH-Tg mice exhibit a normal physical phenotype and survive for months without corticosterone replacement. CRH-Tg mice also show a variety of behavioral features of anxiety that parallel those seen in stress (46-49). The CRH-Tg animal model is particularly well suited to explore neuropathological consequences of altered endocrine and brain CRH system activation because these mice exhibit chronic HPA axis activation and anxiogenic behaviors, features well known to accompany mammalian stress responses.

CRH overproduction in CRH-Tg mice is largely confined to regions in the brain that normally express CRH (e.g. PVN, preoptic area, amygdala, olfactory bulb, lateral septum). Compared to control wild-type mice, CRH expression is increased in these regions with the exception of the PVN where CRH expression matches control levels. This may reflect that endogenous CRH in the PVN is down-regulated due to increased circulating levels of glucocorticoids. Given that the CRH promoter had been replaced by the more broadly expressed metallothionein promoter, we anticipated widespread expression of the transgene in the brain and periphery resulting in marked elevations of CRH in the CNS and circulation. However, this did not occur, suggesting that sequences located within the rat genomic CRH gene contain information that regulates expression in certain CNS sites and peripheral tissues. In keeping with this, peripheral expression of the mMT-CRH transgene follows a pattern similar to that known for endogenous peripheral CRH localization: lung, adrenal, heart and testis. The restricted targeting profile of the CRH transgene results in a model that mimics the physiology of chronic activation of the HPA axis and CRH dysregulation in the brain.

Responses of the HPA Axis in CRH-Tg Mice

CRH-Tg mice show neuroendocrine features similar to chronic HPA axis activation; basal ACTH levels are elevated ~5-fold and corticosterone levels are increased by 10-fold. Circulating levels of CRH are not elevated, consistent with a lack of significant expression of CRH in peripheral sites. However, CRH overproduction in the hypothalamus clearly results in chronic pituitary activation and production of glucocorticoids. This indicates that continuous stimulation of the HPA axis can drive production of ACTH despite chronic elevations in glucocorticoids which are potent negative regulators of ACTH production. Moreover, the CRH-Tg mice retain their ability to further activate the HPA axis in response to exogenous stress (S. Murray and M. Stenzel-Poore, unpublished). Wild-type mice show a rapid increase, within 10 minutes, in corticosterone levels following stress; however, CRH-Tg mice fail to increase their corticosterone levels above basal levels by 10 minutes. By 30 minutes post-stress both wild-type and CRH-Tg mice show similar elevations in corticosterone that are well above their basal levels demonstrating that, in time, stress-responsive activation of hypothalamic neurons does overcome glucocorticoid feedback. Despite chronic HPA axis activation, CRH-Tg mice clearly respond to stress, albeit with altered kinetics. Thus, it appears that CRH-Tg mice have adapted their HPA stress axis to a different set point. This is in keeping with the idea that there are mechanisms that maintain stress-responsiveness in spite of increased corticosterone secretion (50).

This CRH-Tg model of continuous CRH drive provides a unique setting to examine the potential link between chronic CRH input to the pituitary and corticotrope hyperplasia. Excess production of CRH has been implicated in pituitary hyperplasia and microadenoma formation leading to hypersecretion of ACTH; however, CRH-Tg mice do not have larger pituitaries or develop overt pituitary tumors. Furthermore, the corticotrope number is similar or may be slightly increased above that of wild-type mice. This finding is somewhat surprising in light of previous studies showing that chronic overproduction of another hypothalamic peptide, growth hormone-releasing hormone, leads to massive hyperplasia and adenoma formation of somatotropes (51, 52). This suggests that in some cases, the releasing peptide acts as a specific trophic factor for its targets cells. In the case of the CRH-Tg mice, other factors such as elevated glucocorticoids may act to restrain potential trophic effects. Thus, chronic CRH appears to lead to modest increases in corticotropes, but not to adenomas (28, 53). In the CRH-Tg mice, CRH hypersecretion and the resultant increase in ACTH leads to the development of Cushing's syndrome. In humans, abnormal hypothalamic CRH secretion has been suggested as a possible cause of corticotrope hyperplasia associated with Cushing's (54, 55). CRH-Tg mice illustrate that chronically high CRH levels can over-ride normal glucocorticoid negative feedback to the pituitary, resulting in a Cushing's phenotype.

Anxiogenic Behaviors in CRH-Tg Mice
CRH is widely considered to be a critical mediator of stress-related behaviors (56). These effects on anxiety-related behavior occur independent of pituitary-adrenal activation. High level expression of CRH throughout the brain in CRH-Tg mice would be expected to lead to anxiogenic behavior. Indeed, these mice exhibit increased anxiety on the elevated plus maze and enhanced responsiveness to novelty (46). Importantly, anxiogenic behavior is reversed with central administration of the CRH receptor antagonist, α-helCRH9-41. Additionally, deficits in learning and memory processing are seen in CRH-Tg mice consistent with a role for CRH in learning plasticity (48). Thus, chronic elevations in central CRH found in CRH-Tg mice are sufficient to result in increased anxiety and altered learning similar to behavioral responses to stress. Moreover, CRH-Tg mice do not appear to habituate to the effect of CRH despite lifelong elevations in central CRH expression.

Immunological Abnormalities in CRH-Tg Mice
Both Cushing's and corticosteroid-treated patients frequently experience greater susceptibility to infection, presumably due to glucocorticoid-mediated immunosuppression (23). CRH-Tg mice represent an animal model in which specific effects on immune function that arise from modest but unrelenting elevations in glucocorticoids can be examined. It is important to note that these animals also have chronic elevations in peripheral ACTH levels as well as central CRH levels, thus, our observations may not be due exclusively to increased glucocorticoids in the circulation.

CRH-Tg mice have reduced numbers of cells in the bone marrow, thymus, spleen and blood (47). This is primarily due to decreases in the number of T and B lymphocytes. In addition, CRH-Tg mice develop lower antibody titers following immunization consistent with the death of B lymphocytes. In contrast to a global decrease in lymphocytes, the number of neutrophils is actually increased in the blood. This increase in neutrophils represents an important change in host defense particularly since neutrophils are well recognized for their importance in phagocytosis of extracellular microbial pathogens (S. Murray and M. Stenzel-Poore, unpublished).

These findings suggest that chronic HPA axis activation leads to both immunosuppressive as well as immunoenhancing effects. Recently, acute stress has been shown to enhance cell-mediated immunity; this process is glucocorticoid-dependent (57, 58). Furthermore, our findings showing that neutrophils, which play a crucial role in host defense during bacterial infections, are increased in CRH-Tg mice, suggest an additional stress-associated link to immunoenhancement. Taken together, these data support an emerging view that in certain situations, the effects of stress on immune function are adaptive and lead to enhanced survival.

CRH Knock Out Mice (CRH KO)
Mice deficient in CRH were created using targeted gene inactivation in embryonic stem cells (29); thus, CRH is absent in all tissues throughout development. This model has added important new information to our present understanding of CRH roles and in some cases has provided evidence that contradicts current hypotheses. CRH KO mice exhibit severe glucocorticoid deficiency and impaired HPA axis activation—features predicted to occur in the absence of CRH. Surprisingly, although CRH has been shown to be a mitogenic stimulus in pituitary cultures *in vitro* (53), the absence of CRH *in vivo* does not result in any alterations in pituitary structure. Although pituitary ACTH was similar to WT mice, these levels were lower than expected given the presumed lack of negative feedback from glucocorticoids. While pituitary development is normal, the low basal glucocorticoid levels may result from abnormal adrenal development since development of the zona fasciculata (corticosterone-producing region) is impaired in these mice. This defect may be due to altered ACTH input. CRH KO mice do not exhibit the evening rise in ACTH that WT mice do, despite having normal basal levels of this hormone. These findings suggest that CRH is not necessary for pituitary corticotrope development but is essential for adrenal gland maturation.

Low circulating corticosterone levels in this model have been used to further our understanding of the role of glucocorticoids in longevity. Prenatal administration of glucocorticoids to homozygous mothers is required for fetal lung maturation and postnatal survival of homozygous litters (29). Thus, this model has allowed for extensive examination of glucocorticoid influence on fetal lung differentiation and maturation (59). However, glucocorticoid treatment beyond the fetal period is not necessary. CRH KO mice exhibit normal longevity and fertility despite low circulating glucocorticoids (29). In contrast to previous thought, this finding suggests that the many changes associated with adrenal insufficiency such as low body weight, fatigue and decreased fertility may not be related directly to the lack of glucocorticoids.

While CRH KO mice are unable to mount a normal HPA axis response to behavioral stress (ether inhalation and restraint stress) (29), immune activation induced by endotoxin produces a robust (though somewhat lower than WT) rise in plasma ACTH and corticosterone in these mice (60). This indicates that immune stimulation can activate HPA axis pathways independent of CRH and that the abnormally developed zona fasciculata is capable of corticosterone responses under conditions of sufficient stimulation. Cytokines induced during the inflammatory response may act directly at the level of the pituitary or adrenal gland to stimulate glucocorticoid release in the absence of CRH. A similar pattern of HPA axis activation was found in mice lacking CRH-R1 as described below. These findings suggest that immune activation of the HPA axis is selectively preserved in these animals and may underscore an adaptive role for immune-mediated HPA axis activation. Glucocorticoids are necessary to suppress pro-inflammatory cytokines, thereby limiting their deleterious effects. Such a pathway that permits activation of the HPA axis independent of CRH would ensure the ability of the HPA axis to improve survival during immune challenge.

When administered directly into the brain, CRH produces behaviors similar to those observed following exposure to stress (56). Furthermore, many of these behaviors can be inhibited with CRH antagonism. However, measuring release of endogenous CRH from specific brain areas during stress has proven difficult. CRH KO mice provide a unique opportunity to test whether CRH mediates stress-related behaviors. Surprisingly, all behavioral responses to stressors measured to date are normal in mice lacking CRH; freezing, learning and anxiety-like behaviors associated with stress are similar to wild-type controls (61). Interestingly, CRH receptor antagonists inhibit stress-induced behavior in the CRH KO mice (61). In addition, basal feeding as well as feeding responses to hypophagic stimuli is normal (62, 63). These data provide evidence that CRH is not required to initiate these behaviors and leave open the intriguing question as to what CRH-related peptide mediates these behaviors. It has been shown that Ucn is not up-regulated in these animals. Thus, it is not clear whether Ucn normally instigates these behaviors or substitutes for CRH in its absence. Alternatively, these studies could suggest that an unidentified CRH-like molecule is involved in stress-induced behaviors.

CRH Type 1 Receptor Knock Out Mice (CRH-R1 KO)

The recent development of mice lacking functional CRH-R1 lends important insight into the biological effects of CRH that are mediated by this receptor. CRH-R1 KO mice were generated by two independent groups using targeted gene inactivation in embryonic stem cells (30, 31). These two lines of mice generally show similar phenotypes despite differences in the genetic background of the embryonic stem cell lineage. Most notably, these mice have contributed to our understanding of HPA axis regulation. A primary role of hypothalamic CRH is to initiate HPA axis activation by releasing ACTH from pituitary corticotropes, an effect most likely mediated by CRH-R1 since it is the predominant CRH receptor expressed in the pituitary (64). Indeed, CRH-induced release of ACTH is impaired in cultured pituitary cells collected from CRH-R1 KO mice (30, 31). Interestingly, basal ACTH levels *in vivo* are normal in mice lacking CRH-R1. This suggests that other hypothalamic ACTH stimulating hormones can compensate for the lack of CRH input and maintain basal ACTH release. Again, it should be noted that as in the case of CRH KO mice, higher levels of ACTH would be predicted given the presumed lack of glucocorticoid negative feedback; however, such increases were not seen. Turnbull and colleagues (65) have shown that systemic administration of vasopressin antiserum significantly reduced basal levels of ACTH in mutant mice while having negligible effects in wild-type mice. Thus, vasopressin supplies important stimulatory influences on ACTH secretagogues which preserve ACTH levels in the absence of CRH stimulation. These data confirm that CRH-R1 mediates CRH-induced release of pituitary ACTH but also demonstrate that circulating ACTH is conserved by other hormones.

Despite normal circulating ACTH levels, CRH-R1 KO mice exhibit pronounced glucocorticoid deficiency (30, 31). Mutant animals have extremely low circulating corticosterone levels (2-8 ng/ml) with no diurnal rhythm. Similar to CRH KO mice (29), offspring of homozygous matings die within 48 hours of birth (30). Neonatal mortality has been attributed to inadequate lung maturation due to low levels of glucocorticoids. Atrophy of the adrenal gland appears to be responsible for this deficiency. Smith et al. (30) found a marked decrease in the size of the zona fasciculata region of the adrenal gland. Other regions including the zona glomerulus, zona reticularis and medulla were normal. Postnatal treatment (days 10-21) with ACTH was found to prevent this atrophy (30). Thus, adrenal insufficiency appears to be due to lower levels of ACTH during neonatal adrenal maturation in CRH-R1 KO mice. In the other line of CRH-R1 KO mice, Timpl et al. (31) also reported profound glucocorticoid deficiency, but the cause is less clear in this case. The zona fasciculata appeared normal in these mice; however the size of the adrenal medulla was significantly reduced. It is not clear whether low sympatho-medullary drive could lead to reduced corticosterone levels. Alternatively, atrophy of the adrenal medulla could be a secondary effect of low corticosterone. Nonetheless, CRH-R1 KO mice have demonstrated the importance of CRH stimulatory influences on the development of the mature adrenal gland.

CRH-R1 is highly expressed in the pituitary and therefore likely to mediate the stress effects of CRH on pituitary ACTH release. Thus, it was of interest to determine whether CRH-R1 KO mice could respond to stress with HPA axis activation. Vasopressin and CRH act synergistically to release ACTH from the pituitary, raising the possibility that vasopressin may initiate activation of the axis. CRH-R1 KO mice exhibit severely compromised HPA axis activation in response to behavioral stress (30, 31). Acute restraint stress or forced-swim stress did not significantly increase circulating ACTH and corticosterone levels in mutant mice. Thus, as might be predicted, CRH-R1 on pituitary corticotropes is critical for the initiation of the stress response. In contrast to behavioral stress, CRH-R1 KO mice were able to produce a marked activation of the HPA axis in response to turpentine-induced local inflammation (65). Circulating ACTH and corticosterone were both increased in mutant mice following inflammation. However, the mechanism underlying this response appears to be quite different in CRH-R1 deficient mice. Neither antiserum to CRH nor vasopressin prevented the rise in ACTH as it did in wild-type mice. Interestingly, mutant mice responded to inflammation with a more dramatic increase in the cytokine, IL-6, compared to wild-type mice (65). Thus, pronounced increases in IL-6 may contribute to HPA axis activation in mutant mice, independent of CRH or vasopressin pathways as it has been shown that IL-6 is capable of stimulating synthesis and secretion of ACTH in primary pituitary cultures. Robust increases in IL-6 in these mice could reflect low circulating corticosterone or lack of CRH input since both have been shown to restrain cytokine production.

Pharmacological administration of CRH and CRH antagonists into the central nervous system suggests that CRH pathways play an important role in the mediation of anxiety-related behavior (56). In addition, as mentioned above, transgenic mice with CRH overproduction show increased anxiety. Studies with specific CRH-R1 antagonists suggest that CRH-R1 mediates these anxiogenic effects of CRH. In support of this evidence, CRH-R1 KO mice show reduced anxiety-related behaviors under basal conditions and during alcohol withdrawal (30, 31). Thus, it appears that CRH-R1 may indeed mediate anxiety-related behavior induced by CRH.

CRH Type 2 Receptor Knock Out Mice (CRH-R2 KO)

As discussed above, CRH-R1 KO mice have been instrumental in defining central functions for CRH-R1 in anxiety and neuroendocrine stress responses (30, 31). Although our understanding of the role of CRH-R2 is more rudimentary, functions for CRH-R2 are being identified using antagonists specific for this receptor, antisense strategies that reduce receptor expression and the recently developed CRH-R2 KO mice (32-34). The CRH-R2 KO mouse model has provided surprising insights into the role of CRH-R2 in mediating central and peripheral responses to CRH and Ucn. Importantly, studies using these mice suggest a role for CRH-R2 in coordinating responses initiated through CRH-R1. In addition, CRH-R2 may be intimately involved in mediating adaptive responses to stress leading to the re-establishment of stability-a process known as allostasis (66).

CRH-R2 KO mice were generated by us (33) and two other independent groups (32, 34) using targeted gene disruption in embryonic stem cells. Studies thus far have focused on defining the role of CRH-R2 in behavioral and neuroendocrine HPA axis responses to stress and the effect of Ucn in feeding and cardiovascular activities.

HPA Axis Responses to Stress in CRH-R2 KO Mice

The response to stress involves initiation, maintenance and recovery processes. Mice lacking CRH and CRH-R1 have revealed that both CRH and CRH-R1 are critical for initiating the HPA stress hormone cascade (30, 31). The presence of CRH-R2 in the PVN (64) and the observation that icv administration of CRH causes a rapid increase in CRH expression in the PVN (67) led us to postulate that CRH-R2 modifies the stress response following initial activation of the HPA axis. Indeed, studies with CRH-R2 KO mice indicate that following a brief restraint stress, ACTH levels are more robust initially and decline more rapidly compared to wild-type mice (32, 33). More rapid termination of ACTH in CRH-R2 KO mice suggests that CRH-R2 may sustain the early ACTH response possibly through CRH actions on CRH-R2 in the PVN. That such a feed-forward mechanism of regulation exists has been suggested by previous studies showing that pituitary-adrenal activation by stress can be further enhanced with exogenous CRH (68) and the finding that CRH stimulates its own expression in the PVN (67) as mentioned above.

CRH-R2 KO mice also exhibit abnormal recovery from activation of the HPA axis. Corticosterone levels remained significantly elevated post-stress in CRH-R2 KO mice compared to wild-type mice (33). Thus, CRH-R2 may regulate the recovery phase of the stress response, perhaps by influencing negative feedback of the HPA axis—an effect likely to be independent of its feed-forward actions in the hypothalamus. Collectively, the data suggest that CRH-R2 plays an integral part in shaping the HPA axis response to stress.

Behavioral Stress Responses in CRH-R2 KO Mice

As discussed above, CRH KO mice have revealed the unexpected finding that a number of stress-induced behaviors do not depend on the presence of CRH; instead, these behaviors require the presence of CRH-R1. The location of

CRH-R1 in regions thought to be involved in mediating fear and anxiety, particularly the amygdala and the bed nucleus of the stria terminalis, provides additional reinforcement for a central role for CRH-R1 (64). In contrast to CRH-R1, CRH-R2 is located in sites that may have significant involvement with somatosensory processing (nucleus of the solitary tract) and feeding (PVN and ventromedial hypothalamus (VMH)) (64).

We have found no evidence for altered anxiety responses in CRH-R2 KO mice using the elevated plus maze. Our results differ from those obtained with independently generated CRH-R2 KO mice that show increased anxiety-like behavior in the elevated plus maze (32, 34). The discrepancy between our results and these two groups may be due to differences in genetic background that result from use of ES cells derived from different 129 mouse sub strains. Altered anxiety-like behavior was not due to changes in locomotor activity as CRH-R2 KO mice do not differ from wild-type mice in total ambulation when exposed to an open field (32-34). Furthermore, CRH-R2 KO and wild-type mice show similar reductions in locomotor changes following Ucn administration icv indicating that Ucn actions on activity in an open-field do not depend exclusively on CRH-R2 (33).

Enhanced anxiety-like behavior in CRH-R2 KO mice in these paradigms could reflect altered adaptation to stress. It is intriguing to speculate that CRH-R2 may restrain CRH-R1 anxiogenic pathways during stress. Recent studies using anti-sense modulation of CRH-R2 suggested a role for this receptor subtype in stress coping behaviors (69). We explored this possibility further using self-grooming, a behavior thought to reflect dearousal and coping following stress (70). Compared to wild-type mice, CRH-R2 KO mice exhibit significantly reduced grooming behavior in a novel, open-field suggesting that CRH-R2 is involved in adaptations to stress and further supporting a role for this receptor in allostasis.

Feeding Responses in CRH-R2 KO Mice
There is considerable evidence that stress and HPA axis activation influence appetite regulation. The involvement of CRH receptor pathways in regulating appetitive stimuli is suggested by evidence showing that central administration of CRH or Ucn suppresses feeding and increases metabolism thereby blunting energy storage (71, 72). CRH-R2 is a likely candidate to mediate CRH/Ucn-induced anorexia based on its selective predominance in the VMH and PVN (64). CRH-R2 in these locations is ideally situated to interact with neural circuitry regulating feeding (73).

While basal feeding is normal in CRH-R2 KO mice (32, 33), we find that CRH-R2 KO mice and wild-type mice differ in their anorectic responses to Ucn. Both genotypes show similar inhibition of feeding immediately following

Ucn treatment, however, CRH-R2 KO mice recover to normal intake levels more rapidly than wild-type animals. In comparison, Ucn-treated CRH-R1 KO mice are the mirror image-hypophagic only during the late but not early phase (74). Thus, the early phase of Ucn-induced hypophagia occurs via a CRH-R2-independent pathway, perhaps via CRH-R1, but late-phase suppression critically depends on CRH-R2. In addition to Ucn stimulated feeding, food intake is altered in CRH-R2 KO mice following metabolic (32) and isolation stress (S. Coste and M. Stenzel-Poore, unpublished). Collectively, CRH-R2 plays a critical role in mediating the actions of CRH related pathways on feeding behavior.

Cardiovascular Responses in CRH-R2 KO Mice
CRH is well recognized for its ability to modulate cardiovascular function. CRH delivered icv increases arterial blood pressure and heart rate similar to effects of stress (75). In contrast, Ucn or CRH delivered systemically induces a marked decrease in blood pressure (6) due to vasodilation in specific vascular beds (75). In addition, several recent studies indicate that CRH or Ucn may have direct actions on the heart leading to increased contractile function (76, 77). The presence of CRH-R2 in the heart and vasculature suggests that this receptor subtype mediates the peripheral actions of CRH or Ucn. We find that CRH-R2 KO mice show significantly elevated mean arterial pressure compared with wild-type mice (33). These findings indicate a role for CRH-R2 in basal blood pressure homeostasis. In addition, systemic Ucn administration fails to decrease mean arterial pressure in CRH-R2 KO, whereas wild-type mice show a marked drop demonstrating that CRH-R2 mediates the hypotensive effect of systemically administered Ucn (32, 33). These studies clearly implicate CRH-R2 in modulating these effects on blood pressure; however, the source of CRH or Ucn that mediates these effects is less obvious. It is possible that basal blood pressure is influenced by Ucn located in the CNS (6) or periphery (6, 78).

In addition to effects on blood pressure regulation, an intriguing role is emerging for CRH or Ucn concerning direct actions for these peptides in cardiac function. Both CRH and Ucn have been shown recently to increase cardiac contractility *in vitro* (77) and *in vivo* following systemic administration (76). That cardiomyocytes express CRH-R2 and respond to CRH and Ucn with robust increases in cAMP production (79) suggests that cardiac contractile responses to Ucn are CRH-R2 dependent. CRH-R2 KO mice do not exhibit detectable cardiac responses to Ucn whereas wild-type mice show a pronounced increase in cardiac function (33). Increased cardiac function in wild-type mice is likely due to direct actions of Ucn on cardiomyocytes since CRH-R2 activation in these cells increases cAMP (79), which is known to stimulate cardiac contractility (80). However, decreased blood pressure following Ucn injection may also contribute to the increase in contractile function.

The precise mechanism(s) involved in Ucn actions on the heart are not yet clear; however, these findings show that changes in cardiac function and blood pressure critically depend on CRH-R2. It is noteworthy that stress-induced effects and those induced by injection of CRH into the CNS lead to similar cardiovascular changes; namely, elevation in arterial pressure and heart rate and a marked change in regional blood flow resulting in shunting from mesentery to skeletal muscle (75). These cardiovascular changes are favorable during the "fight or flight" response. It is possible that systemic or paracrine actions of Ucn may oppose these CNS effects by re-directing local blood flow thereby bringing regional hemodynamics closer to a basal state while maintaining increased cardiac function.

It is clear from studies using CRH-R1 KO mice that CRH-R1 plays a pivotal role in neuroendocrine and behavioral responses to stress. In contrast, our findings in CRH-R2 KO mice suggest that CRH-R2 may mediate specific adaptations to stress. CRH-R2 is the dominant CRH receptor in the PVN and VMH (64); thus it is particularly well-poised to co-ordinate effects of CRH/Ucn in neuroendocrine, cardiovascular and feeding responses to stress. In addition, the location of CRH-R2 in peripheral systems such as the heart and vasculature may provide a vital link to stress-responsive pathways operating outside of the CNS. It is tempting to speculate that a common theme exists wherein neuroendocrine, behavioral and cardiovascular responses to stress are initiated through CRH-R1 and are then specifically tailored within various stress pathways through the actions of CRH-R2.

ALTERED REGULATION OF CRH PATHWAYS

CRH-BP Transgenic Mice
CRH-BP binds CRH and Ucn with an affinity equal to or greater than does CRH-R1 (13-15). Therefore, it is believed that this molecule may regulate CRH actions *in vivo* by clearing CRH and thereby reducing CRH levels available for binding to CRH receptors. *In vitro*, CRH-BP has been shown to antagonize CRH-induced secretion of ACTH from pituitary cells (14, 16). However, while the affinity of CRH for CRH-R1 and CRH-BP is similar, CRH interacts with CRH-BP with slower kinetics. Thus, it has been hypothesized that an interaction between CRH and pituitary CRH-BP may not occur rapidly enough to affect CRH-R1 signaling (16). Two CRH-BP transgenic models have been created to address the physiologic role of CRH-BP. One transgenic model (α-GSU-CRH-BP) was created using the CRH-BP cDNA linked to the pituitary glycoprotein hormone α-subunit (α-GSU) promoter to specifically enhance anterior pituitary expression (35).

Accordingly, the CRH-BP trangene is highly expressed in gonadotropes and thyrotropes. In this model, CRH-BP secretion from these cells was postulated to bind CRH in extracellular regions surrounding corticotropes (17). The second transgenic model, mMT-CRH-BP, was created using the rat CRH-BP cDNA under the control of the mouse metallothionein promoter (mMT-1) (36). These mice express the transgene in the liver, heart, lung, kidney, spleen, adrenals, and testes, and in several brain regions including olfactory lobes, forebrain, brain stem, and pituitary. In addition, CRH-BP is detectable in the blood.

At first glance, HPA axis activity appears normal in these transgenic models as basal levels of circulating ACTH and corticosterone were similar to wild-type mice (35, 36). However, α-GSU CRH-BP Tg mice exhibited significantly higher CRH and AVP mRNA in the PVN, suggesting that the CRH-BP transgene was able to reduce available CRH but compensatory mechanisms were rapidly initiated to increase CRH synthesis (35). This conclusion is somewhat speculative since free CRH was not measured in the pituitary. Nonetheless, these findings suggest that endogenous CRH-BP may play a role in regulating basal HPA axis tone; however such effects are masked by tight feedback control. Conversely, stress responses do not appear to be modified by CRH-BP since α-GSU-CRH-BP mice show normal stress-induced increases in ACTH and corticosterone.

It is conceivable that CRH-BP may modulate CRH-induced behaviors because of its co-localization with CRH and CRH receptors in several brain regions that subserve behavioral function, including the amygdala and the pre-optic nucleus (17, 81). Indeed, some behavioral changes were observed in α-GSU-CRH-BP mice. Specifically, these mice display increased activity in standard behavioral tests and an altered circadian pattern of food intake. Although total food intake was similar to wild-type mice, feeding behavior was increased during the light phase but diminished during the dark phase (35). Similarly, food intake may have been altered in the mMT-CRH-BP mice since they gained weight more quickly than wild-type mice (36). Taken together, these findings suggest that CRH-BP may modulate certain behaviors associated with CRH/Ucn although it is unclear at this time where in the brain such interactions occur.

CRH-BP Knock Out Mice (CRH-BP KO)

In addition to CRH-BP transgenic models, a CRH-BP KO mouse was created by targeted gene deletion (37). These mice would be expected to exhibit overactive CRH pathways. Although no obvious changes in baseline or stress-induced HPA hormones were seen, CRH-BP KO mice display a modest increase in anxiety-related behavior. In addition, CRH-BP KO mice showed alterations in feeding behavior. Male CRH-BP KO mice ate less during the light and dark cycles and did not gain weight as rapidly as wild-type males (37). These findings complement those observed with mMT-CRH-BP mice wherein increased weight gain accompanied global excess of CRH-BP (36).

Collectively, CRH-BP transgenic and KO models suggest that CRH-BP may be important in regulating behavioral effects of CRH/Ucn, yet plays a minor role in HPA axis homeostasis. That food intake was altered in these models despite normal feeding behavior in CRH KO mice strongly suggests that these effects of CRH-BP may be mediated through an alternate ligand, possibly Ucn. These models demonstrate that HPA axis and central CRH pathways are regulated independently. Furthermore, when regulatory components are altered, the HPA axis seems to be exquisitely capable of readjusting setpoints in order to maintain allostasis.

GLUCOCORTICOID FEEDBACK DYSREGULATION

Several animal models have been created to examine glucocorticoid actions *in vivo*. In these models, glucocorticoid activity is diminished by inactivating type II glucocorticoid receptors (GR) through various means of disrupting the GR gene.

Glucocorticoid Receptor Knock Out Mice

Complete disruption of the GR gene results in perinatal lethality (82, 83). The loss of glucocorticoid signaling via GR leads to respiratory, distress caused by severe lung atelectasis. This indicates that glucocorticoid signaling through GR is indeed necessary for lung maturation and neonatal survival. Since these mice die shortly after birth, the role of GR in physiology has only been addressed in embryos and newborns (82, 83). Unfortunately, early mortality in this model has precluded extensive investigation into specific mechanisms of glucocorticoid modulation of lung development. CRH KO mice exhibit a similar phenotype but are a more useful model for examining this issue since mortality can be rescued with prenatal glucocorticoid treatment (29, 59).

Glucocorticoid Receptor Antisense Knock Down

Partial knock-down and tissue specific mutations of the glucocorticoid receptor have circumvented perinatal mortality and allowed for examination of glucocorticoid pathways in adult physiology. Pepin et al. (38) created a transgenic mouse model that harbors a transgene that constituitively expresses antisense RNA against type II GR. The transgene is under the control of a neurofilament promoter and thus antisense expression impairs production of GR mRNA predominantly in neural tissue, but also pituitary and several peripheral tissues. Type II GR signaling is only partially impaired in transgenic mice as GR mRNA levels are reduced 50-70% in hypothalamus and only 30-55% in peripheral organs (38). At the onset, it appeared that this model may closely resemble clinical depression in humans, in terms of neuroendocrine function. Transgenic mice display decreased negative feedback efficacy to both corticosterone and dexamethasone; a 10-fold higher dosage of dexamethasone is required to suppress plasma ACTH and corticosterone levels (84, 85). This resistance to the suppressive effects of dexamethasone is similar to human depression, where 60-70% of severe clinical cases are nonsuppressors (86). Using *in vitro* methods with hypothalamic-pituitary complexes from transgenic mice, Karanth et al. (87) have subsequently shown that impaired glucocorticoid negative feedback occurs primarily at the level of the hypothalamus and not the pituitary. In addition, these mice display exaggerated ACTH responses to stress and exogenously administered CRH (84, 87), whereas corticosterone responses are reduced due to hyposensitivity of the adrenal gland (84). However, unlike human depression which typically presents with elevated urinary free cortisol levels and elevated levels of cortisol and CRH in cerebrospinal fluid, transgenic mice show normal or elevated corticosterone levels only in the morning (38, 87). Furthermore, recent studies indicate that transgenic mice show reduced hypothalamic CRH activity rather than CRH overactivity (88). These findings challenge the notion that CRH hypersecretion in human depression is a consequence of impaired glucocorticoid function. It should be cautioned that this model represents partial disruption of GR signaling and therefore changes seen here may be unique. Nevertheless, behavioral studies with this model have revealed complex pathways of GR regulation (reviewed in (89)).

Brain-Specific Glucocorticoid Receptor Knock Out Mice (Gr^{NesCre})

Recently, Tronche et al. (40) developed a mouse model in which GR was selectively inactivated in the central nervous system using a Cre/loxP-recombination system in which Cre is under the control of the nestin promoter/enhancer. Thus, GR protein is absent in the brains of mutant mice but is normally distributed in the anterior pituitary and other peripheral tissues. Gr^{NesCre} mice display pronounced alterations in HPA axis equilibrium despite intact negative feedback on pituitary cells. Hypothalamic CRH expression is elevated, leading to increased POMC transcription in the pituitary and hence elevated plasma corticosterone. Because glucocorticoids are still able to act on non-neuronal cells with intact GR, these mice display several symptoms of Cushing's syndrome, including growth retardation, altered fat distribution and osteoporosis. Behaviorally, these mice show reduced anxiety-related behavior in the dark-light box and elevated zero maze paradigms. These findings suggest a new role for glucocorticoids in modulating anxiogenic behavior. Through HPA axis regulation, glucocorticoids may serve to augment CRH-induced anxiety during stress. The *selective* targeting of brain GR in the Gr^{NesCre} mouse is therefore a very important model in examining central effects of GR on anxiety and stress coping.

Glucocorticoid Receptor DNA Binding Domain Knock Out Mice (GR^{dim})
Tronche et al. also developed a mouse model with a point mutation within the
D loop of the second zinc finger of the receptor (39). This mutation caused loss
of dimerization and binding of the receptor to DNA targets, thereby eliminating
one mechanism that GR utilizes to alter transcription rates. However, due to
the precise mutation, modulation of transcription by protein-protein
interactions remains intact. Thus, this model has the potential to relate
molecular signaling to overt glucocorticoid actions. Interestingly, GR^{dim} mice
show normal viability and lung development, indicating that protein-protein
interactions alone may be sufficient for lung maturation. With regard to HPA
axis regulation, feedback inhibition by glucocorticoids appears to require DNA
binding of GR at the pituitary but not the hypothalamus. Newborn GR^{dim}
mice display normal CRH immunoreactivity in the median eminence; however,
POMC mRNA and ACTH immunostaining were greatly elevated in the
anterior pituitary. Thus, the authors suggest that glucocorticoids act through
different regulatory mechanisms within the same physiological system.
Although intriguing, this interpretation should be considered with caution since
regulation of the developing HPA axis may differ from regulation in adult
animals.

PITUITARY DYSREGULATION

Models of dysregulated pituitary function provide valuable information
regarding normal processes of pituitary development and hormone regulation.
Here we focus on two specific models that result in aberrant hormone
processing and secretion: dopamine receptor knock out mice and mice lacking
7B2, a neuroendocrine binding protein. We also discuss a striking example of
pituitary regulation and differentiation under control of the immune molecule,
leukemia inhibitory factor (LIF). These models highlight diverse mechanisms
leading to similar disease phenotypes: increased ACTH production arising
from altered POMC processing, insufficient inhibitory input and altered
immune system regulation. In each case regulatory mechanisms fail to achieve
HPA axis balance and a Cushing's-like syndrome ensues.

Dopamine D2 Receptor Knock Out Mice (D2R KO)
Dopamine is the primary inhibitory transmitter controlling secretion from the
pituitary. Dopamine D2 receptor knock out mice (D2R KO) exhibiting a
marked pituitary phenotype have recently been generated by two independent
groups (41, 42, 90, 91). D2R KO mice exhibit hyperprolactinemia resulting
from de-repression of target cells in the pituitary. In addition, these mice
develop lactotrope hyperplasia and subsequent pituitary adenomas composed
of lactotropes indicating that lack of dopamine inhibition leads to pituitary

oncoplastic development. Previous pharmacological studies suggested such an inhibitory role for dopamine in the development and proliferation of anterior lobe mammotropes (92, 93). Findings with the D2R KO mice are partially corroborated by dopamine transporter deficient mice that exhibit a dopaminergic gain-of-function and in these animals severe *hypo*plasia of the anterior lobe develops (94).

Interestingly, the D2R KO mouse line created by Saiardi and Borrelli unexpectedly exhibited increased ACTH production, as well as excess α-MSH and β-endorphin (41). This effect presumably resulted from the absence of dopaminergic control on melanotropes and subsequent intermediate lobe hyperplasia (90). Consistent with this expansion, the levels of mRNA encoding the POMC proteolytic processing enzymes, PC1 and PC2 were increased in the intermediate lobes of these animals. In addition, there was a two-fold increase in POMC levels consistent with increased levels of circulating α-MSH, β-endorphin and intact ACTH. Adrenal cortical hypertrophy was also observed, and circulating corticosterone was found to be elevated by 1.5-fold, leading to a Cushing's-like syndrome. The fact that only one of the two D2R KO models resulted in changes in intermediate lobe morphology and secretion of hormones suggests that background mouse strain differences may play a role in the phenotype of dopaminergic null animals. Indeed, Kelly et al. (42) have documented that 129 mice possess measurably larger intermediate lobes than C57BL/6 mice; these authors suggest that the 129 mouse intermediate lobe has a high growth potential. Furthermore, backcrossing the D2R KO mice onto the C57BL/6 background results in normal intermediate lobe size (42) suggesting that loss of D2R may be subject to the influence of modifier genes in the 129 strain. Thus the intermediate lobe hyperplasia and resultant Cushing's-like syndrome observed by Saiaridi and Borrelli (41) may be due to a combination of factors not well understood at this time.

Collectively, findings in the D2R KO mice indicate that dopaminergic systems play an important role in regulation of pituitary growth and lactotrope hyperplasia. In addition, they suggest an intriguing role for D2R in the development of some forms of Cushing's disease via abnormal regulation of POMC expression and processing in intermediate lobe melanotropes.

7B2 Knock Out Mice (7B2 KO)
The cleavage of prohormone precursors such as POMC is accomplished by the prohormone convertases, PC1 and PC2 (as discussed above). The intermediate lobe specific enzyme, PC2, operates in conjunction with the neuroendocrine binding protein known as 7B2, which is exclusively localized to neuroendocrine tissue. 7B2 is absolutely required for the production of active PC2 in AtT-20 cells (95). Mice null for the PC2 gene exhibit impairment in prohormone processing but overall show a remarkably mild phenotype (96).

By contrast, 7B2 KO mice exhibit signs of severe illness, including aberrant distribution of fat, hypoglycemia, altered skeletal development, mineral deposits in fat pads, splenic atrophy, and vacuolization of the liver leading to death prior to 9 weeks of age (43). The aberrant sugar and fat metabolism likely stems from disturbances in HPA axis regulation. In particular, pituitary and plasma ACTH levels are markedly elevated in 7B2 KO mice, leading to adrenocortical hyperplasia (43). Interestingly, pituitary ACTH is localized exclusively to the intermediate lobe with undetectable levels of POMC mRNA in anterior pituitary corticotropes (B. Peng and J. Pintar, unpublished). Lack of ACTH in corticotropes may result from transcriptional repression due to chronically elevated corticosterone.

As described above, ACTH is normally synthesized by intermediate lobe melanotropes; however, in the intermediate lobe, the expression of PC2 results in internal cleavage of intact ACTH generating the relatively non-corticotropic product, α-MSH (20). Therefore, the lack of PC2 should result in ACTH accumulation similar to findings in the 7B2 KO model. However, there is no evidence for a Cushing's-like syndrome in PC2 KO mice; circulating corticosterone levels are not elevated, nor is anterior lobe POMC suppressed (D. Steiner, pers. commun.). Although PC2 KO mice do exhibit high levels of intact ACTH in the intermediate lobe (I. Lindberg, unpublished), this accumulation does not result in hypersecretion as was observed in the 7B2 KO. This phenotypic divergence is unlikely to be due to differences in genetic background between the two mouse models because the 7B2 KO mouse line exhibits a severe Cushing's phenotype when bred onto the same background as the PC2 KO mice (I. Lindberg, unpublished).

These data suggest that 7B2 plays a role in pituitary development or hormone release which is unrelated to its known role as a helper protein for the enzymatic maturation of PC2. It is interesting that the anterior lobes of the 7B2 KO show a marked hypotrophy which is unexpected given the naturally low abundance of PC2 in this lobe. In addition, 7B2 is present in brain areas where PC2 is absent (97) suggesting alternative roles for this neuroendocrine protein. Examination of general biochemical defects in hormone synthesis and regulated secretion in 7B2 KO mice may lead to new insights into the contribution of 7B2 to secretory function. It is possible that 7B2 contributes to the maintenance of the regulated secretory phenotype, perhaps by regulating specific genes or proteins involved in the formation of secretory granules. This rather speculative idea is based on the following: there are no known neuroendocrine tissues or cell lines lacking 7B2 and neurally-derived cells that have lost 7B2 expression (such as SK-N-MC; (97)) contain few if any secretory granules (R. Moses and I. Lindberg, unpublished). However, it should be emphasized that as yet, there are no data generalizing the effect of

7B2 loss to increased basal secretion of other hormones or tissues beyond the intermediate lobe.

The 7B2 KO model is useful primarily as a model for a severe form of Cushing's disease in which both ACTH and corticosterone circulate at elevated levels, most likely from early developmental times. In this regard, 7B2 KO mice share numerous features with the CRH-Tg mice (28). However, unlike the CRH-Tg mice, the entire complement of circulating ACTH in the 7B2 null arises from the intermediate lobe whereas the majority of ACTH in the CRH-Tg likely originates from anterior lobe corticotropes. That a number of distinct mechanisms of disease can lead to a Cushing's phenotype underscores the importance of using several models with different defects in pituitary regulation in testing novel therapeutic strategies aimed at the treatment of Cushing's syndrome.

Leukemia Inhibitory Factor Mutant Mice
Leukemia inhibitory factor (LIF) is a pleiotropic cytokine that affects both the immune and nervous systems. LIF is produced by many cell types including corticotropes and immune cells. This cytokine has recently been ascribed a role in the regulation of the HPA axis. LIF stimulates POMC transcription and ACTH secretion by corticotropes and is able to potentiate the effect of CRH on POMC expression and ACTH release (98, 99). Recent evidence indicates inflammatory signals such as LPS and IL-1β induce peripheral and pituitary LIF, which may synergize with CRH to induce POMC transcription (99-101). Thus, LIF has been suggested to act at the interface of neuroendocrine and inflammatory pathways.

Although the evidence showing inflammation-induced LIF expression in the pituitary was intriguing, the precise function of LIF in HPA axis regulation *in vivo* was difficult to test rigorously until the development of LIF transgenic (LIF Tg) and LIF knockout (LIF KO) mice. Several important points have now been clarified using LIF KO mice. LIF appears to be involved in stress-induced ACTH release because LIF KO mice show a severely blunted ACTH response to restraint stress (44, 102). In addition, adjuvant treatment (immune activation) increased corticosterone levels in wild-type but not LIF KO mice indicating a role for LIF in mediating HPA axis responses to inflammatory stimulation. The importance of LIF in activation of the HPA axis was confirmed further by exogenous administration of LIF to LIF KO mice which caused a marked increase in POMC expression and ACTH. These findings provide strong evidence that LIF regulates HPA axis responses to both psychological and inflammatory stress.

LIF also appears to regulate pituitary development. Previous studies using the mouse corticotrope cell line, AtT20, suggested that LIF negatively affects corticotrope differentiation (98). Further clarification of these observations has risen from recent results obtained in mice expressing a LIF transgene under the control of the α-GSU promoter (45). Interestingly, LIF overexpression under the control of this promoter alters pituitary development. Pituitaries from these animals are smaller with fewer somatotropes, thyrotropes, and gonadotropes, but significantly more corticotropes, implying that early pituitary overexpression of LIF favors the development of corticotropes. This is consistent with effects of LIF known to induce differentiation and de-differentiation of various cell types. As a result of increased corticotropes, these mice have elevated basal levels of ACTH and corticosterone (45). LIF-Tg mice exhibit features of Cushing's syndrome such as truncal obesity, thin skin, and enlarged adrenals at the time of death, similar to other models with ACTH and glucocorticoid excess. In addition, these animals show a range of growth abnormalities due to hypoplasia of non-corticotrope cells in the pituitary.

LIF-Tg mice are particularly interesting because they demonstrate that dysregulation of pituitary development and HPA axis responses can occur through immune system effectors. This model shows that abnormal expression of a cytokine such as LIF leads to chronic HPA activation at the level of the pituitary. In addition, we learn that LIF may alter the normal developmental program of the HPA axis through its positive actions on corticotrope growth and negative effects on other pituitary cells types. Admittedly, it is difficult to know from these studies whether constitutive pituitary activation results from chronic overexpression of LIF or is a consequence of inappropriate timing of LIF during ontogeny. Nevertheless, this model shows a novel interaction between cytokines and HPA axis activity.

CONCLUSIONS

We have described here various genetic models in which components at all levels of the HPA axis have been individually compromised, ranging from key hormones and receptors to enzymatic cleavage proteins. This collection of mutant models provides a powerful dissection of corticotropin pathways-a view not previously achieved with pharmacological methods. Most notably, these models have strengthened our understanding of the numerous mechanisms that achieve HPA axis regulation. Several models demonstrate the diverse nature of tissue specific regulation at the level of the pituitary. The 7B2 KO mouse has revealed that proteins involved primarily in prohormone precursor processing also have alternative means of regulating production or secretion of active peptides such as ACTH. The D2R KO mouse demonstrates the importance of dopaminergic inhibitory input to restrain pituitary proliferation and expression of prohormone precursors, particularly POMC. Conversely, stimulatory influences of the immune system were revealed in mice with LIF overexpression, where pituitary differentiation was biased towards corticotropes, leading to elevated HPA axis activity. Other models lend insight into HPA axis regulation that occurs through distinct pathways directed by different receptors. CRH-R1 and CRH-R2 KO mice have provided an excellent opportunity to examine CRH actions that are mediated by these receptors. CRH-R1 KO mice confirm that CRH-R1 via its presence on pituitary corticotropes mediates initiation of the HPA axis. CRH-R2 KO mice have revealed that CRH-R2 restrains behavioral and HPA axis responses to stress; in certain cases, CRH-R2 may oppose the actions of CRH-R1. We anticipate that in the near future, CRH-R1 and CRH-R2 double KO models will offer extreme promise by providing unequivocal removal of conventional CRH pathways.

In addition, this collection of models has highlighted the role of CRH pathways in promoting stability through adaptive changes. In each of these models, allostatic processes are altered to varying degrees. Certain models show overt changes in basal functioning as a consequence of HPA axis impairment. In these instances, compensatory mechanisms are not able to over-ride the deficiency resulting in allostatic load and subsequent pathophysiology. The CRH transgenic mouse reflects a failure to turn off allostatic processes. Thus, adaptive hormones such as glucocorticoids are unable to negatively regulate unmitigated CRH overexpression leading to an elevation in hormones at all levels of the HPA axis. Such allostatic load is readily apparent in the Cushing's phenotype of these mice. In other models, compensatory mechanisms are more effective in maintaining normal function and allostatic load is not evident in the basal state. However, altered allostatic processes become apparent when systems are challenged. These models have contributed greatly to our understanding of the plasticity of neurohormonal regulation. For example, both CRH and CRH-R1 KO mice are unable to elevate glucocorticoids in response to behavioral stress. Yet, robust HPA axis activation is seen following immune challenge, showing that the immune system can directly signal pituitary corticotropes, circumventing hypothalamic input. Finally, these models have shown that different mechanisms of molecular signaling, in part, confer plasticity to HPA axis function. Using these models and others which are sure to follow, we can anticipate increasing clarification of the complexities of the HPA axis and its essential role in maintaining allostasis.

REFERENCES

1. Vale W, Spiess J, Rivier C, Rivier J. Characterization of a 41-residue ovine hypothalamic peptide that stimulates secretion of corticotropin and β-endorphin. Science 1981;213:1394-1397.
2. Jingami H, Matsukura S, Numa S, Imura H. Effects of adrenalectomy and dexamethasone administration on the level of prepro-corticotropin-releasing factor

messenger ribonucleic acid (mRNA) in the hypothalamus and adrenocorticotropin/β-lipotropin precursor mRNA in the pituitary in rats. Endocrinology 1985;117:1314-1320.

3.   Vale W, Vaughan J, Perrin M. Corticotropin-releasing factor (CRF) family of ligands and their receptors. Endocrinologist 1997;7:3S-9S.

4.   Swanson LW, Sawchenko PE, Rivier J, Vale WW. Organization of ovine corticotropin-releasing factor immunoreactive cells and fibers in the rat brain: an immunohistochemical study. Neuroendocrinol1983;36:165-186.

5.   Bittencourt JC, Vaughan J, Arias C, Rissman RA, Vale WW, Sawchenko PE. Urocortin expression in rat brain: evidence against a pervasive relationship of urocortin-containing projections with targets bearing type 2 CRF receptors. J Comp Neurol. 1999;415:285-312.

6.   Vaughan J, Donaldson C, Bittencourt J, Perrin MH, Lewis K, Sutton S, Chan R, Turnbull AV, Lovejoy D, Rivier C, et al. Urocortin, a mammalian neuropeptide related to fish urotensin I and to corticotropin-releasing factor. Nature 1995;378:287-292.

7.   Stenzel P, Kesterson R, Yeung W, Cone RD, Rittenberg MB, Stenzel-Poore MP. Identification of a novel murine receptor for corticotropin-releasing hormone expressed in the heart. Mol Endocrinol.1995;9:637-645.

8.   Chen R, Lewis KA, Perrin MH, Vale WW. Expression cloning of a human corticotropin-releasing-factor receptor. 1993;90:8967-8971.

9.   Chang C-P, Pearse RV 2d, O'Connell S, Rosenfeld MG. Identification of a seven transmembrane helix receptor for corticotropin-releasing factor and sauvagine in mammalian brain. Neuron 1993;11:1187-1195.

10. Lovenberg TW, Liaw CW, Grigoriadis DE, Clevenger W, Chalmers DT, De Souza EB, Oltersdorf T. Cloning and characterization of a functionally distinct corticotropin-releasing factor receptor subtype from rat brain. Proc Natl Acad Sci USA 1993;92:836-840.

11. Perrin M, Donaldson C, Chen R, Blount A, Berggren T, Bilezikjian L, Sawchenko P, Vale W. Identification of a second corticotropin-releasing factor receptor gene and characterization of a cDNA expressed in heart. Proc Natl Acad Sci USA 1995;92:2969-2973.

12. Kostich WA, Chen A, Sperle K, Largent BL. Molecular identification and analysis of a novel human corticotropin-releasing factor (CRF) receptor: the CRH2γ receptor. Mol Endocrinol 1998;12:1077-1085.

13. Behan DP, Linton E A, Lowry P J. Isolation of the human plasma corticotrophin-releasing factor-binding protein. J Endocrinol 1989;22:23-31.

14. Cortwright DN, Nicoletti A, Seasholtz AF. Molecular and biochemical characterization of the mouse brain corticotropin-releasing hormone-binding protein. Mol. Cell. Endocrinol 1995;111:147-157.

15. Potter E, Behan DP, Fischer WH, Linton EA, Lowry PJ, Vale WW. Cloning and characterization of the cDNAs for human and rat corticotropin releasing factor-binding proteins. Nature 1991;349:423-426.

16. Linton EA, Behan DP, Saphier PW, Lowry PJ. Corticotropin-releasing hormone (CRH)-binding protein: reduction in the adrenocorticotropin-releasing activity of placental but not hypothalamic CRH. J Clin Endocrinol Metab 1990;70:1574-1580.

17. Potter E, Behan DP, Linton EA, Lowry PJ, Sawchenko PE, Vale WW. The central distribution of a corticotropin-releasing factor (CRF)-binding protein predicts multiple sites and modes of interaction with CRF. Proc Natl Acad Sci USA 1992;89:4192-4196.

18. de Kloet ER, Vreugdenhil E, Oitzl MS, Joels M. Brain corticosteroid receptor balance in health and disease. Endocr Rev 1998;19:269-301.

19. Birnberg NC, Lissitzky JC, Hinman M, Herbert E. Glucocorticoids regulate proopiomelanocortin gene expresssion in vivo at the levels of transcription and secretion. Proc Natl Acad Sci USA 1983;80:6982-6986.

20. Zhou A, Bloomquist B T, Mains R E The prohormone convertases PC1 and PC2 mediate distinct endoproteolytic cleavages in a strict temporal order during proopiomelanocortin biosynthetic processing. J Biol Chem 1993;268:1763-1769.
21. Autelitano DJ, Lundblad JR, Blum M, Roberts JL. Hormonal regulation of POMC gene expression. Ann Rev Physiol 1989;51:715-726.
22. Behan DP, Khongsaly O, Owens MJ, Chung HD, Nemeroff CB, De Souza EB. Corticotropin-releasing factor (CRF), CRF-binding protein (CRF-BP), and CRF/CRF-BP complex in Alzheimer's disease and control postmortem human brain. J Neurochem 1997;68:2053-2060.
23. Nelson DH. Cushing's syndrome, in endocrinology, (Ed: L.J. DeGroot) pp1660-1675 W.B. Saunders Company: Philadelphia: W.B. Saunders, 1989.
24. Gold P W, Gwirtsman H, Avgerinos PC, Nieman LK, Gallucci WT, Kaye W, Jimerson D, Ebert M, Rittmaster R, Loriaux DL, et al. Abnormal hypothalamic-pituitary-adrenal function in anorexia nervosa. Pathophysiologic mechanisms in underweight and weight-corrected patients. N Engl J Med 1986;314:1335-1342.
25. Nemeroff CB, Widerlov E, Bissette G, Walleus H, Karlsson I, Eklund K, Kilts CD, Loosen PT, Vale W. Elevated concentrations of CSF corticotropin-releasing factor-like immunoreactivity in depressed patients. Science 1984;226:1342-1344.
26. Behan DP, Heinrichs SC, Troncoso JC, Liu XJ, Kawas CH, Ling N, De Souza EB. Displacement of corticotropin releasing factor from its binding protein as a possible treatment for Alzheimer's disease. Nature 1995;378:284-287.
27. Heilig M, Sjogren M, Blennow K, Ekman R, Wallin A. Cerebrospinal fluid neuropeptides in Alzheimer's disease and vascular dementia. Biol Psychiatry 1995;38:210-216.
28. Stenzel-Poore MP, Cameron VA, Vaughan J, Sawchenko PE, Vale W. Development of Cushing's syndrome in corticotropin-releasing factor transgenic mice. Endocrinology 1992;130:3378-3386.
29. Muglia L, Jacobson L, Dikkes P, Majzoub JA. Corticotropin-releasing hormone deficiency reveals major fetal but not adult glucocorticoid need. Nature 1995;373:427-432.
30. Smith GW, Aubry JM, Dellu F, Contarino A, Bilezikjian LM, Gold LH, Chen R, Marchuk Y, Hauser C, Bentley CA, Sawchenko PE, Koob GF, Vale W, Lee KF. Corticotropin releasing factor receptor 1-deficient mice display decreased anxiety, impaired stress response, and aberrant neuroendocrine development. Neuron 1998;20:1093-1102.
31. Timpl P, Spanagel R, Sillaber I, Kresse A, Reul JM, Stalla GK, Blanquet V, Steckler T, Holsboer F, Wurst W. Impaired stress response and reduced anxiety in mice lacking a functional corticotropin-releasing hormone receptor. Nature Gen 1998;19:162-166.
32. Bale TL, Contarino A, Smith GW, Chan R, Gold LH, Sawchenko PE, Koob GF, Vale WW, Lee KF. Mice deficient for corticotropin-releasing hormone receptor-2 display anxiety-like behavior and are hypersensitive to stress. Nature Gen 2000;24:410-414.
33. Coste SC, Kesterson RA, Heldwein KA, Stevens SL, Heard AD, Hollis JH, Murray SE, Hill JK, Pantely GA, Hohimer AR, Hatton DC, Phillips TJ, Finn DA, Low MJ, Rittenberg MB, Stenzel P, Stenzel-Poore MPl. Abnormal adaptations to stress and impaired cardiovascular function in mice lacking corticotropin-releasing hormone receptor-2. Nature Gen 2000;24:403-409.
34. Kishimoto T, Radulovic J, Radulovic M, Lin CR, Schrick C, Hooshmand F, Hermanson O, Rosenfeld MG, Spiess J. Deletion of Crhr2 reveals an anxiolytic role for corticotropin-releasing hormone receptor-2. Nature Gen 2000;24:415-419.
35. Burrows HL, Nakajima M, Lesh JS, Goosens KA, Samuelson LC, Inui A, Camper SA, Seasholtz AF. Excess corticotropin-releasing hormone-binding protein in the hypothalamic-pituitary-adrenal axis in transgenic mice. J Clin Invest 1998;101:1439-1447.

36. Lovejoy DA, Aubry JM, Turnbull A, Sutton S, Potter E, Yehling J, Rivier C, Vale WW. Ectopic expression of the CRF-binding protein: minor impact on HPA axis regulation but induction of sexually dimorphic weight gain. J Neuroendocrinol 1998;10:483-491.

37. Karolyi IJ, Burrows HL, Ramesh TM, Nakajima M, Lesh JS, Seong E, Camper SA, Seasholtz AF. Altered anxiety and weight gain in corticotropin-releasing hormone-binding protein-deficient mice. Proc Natl Acad Sci USA 1999;96:11595-11600.

38. Pepin M, Pothier F, Barden N Impaired type II glucocorticoid-receptor function in mice bearing antisense RNA transgene. Nature 1992;355:725-728.

39. Reichardt HM, Kaestner KH, Tuckermann J, Kretz O, Wessely O, Bock R, Gass P, Schmid W, Herrlich P, Angel P, Schutz G. DNA binding of the glucocorticoid receptor is not essential for survival. Cell 1998;93:531-541.

40. Tronche F, Kellendonk C, Kretz O, Gass P, Anlag K, Orban PC, Bock R, Klein R, Schutz G. Disruption of the glucocorticoid receptor gene in the nervous system results in reduced anxiety. Nature Gen 1999;23:99-103.

41. Saiardi A, Bozzi Y, Baik JH, Borrelli E. Antiproliferative role of dopamine: loss of D2 receptors causes hormonal dysfunction and pituitary hyperplasia. Neuron 1997;19:115-126.

42. Kelly MA, Rubinstein M, Asa SL, Zhang G, Saez C, Bunzow JR, Allen RG, Hnasko R, Ben-Jonathan N, Grandy DK, Low MJ. Pituitary lactotroph hyperplasia and chronic hyperprolactinemia in dopamine D2 receptor-deficient mice. Neuron 1997;19:103-113.

43. Westphal CH, Muller L, Zhou A, Zhu X, Bonner-Weir S, Schambelan M, Steiner DF, Lindberg I, Leder P l. The neuroendocrine protein 7B2 is required for peptide hormone processing in vivo and provides a novel mechanism for pituitary Cushing's disease. Cell 1999;96:689-700.

44. Akita S, Malkin J, Melmed S Disrupted murine leukemia inhibitory factor (LIF) gene attenuates adrenocorticotropic hormone (ACTH) secretion. Endocrinology 1996;137:3140-3143.

45. Yano H, Readhead C, Nakashima M, Ren SG, Melmed S. Pituitary-directed leukemia inhibitory factor transgene causes Cushing's syndrome: neuro-immune-endocrine modulation of pituitary development. Mol Endocrinol 1998;12:1708-1720.

46. Stenzel-Poore MP, Heinrichs SC, Rivest S, Koob GF, Vale WW. Overproduction of corticotropin-releasing factor in transgenic mice: a genetic model of anxiogenic behavior. J Neurosci 1994;14:2579-2584.

47. Stenzel-Poore MP, Duncan JE, Rittenberg MB, Bakke AC, Heinrichs SC. CRH overproduction in transgenic mice: behavioral and immune system modulation. Ann NY Acad Sci 1996780:36-48.

48. Heinrichs SC, Stenzel-Poore MP, Gold LH, Battenberg E, Bloom FE, Koob GF, Vale WW, Pich EM. Learning impairment in transgenic mice with central overexpression of corticotropin releasing hormone. Neurosci 1996;74:303-311.

49. Heinrichs SC, Min H, Tamraz S, Carmouche M, Boehme SA, Vale WW. Anti-sexual and anxiogenic behavioral consequences of corticotropin-releasing factor overexpression are centrally mediated. Psychoneuroendocrinol 1997;22:215-224.

50. Dallman MF, Jones MT Corticosteroid feedback control of ACTH secretion; effects of stress-induced corticosterone secretion on subsequent stress responses in the rat. Endocrinology 1973;92:1367-1375.

51. Asa SL, Kovacs K, Stefaneanu L, Horvath E, Billestrup N, Gonzalez-Manchon C, Vale W. Pituitary adenomas in mice transgenic for growth hormone-releasing hormone. Endocrinology 1992;131:2083-2089.

52. Mayo KE, Hammer RE, Swanson LW, Brinster RL, Rosenfeld MG, Evans RM. Dramatic pituitary hyperplasia in transgenic mice expressing a human growth hormone-releasing factor gene. Mol Endocrinol 1988;2:606-612.

53. Gertz BJ, Contreras LN, McComb DJ, Kovacs K, Tyrrell JB, Dallman MF. Chronic administration of corticotropin-releasing factor increases pituitary corticotroph number. Endocrinology 1987;120:381-388.

54. Carey RM, Varma SK, Drake CR Jr, Thorner MO, Kovacs K, Rivier J, Vale W. Ectopic secretion of corticotropin-releasing factor as a cause of Cushing's syndrome. N Engl J Med 1984;311:13-20.

55. Schteingart DE, Lloyd RV, Akil H, Chandler WF, Ibarra-Perez G, Rosen SG, Ogletree R. Cushing's syndrome secondary to ectopic corticotropin-releasing hormone-adrenocorticotropin secretion. J Clin Endocrinol Metab 1986;63:770-775.

56. Dunn AJ, Berridge CW. Physiological and behavioral responses to corticotropin-releasing factor administration: is CRF a mediator of anxiety or stress responses? Brain Res Rev 1990;15:71-100.

57. Dhabhar FS, McEwen BS. Stress-induced enhancement of antigen-specific cell-mediated immunity. J Immunol 1996;156:2608-2615.

58. Dhabhar FS, McEwen BS. Enhancing versus suppressive effects of stress hormones on skin immune function. Proc Natl Acad Sci USA1999;96:1059-1064.

59. Muglia LJ, Bae DS, Brown TT, Vogt SK, Alvarez JG, Sunday ME, Majzoub JA. Proliferation and differentiation defects during lung development in corticotropin-releasing hormone-deficient mice. Am J Respir Cell Mol Biol 1999;20:181-188.

60. Karalis K, Muglia LJ, Bae D, Hilderbrand H, Majzoub JA. CRH and the immune system. J Neuroimmunol 1997;72:131-136.

61. Weninger SC, Dunn AJ, Muglia LJ, Dikkes P, Miczek KA, Swiergiel AH, Berridge CW, Majzoub JA Stress-induced behaviors require the corticotropin-releasing hormone (CRH) receptor, but not CRH. Proc Natl Acad Sci USA 1999;96:8283-8288.

62. Weninger SC, Muglia LJ, Jacobson L, Majzoub JA. CRH-deficient mice have a normal anorectic response to chronic stress. Regul Pep. 1999;84:69-74.

63. Swiergiel AH, Dunn AJ. CRF-deficient mice respond like wild-type mice to hypophagic stimuli. Pharmacol Biochem Behav 1999;64:59-64.

64. Chalmers DT, Lovenberg TW, DeSouza EB. Localization of novel corticotropin-releasing factor receptor (CRF2) mRNA expression to specific subcortical nuclei in rat brain: comparison with CRF1 receptor mRNA expression. Neurosci 1995;15:6340-6350.

65. Turnbull AV, Smith GW, Lee S, Vale WW, Lee KF, Rivier C. CRF type I receptor-deficient mice exhibit a pronounced pituitary-adrenal response to local inflammation. Endocrinology 1999;140:1013-1017.

66. McEwen BS. Stress, adaptation and disease: allostasis and allostatic load. Ann N Y York Acad Sci 1998;840:33-44.

67. Parkes D, Rivest S, Lee S, Rivier C, Vale W. Corticotropin-releasing factor activates c-fos, NGFI-B, and corticotropin-releasing factor gene expression within the paraventricular nucleus of the rat hypothalamus. Mol Endocrinol. 1993;7:1357-1367.

68. Ono N, Bedran de Castro JC, McCann SM. Ultrashort-loop positive feedback of corticotropin (ACTH)-releasing factor to enhance ACTH release in stress. Proc Natl Acad Sci USA 1985;82:3528-3531.

69. Liebsch G, Landgraf R, Engelmann M, Lorscher P, Holsboer F. Differential behavioral effects of chronic infusion of CRH 1 and CRH 2 receptor antisense oligonucleotides into the rat brain. J Psychiatr Res 1999;33:153-163.

70. Spruijt BM, van Hooff J A, Gispen W H Ethology and neurobiology of grooming behavior. Physiol Rev 199272:825-852.

71. Spina M, Merlo-Pich E, Chan RK, Basso AM, Rivier J, Vale W, Koob GF. Appetite-suppressing effects of urocortin, a CRF-related neuropeptide. Science 1996;273:1561-1564.

72. Arase K, York DA, Shimizu H, Shargill N, Bray GA. Effects of corticotropin-releasing factor on food intake and brown adipose tissue thermogenesis in rats. Am J Physiol 1988;255:E255-259.

73. Woods SC, Seeley RJ, Porte D Jr, Schwartz MW. Signals that regulate food intake and energy homeostasis. Science 1998;280:1378-1383.

74. Bradbury MJ, et al. Divergent effects of CRF receptors on food intake and weight gain: acute vs. chronic urocortin administration in WT and CRFR1-/- mice. Endocrine Soc. Abstr. 1999;81:224.
75. Overton JM, Fisher LA. Differentiated hemodynamic responses to central versus peripheral administration of corticotropin-releasing factor in conscious rats. J Auton Nerv Syst 1991;35:43-52.
76. Parkes DG, Vaughan J, Rivier J, Vale W, May CN. Cardiac inotropic actions of urocortin in conscious sheep. Am J Physiol 1997;272:H2115-2122.
77. Grunt M, Haug C, Duntas L, Pauschinger P, Maier V, Pfeiffer EF. Dilatory and inotropic effects of corticotropin-releasing factor (CRF) on the isolated heart. Effects on atrial natriuretic peptide (ANP) release. Horm Metab Res 1992;24:56-59.
78. Okosi A, Brar BK, Chan M, D'Souza L, Smith E, Stephanou A, Latchman DS, Chowdrey HS, Knight RA. Expression and protective effects of urocortin in cardiac myocytes. Neuropeptides 1998;32:167-171.
79. Heldwein KA, Redick DL, Rittenberg MB, Claycomb WC, Stenzel-Poore MP. Corticotropin-releasing hormone receptor expression and functional coupling in neonatal cardiac myocytes and AT-1 cells. Endocrinology 1996;137:3631-3639.
80. Miyakoda G, Yoshida A, Takisawa H, Nakamura T. β-Adrenergic regulation of contractility and protein phosphorylation in spontaneously beating isolated rat myocardial cells. J Biochem 1987;102:211-224.
81. Kemp CF, Woods RJ, Lowry PJ. The corticotrophin-releasing factor-binding protein: an act of several parts. Peptides 1998;9:1119-1128.
82. Cole TJ, Blendy JA, Monaghan AP, Krieglstein K, Schmid W, Aguzzi A, Fantuzzi G, Hummler E, Unsicker K, Schutz G. Targeted disruption of the glucocorticoid receptor gene blocks adrenergic chromaffin cell development and severely retards lung maturation. Genes Dev 1995;9:1608-1621.
83. Kellendonk C, Tronche F, Reichardt HM, Schutz G. Mutagenesis of the glucocorticoid receptor. J Steroid Biochem Mol Biol 1999;69:253-259.
84. Barden N, Stec IS, Montkowski A, Holsboer F, Reul JM. Endocrine profile and neuroendocrine challenge tests in transgenic mice expressing antisense RNA against the glucocorticoid receptor. Neuroendocrinol 1997;66:212-220.
85. Stec I, Barden N, Reul JM, Holsboer F. Dexamethasone nonsuppression in transgenic mice expressing antisense RNA to the glucocorticoid receptor. J Psychiatr Res 1994;28:1-5.
86. Carroll BJ. The dexamethasone suppression test for melancholia. Br J Psychiatry 1982;140:292-304.
87. Karanth S, Linthorst AC, Stalla GK, Barden N, Holsboer F, Reul JM. Hypothalamic-pituitary-adrenocortical axis changes in a transgenic mouse with impaired glucocorticoid receptor function. Endocrinology 1997;138:3476-3485.
88. Dijkstra I, Tilders FJ, Aguilera G, Kiss A, Rabadan-Diehl C, Barden N, Karanth S, Holsboer F, Reul JM Reduced activity of hypothalamic corticotropin-releasing hormone neurons in transgenic mice with impaired glucocorticoid receptor function. J Neurosci 1998;18:3909-3918.
89. Holsboer F, Barden N Antidepressants and hypothalamic-pituitary-adrenocortical regulation. Endocr Rev 1996;17:187-205.
90. Saiardi A, Borrelli E Absence of dopaminergic control on melanotrophs leads to Cushing's-like syndrome in mice. Mol Endocrinol 1998;12:1133-1139.
91. Asa SL, Kelly MA, Grandy DK, Low MJ. Pituitary lactotroph adenomas develop after prolonged lactotroph hyperplasia in dopamine D2 receptor-deficient mice. Endocrinology 1999;140:5348-5355.
92. Gehlert DR, Bishop JF, Schafer MP, Chronwall BM. Rat intermediate lobe in culture: dopaminergic regulation of POMC biosynthesis and cell proliferation. Peptides 1988;9:161-168.

93. Chronwall BM, Hook GR, Millington WR Dopaminergic regulation of the biosynthetic activity of individual melanotropes in the rat pituitary intermediate lobe: a morphometric analysis by light and electron microscopy and in situ hybridization. Endocrinology 1988;123:1992-2002.

94. Bosse R, Fumagalli F, Jaber M, Giros B, Gainetdinov RR, Wetsel WC, Missale C, Caron MG. Anterior pituitary hypoplasia and dwarfism in mice lacking the dopamine transporter. Neuron 1997;19:127-138.

95. Zhu X, Lindberg I. 7B2 facilitates the maturation of proPC2 in neuroendocrine cells and is required for the expression of enzymatic activity. J Cell Biol 1995;129:1641-1650.

96. Furuta M, Yano H, Zhou A, Rouille Y, Holst JJ, Carroll R, Ravazzola M, Orci L, Furuta H, Steiner DF. Defective prohormone processing and altered pancreatic islet morphology in mice lacking active SPC2. Proc Natl Acad Sci USA 1997;94:6646-6651.

97. Seidel B, Dong W, Savaria D, Zheng M, Pintar JE, Day R. Neuroendocrine protein 7B2 is essential for proteolytic conversion and activation of proprotein convertase 2 in vivo. DNA Cell Biol 1998;17:1017-1029.

98. Stefana B, Ray DW, Melmed S. Leukemia inhibitory factor induces differentiation of pituitary corticotroph function: an immuno-neuroendocrine phenotype switch. Proc Natl Acad Sci USA 1996;93:12502-12506.

99. Ray DW, Ren S, Melmed S. Leukemia inhibitory factor (LIF) stimulates proopiomelanocortin (POMC) expression in a corticotroph cell line: role of STAT pathway. J Clin Invest 1996;97:1852-1859.

100. Bousquet C, Ray DW, Melmed S. A common pro-opiomelanocortin-binding element mediates leukemia inhibitory factor and corticotropin-releasing hormone transcriptional synergy. J Biol Chem 1996;272:10551-10557.

101. Auernhammer CJ, Chesnokova V, Melmed S. Leukemia inhibitory factor modulates interleukin-1β-induced activation of the hypothalamo-pituitary-adrenal axis. Endocrinology 1998;139:2201-2208.

102. Chesnokova V, Auernhammer CJ, Melmed S Murine leukemia inhibitory factor gene disruption attenuates the hypothalamo-pituitary-adrenal axis stress response. Endocrinology 1998;139:2209-2216.

9

SPONTANEOUS AND INDUCED GENETIC MUTATIONS OF THE POMC SYSTEM

James L. Smart and Malcolm J. Low
Vollum Institute, Oregon Health Sciences University, Portland OR 97201, U.S.A.

INTRODUCTION

Two decades ago a series of biochemical experiments conclusively demonstrated that adrenocorticotropin (ACTH) and β-lipotropin (βLPH) peptides were stoichiometrically generated from a common precursor (1). This result was heralded by one of the landmark studies in the nascent field of eukaryotic cloning, the sequencing of a pituitary cDNA whose deduced amino acids encoded both peptides separated by a pair of basic amino acids (Lys-Arg) (2). Subsequent work demonstrated that the proopiomelanocortin (POMC) prohormone is posttranslationally processed into several biologically active peptides in addition to ACTH and βLPH. Among these are the melanocortin melanocyte-stimulating hormone (αMSH) and the potent opioid β–endorphin (βEND). The mechanism of POMC processing became a paradigm in cellular biology for the elucidation of vesicular trafficking and enzymatic cleavage of secreted peptides from prohormone precursors. A renaissance of scientific interest in POMC has occurred in more recent years with the cloning of a gene family of melanocortin receptors and the demonstration of central melanocortin effects on energy homeostasis. This chapter will review the ongoing role that spontaneous and induced (transgenic and "knockout") mutant mouse models have played in the analysis of POMC gene regulation and biological function.

Biosynthesis and Post-translational Processing of POMC
Several comprehensive reviews of POMC processing have been published (3, 4). This section will serve only to introduce major points relevant to the mouse models under consideration and is summarized in Table 1. The principal sites

of POMC expression are corticotrophs and melanotrophs of the pituitary gland and subsets of neurons in the arcuate nucleus (ARC) of the hypothalamus and in the caudal nucleus of the solitary tract (NTS) within the dorsal vagal complex of the medulla (5-7). Endoproteolytic cleavage of POMC by a prohormone convertase (PC1/PC3) in corticotrophs yields ACTH, βLPH, and an NH$_2$-terminal peptide (Pro-γMSH). The latter peptide and/or further proteolytic derivatives have been postulated to mediate paracrine growth and secretory effects within the anterior lobe and to stimulate mitogenesis in the adrenal cortex (8, 9). A small proportion of βLPH is cleaved to yield γLPH and βEND1-31. However, in melanotrophs of the intermediate lobe virtually all βLPH is cleaved by PC2 to γLPH and βEND1-31. The latter peptide is acetylated on its NH$_2$-tyrosine and truncated by the action of a carboxypeptidase to yield Ac–βEND1-27 and Ac–βEND1-26. The acetylated opioid peptides have a greatly reduced affinity for μ and δ opioid receptors and their biological function has not been satisfactorily determined. αMSH is acetylated on its NH$_2$-serine and amidated on its COOH-terminal to yield a fully bioactive peptide. ßMSH is an additional endoproteolytic product of human γLPH but is not generated from rodent γLPH because it lacks the necessary consensus amino acid sequence for cleavage by a PC. Endoproteolytic processing of POMC in ARC neurons closely mirrors the processing in melanotrophs. The neuronal αMSH is fully amidated, but there is minimal NH$_2$acetylation of peptides or COOH-terminal shortening of βEND1-31. γMSH and/or γ$_3$MSH are processed from the NH$_2$-POMC domain in ARC.

Table 1. POMC Peptides and their Cognate G-Protein Coupled Receptors

Peptide	Cell of origin	Receptor	Target site	Function
ACTH	Corticotrophs	MC2-R	Adrenal cortex	Cort. secretion
Pro-γMSH	Corticotrophs	MC3-R, ?	Pituitary cells	Paracrine action
Ac-αMSH-NH$_2$	Melanotrophs	MC1-R	Melanocytes	Pigmentation
Ac–βEND1-27	Melanotrophs	μ,δ ↑ K$_d$	Peripheral sites	Modulation ?
DesAc-αMSH	Pituitary	MC5R	Exocrine glands	Secretion
γMSH, γ$_3$MSH	ARC neurons	MC3-R	Limbic system	Homeostasis
DesAc-αMSH-NH$_2$, ACTH	ARC neurons	MC4-R	Hypothalamus	Homeostasis
βEND1-31	ARC neurons	μ = δ > κ	Diffuse CNS	Analgesia, reward

Five distinct members of a G-protein coupled, melanocortin receptor gene family have been cloned (reviewed in (10)). MC1-R is expressed on melanocytes within the skin and its activation by Ac-αMSH-NH$_2$ leads to increased membrane adenylyl cyclase activity by interaction with G$_{s\alpha}$ and subsequently greater production of the brown-black eumelanin pigment and less of the yellow-red phaeomelanin. MC2-R is the classical ACTH receptor expressed on zona glomerulosa and fasiculata cells of the adrenal cortex and it has very low affinity for the various melanocortin peptides other than ACTH and ACTH1-24 (11). Two additional members of the gene family, MC3-R and MC4-R are expressed exclusively in the CNS in partially overlapping but distinct neuroanatomical areas with particularly high abundance in the hypothalamus and limbic system (12, 13). The MC3-R has the highest affinity for γMSH, although it is also activated by other melanocortins, while the MC4-R is activated most selectively by αMSH and ACTH. Together these receptor subtypes mediate most of the described melanocortin effects on appetite, metabolism, thermoregulation, autonomic outflow, and behavior. The MC5-R is the most ubiquitously expressed of the melanocortin receptors and its best-documented function is the stimulation of exocrine function including the sebaceous, preputial, Harderian, and lacrimal glands (14).

Three opioid receptor genes have been cloned corresponding to the μ, δ, and κ receptors originally defined by their pharmacological profiles (15, 16). βEND1-31 has highest and nearly equivalent affinity to both μ and δ receptors (17). However, each of these subtypes has significant affinity for other opioid peptides derived from the proenkephalin and prodynorphin precursors.

POMC Peptides and Energy Homeostasis
Classical ablation or disconnection experiments utilizing electrolytic and chemically generated lesions or knife cuts of the mammalian brain revealed a number of discrete areas within the medial-basal hypothalamus to be nodal-points in the regulation of energy homeostasis (reviewed in (18)). One principal area characterized in this fashion is the ARC located at the base of the hypothalamus and adjacent to the third ventricle. The ARC extends rostrocaudally from the posterior border of the optic chiasm to the anterior border of the mammillary nuclei. A subset of hypothalamic neurons distributed throughout the ARC express the POMC gene. These POMC neurons densely innervate other brain areas implicated in energy homeostasis, motivated behavior, and autonomic control including the vetromedial nucleus of the hypothalamus (VMN), dorsomedial nucleus of the hypothalamus (DMH), paraventricular nucleus of the hypothalamus (PVN), lateral hypothalamic area (LH), ventral tegmental area (VTA), parabrachial nucleus, and NTS.

The lesion studies showed that disruption of the mammalian ARC correlated with an obese phenotype and led to the concept that this area of the hypothalamus contained a satiety center. Later efforts focused on determining the specific neuronal phenotype in the ARC that was involved in the regulation of energy homeostasis. An early indication of the importance of POMC peptides was the detection of MSH binding by autoradiography in the VMN, another hypothalamic nucleus involved in regulating appetite and feeding (19). In addition, central or peripheral injections of melanocortin peptides were shown to affect food intake. Melanocortins appear to have dual roles in energy balance, depending on the site of action. Administration of ACTH and Ac-αMSH directly into adipose tissue increases lipolysis whereas intramuscular injection of DesAc-αMSH increases food intake. Furthermore, peripheral DesAc-α-MSH inhibits pituitary production of ACTH and Ac-αMSH, which may lead to increased lipogenesis (20). MSH analogues administered i.c.v. or injected stereotaxically directly into the PVN of rodents have potent anorexigenic effects and stimulate metabolic rate (21). Exogenous opiates and endogenous opioid peptides have also been shown to affect food consumption and food preference, suggesting a role for POMC-derived β-endorphin, in addition to the melanocortin peptides, as a modulator of energy homeostasis (22-24). The physiology of this complex neuroendocrine system is virtually impossible to replicate in *in vitro* models. Thus, transgenic mice and mice with naturally occurring mutants or targeted genetic alterations have been extremely informative tools for the analysis of POMC peptides and their receptors in the control of food intake, energy partitioning, and body weight.

One of the first mutant mouse models of obesity to be studied was the lethal yellow (A^Y/a) mouse (25). The mutation arose spontaneously in the C57BL/6 inbred strain and is associated with a striking yellow coat color and the development of obesity and hyperphagia. Genetically, the mutation was shown to be dominant to the extension (E) locus, later identified as the gene encoding the MC-1 receptor (26). Cloning of the agouti gene revealed that it encodes a secreted protein characterized by multiple pairs of intrachain disulfide bonds in its C-terminal domain. Agouti has high affinity binding for the MC1-R, antagonizes receptor activation by MSH (27), and is normally transiently expressed only in the hair follicles accounting for the yellow banding of individual black hairs in the fur of many mammals. The mutant A^Y allele results from a 120 kb genomic deletion and juxtaposition of the agouti coding sequences to the Raly gene locus and subsequent ubiquitous expression of agouti protein from the Raly promoter (25). These observations led to the agouti-obesity hypothesis, which states that agouti protein expressed within the brain antagonizes the action of central MSH to a neural MC receptor involved in regulation of energy balance. Some of the experimental models reviewed in this chapter provide essential data in support of this hypothesis, which has been further validated by the discovery and cloning of an agouti homologue, agouti gene-related protein (AGRP).

AGRP is expressed in the nervous system and functions as an antagonist of central MC receptors (28).

Genetic linkage studies and quantitative trait loci analyses have strongly implicated the POMC gene locus as an important determinant of weight homeostasis in humans of many different ethnic populations, although specific alleles associated with obesity have not yet been demonstrated (29, 30). Because no mutations within the coding region of the POMC gene that alter peptide activity have been identified in these populations, a current hypothesis is that mutations in regulatory regions of the gene decrease the level of POMC expression in the brain. However, a small number of children from cosanguineous parents have been found to have null mutations in the POMC gene resulting in the absence of detectable circulating ACTH (31). These children presented with a syndrome of red hair, adrenal insufficiency and severe, early-onset obesity. In addition, both dominant and recessive mutations in the MC4-R gene have been found in the human population, and MC4-R mutations have been proposed to play a role in as many as 5% of pediatric obesity cases (32, 33).

POMC GENE REGULATION

A standard method to characterize the *cis* DNA elements and transcription factors that control POMC gene transcription has been the expression of fusion genes, consisting of POMC promoter sequences and a reporter sequence, in transfected cell lines. The most commonly used cell line is AtT20, originally derived from a radiation-induced pituitary tumor (34). These cells have many characteristics of primary corticotrophs. Small cell lung cancers often ectopically express POMC and cell lines derived from these tumors have also been used for the analysis of POMC gene expression (35). It is clear, however, that neither of these cell lines are adequate models for elucidation of the mechanisms governing neural-specific expression of the POMC gene. For this reason, our laboratory has focused on a transgenic mouse approach for the identification of cell-type specific regulatory elements in the POMC gene.

Promoter Analysis in Transgenic Mice
The Drouin laboratory was the first to report the generation of transgenic mice expressing a neomycin-resistance coding sequence (*neo*) from 770 bases of the rat POMC promoter (36). Although the expression pattern in pituitary and regulation by adrenalectomy and dexamethasone were consistent with accurate cellular expression of the transgene, the low levels of transcript from the *neo* reporter prevented the direct demonstratation of corticotroph and melanotroph specific transcription. Our laboratory initially

approached this problem using an *E. coli lacZ* reporter gene encoding ß-galactosidase, which was useful for both the precise cellular identification of transgene expression and quantitation of transcriptional activity by an enzymatic assay utilizing a fluorigenic substrate. These studies demonstrated that 770 bases of the rat POMC promoter from nucleotide -706 in the 5' flanking sequences to +64 in the 5' untranslated region of exon 1 were sufficient to direct *lacZ* expression selectively to pituitary corticotrophs and melanotrophs (Figure 1, POMC-LacZ) (37). The ontogeny of transgene expression also closely approximated the time course of endogenous POMC gene expression in the two lobes of the pituitary gland. However, there was no detectable expression of the -706POMC-*lacZ* transgene in the brain suggesting that additional genomic sequences were necessary for neuronal transcription.

Additional studies from our laboratory analyzed the rat POMC promoter in more detail to define the minimal elements essential for pituitary expression (38, 39). Accurate cell-specific colocalization was verified in all instances by double-label immunofluoresence histochemistry. Truncations of the rat POMC promoter to nucleotide -323 or -234 had minimal effects on both the ratio of expression-positive to total transgenic pedigrees and qualitative levels of reporter expression in the positively expressing lines. Further deletion of the 5' flanking sequences to nucleotide -160 abolished expression in the pituitary gland. Transfection studies in AtT20 cells had identified important regulatory elements in the core promoter region of the POMC gene. However, we demonstrated that the enhancerless promoter and TATA box from the HSV1 thymidine kinase gene were equivalent to the native POMC promoter sequences between nucleotides -34 and +64, in the context of upstream flanking sequences of POMC, to support pituitary cell-specific expression and normal upregulation by adrenalectomy. Based on these promoter truncation studies in transgenics and our data from DNAse I protection assays and gel-shift assays utilizing POMC oligonucleotide probes and fractionated nuclear proteins from AtT20 cells, we designed a series of more discrete mutations in the rat POMC promoter. This second generation of transgenic experiments identified two functionally important protein binding sites at nucleotide positions -262/-253 and -202/-193 of the POMC gene (39). At least one of these sites had to be intact to support detectable pituitary expression of the transgenes. In addition, the ubiquitous transcription factor SP1 appeared to play a supportive role in POMC transgene expression through its interaction with DNA sites at positions -201/-192 and -146/-136.

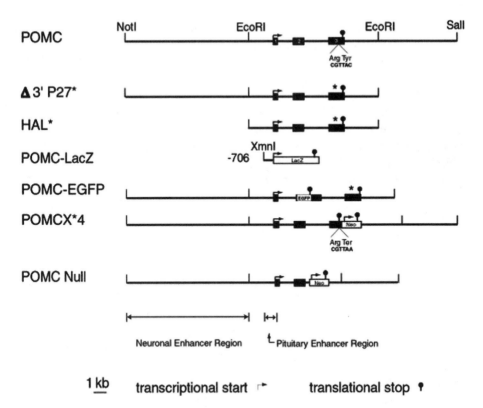

Fig 1. Restriction maps of the mouse POMC gene locus, transgene constructs utilized to map the location of tissue-specific regulatory elements, and targeted POMC alleles that either selectively prevent ß-endorphin production or result in the absence of all POMC peptides. See text for details of the individual alleles.

None of the transgene constructs that we studied selectively directed expression to only pituitary melanotrophs or corticotrophs, suggesting that both cell types share one or more essential components for POMC gene transcription. In common with most published transgenic promoter analyses, we observed a large quantitative and qualitative range of expression levels among independent pedigrees for each construct. It has been suggested that these mosaic patterns of gene expression are influenced by repressive effects of multiple-copy transgene integration in some cases (40). In general, basal expression of the transgenes appeared to be greater in individual melanotrophs than corticotrophs, regardless of the type of reporter molecule.

The data from our transgenic promoter analyses revealed a number of inconsistencies with the mapping of functional POMC promoter elements by transfection studies in AtT20 cells. These differences may be due in part to the more stringent requirement in transgenic mice for the definition of cell-specific expression and the importance of chromatin remodeling during *in vivo* development. However, there is good agreement between both kinds of studies for the key importance of DNA elements between nucleotides -160 and −323 in the 5' flanking domain of the POMC gene. This region contains a site between -173/-160 that appears to mediate the transcriptional activating effects of both corticotroph releasing hormone (CRH) and the cytokine leukemia inhibitory factor (LIF), although it is neither a binding site for CREB, AP-1 or STAT proteins (41). In contrast, the POMC TATA box, a putative AP-1 binding site in exon 1, and more distal sites in the POMC 5' flanking sequences that bind HLH proteins and confer a synergistic activation of POMC expression in AtT20 cells do not appear to be essential for pituitary cell-specific expression in the *in vivo* paradigm (42).

Generation of Novel POMC Expressing Cell Lines
Targeted cellular expression of oncogenes with tissue-specific promoters has led to new insights into the molecular basis of tumorigenesis. Several laboratories have used this technique to isolate novel differentiated cell lines (43-46). We used the rat POMC promoter elements to express two isoforms of the simian virus SV40-Large T antigen in pituitary corticotrophs and melanotrophs. Wild-type SV40Tag consistently induced tumors in the pituitary gland, as predicted, but also in the thymus gland with incomplete penetrance of the phenotype (47). However, the incidence of thymic tumors decreased and eventually disappeared after 15 generations of backcrossing to an outbred CD-1 genetic background, suggesting that epistatic interactions with gene alleles from the original B6D2 genetic background were necessary for expression of the transgene in thymus. In contrast, the pituitary tumor phenotype remained stable after 27 generations of breeding (48). The variable latency to tumor development in these mice is consistent with the idea that stochastic secondary events likely occur in individual T-antigen expressing cells before they are transformed and lose their inhibitory growth controls. Other transgenic mice expressing the temperature-sensitive tsA58 mutant form of SV40Tag from the same POMC promoter sequences only developed pituitary tumors after 1 year of age (our unpublished data). The longer latency was presumably due to the expression of minimal amounts of biologically stable tsA58-Tag at the partially permissive core body temperature of the mice.

Although the POMC-Tag transgene is expressed in both pituitary corticotrophs and melanotrophs, every pituitary tumor that has developed in the mice has been of melanotroph origin based on visual inspection and biochemical and molecular studies. The melanotroph tumor cells are fully transformed because they will produce secondary tumors after subcutaneous passage in nude mice and will freely grow in suspension culture (47, 49). Unlike the strict dependence for IGF-II in the induction and progression of pancreatic islet ß-cell tumors from an insulin promoter-Tag transgene (50), we found no evidence for a critical role of the IGF system in the POMC-Tag induced pituitary tumors (51). Posttranslational processing of POMC in the primary tumors was indistinguishable from normal melanotrophs and was characterized by high proportions of NH_2-acetylated and COOH-truncated forms of ß-endorphin and Ac-αMSH with virtually no ACTH (47). Even though the tumors expressed high levels of PC2, which is characteristic of melanotrophs, there was frequently a sufficient quantity of an ACTH-like peptide secreted from the neoplastic glands to induce adrenal cortical hyperplasia and elevated serum corticosterone. A syndrome of fatal melanotroph-dependent Cushing's disease was recently reported in mice with a null mutation in the gene encoding the neuroendocrine protein 7B2 (52) (see Chapter 8). 7B2 has been implicated in the activation of PC2, accounting for the lack of functional PC2 production and generation of ACTH instead of Ac-αMSH and CLIP in melanotrophs of the 7B2 null mice. However, it remains unexplained why ACTH is constitutively secreted from the intermediate lobe of these mice because a knockout of the PC2 gene results in similarly high ACTH production in melanotrophs without the unregulated secretion of the hormone (53). Therefore, 7B2 may have an additional function in vesicular trafficking or secretion.

Two cell lines with distinct phenotypes were isolated from cultures of the mouse melanotroph tumors (54). mIL39 cells are melanotrophs based on their expression of both POMC and the dopamine D2 receptor. They are small, bipolar cells and rapidly produce secondary tumors after transplantation into nude mice. In contrast, mIL5 cells fail to express either POMC or detectable levels of SV40Tag. The exact origin of these latter cells is unknown but they are of interest because they secrete a prolactin-regulating factor. The lack of concordance between this prolactin-regulating activity and POMC gene expression suggests that the activity is not due to a peptide product of POMC. More recently our laboratory has isolated another melanotroph-like cell line from a pituitary tumor induced by the tsA58SV40Tag (Chronwall et al, submitted). These cells express high levels of POMC mRNA and POMC prohormone but they have few secretory granules and limited posttranslational processing of POMC. Preliminary studies have demonstrated the expression of several G-protein coupled receptors including dopamine D2, CRH R1, and $GABA_B$ that are normally expressed on melanotrophs. These receptors are functional and therefore this cell line may be useful for analyzing the biochemical cross-talk in signaling pathways that are unique to pituitary melanotrophs.

Targeting POMC Neurons *In Vivo*

Because neither 770 or 4000 bp of the 5' flanking sequence from the rat POMC gene were sufficent to produce appropriate neuronal expression in transgenic mice (55), longer mouse genomic POMC clones were isolated from a cosmid library. Exon 3 was marked with an oligonucleotide sequence to distinguish transgenic mRNA from endogenous POMC mRNA and to enable the simultaneous quantitation of both mRNA species by an RNAse protection assay. A new series of transgenic mice was generated and analyses of mRNA expression in the mice clearly established the location of a putative neuron-specific enhancer within an 11 kb stretch of 5' flanking DNA sequences between −13 kb and -2 kb relative to the transcriptional start site (Figure 1, Δ3'P27* and HAL*) (56). Transgene constructs containing these sequences have had a 100% penetrance of correctly targeted expression in neurons of both the ARC and NTS, consistent with the existence of a locus-control type regulatory element within the POMC gene. There is also total concordance of pituitary expression with neural expression from the longer genomic constructs indicating that an overlapping subset of POMC regulatory elements and cognate transcription factors may be required for transcription in both cell types. The recent identification of a novel neuropeptide precursor gene, cocaine and amphetamine responsive-transcript (CART), as a cotransmitter in POMC arcuate neurons suggests that these two distinct genes may also share a subset of regulatory factors (57, 58). Ongoing functional analyses of the distal POMC genomic sequences combined with recent advances in the human and mouse genome projects should rapidly lead to the identification of the minimal DNA elements that are responsible for neural-specific expression. It will then be possible to compare these sequences with the corresponding regions of DNA amplified from obese subjects to test the hypothesis that mutations in POMC regulatory sites are the molecular basis for the genetic linkage established by QTL analyses (29). Ultimately, the functional significance of nucleotide differences identified by this type of horizontal genetic comparison can be tested in mice by targeted mutagenesis and analyses of POMC expression in hypothalamus, feeding behavior, and energy partitioning.

Our characterization of neural-specific regulatory regions in the POMC gene provided a means to identify viable POMC neurons using a nondestructive, green fluorescent protein (enhanced-GFP) reporter (Cowley et al., submitted). Previously, the analysis of cellular events in POMC neurons has been limited because of their small number and sparse distribution. Bright green fluorescence was readily identified in the basal hypothalamus and in the NTS of transgenic mice expressing a POMC-EGFP construct (Figure 1). Double-labeling immunofluorescence with a ß-endorphin antisera revealed >99% cellular colocalization of EGFP and endogenous POMC peptide expression within the ARC. The developmental onset of EGFP expression in both hypothalamic and pituitary cells also matched the ontogeny of native POMC.

POMC neurons are an important site of leptin receptor expression in the hypothalamus (59). The adipocyte derived hormone leptin regulates long-term aspects of energy homeostasis by signaling the levels of fat mass to the brain. Leptin also has acute actions to stimulate *c-fos* and *Socs3* expression in POMC neurons and can cause the release of ß-endorphin from hypothalamic fragments *in vitro* (60). Therefore, the POMC-EGFP transgenic mice will be a useful tool to test the hypothesis that leptin directly activates POMC neurons. Preliminary electrophysiological experiments in hypothalamic slices suggest that leptin has both pre- and postsynaptic actions on POMC neurons to increase the frequency of spontaneous action potentials. Additional studies using this transgenic mouse line will decipher the role of other peptidergic modulators, cytokines, and amino acid neurotransmitters on POMC neuronal activity. EGFP-labeled POMC neurons can also be used for investigation of neuronal migration during development and as a unique source of mRNA for differential gene expression analyses. Similar techniques to identify rare neuronal populations *in vivo* are increasingly being used by other laboratories. A prime example is the identification of GnRH neurons as described in Chapter 3 of this volume and by other investigators (61).

In principle, the neural-specific regulatory sequences of the POMC gene can be used to target expression of foreign or mutated proteins to POMC neurons to investigate intracellular signaling pathways *in situ*. For example, leptin resistance may be associated with alterations in the capacity for leptin signaling through either JAK/Stat or MAPK cascades. Individual molecular steps in these pathways could be perturbed by the use of dominant-negative factors or expression of natural inhibitor proteins. Finally, targeted oncogene expression to POMC neurons may lead to the isolation of a differentiated POMC neuronal cell line, which would greatly facilitate molecular studies of POMC gene regulation and the membrane properties of the neurons themselves.

MUTATIONAL ANALYSIS OF POMC FUNCTION

TARGETED MUTAGENESIS OF THE POMC GENE

ß-endorphin-Deficient Mice
ß-endorphin-deficient mice were generated by the introduction of a point mutation into the POMC gene by homologous recombination in embryonic

stem cells (62). The nucleotide insertion at the amino-terminal tyrosine of ßEND introduced a premature translational stop-codon into exon 3 (Figure 1, construct POMCX*4). Homozygous mice carrying this mutated allele express normal levels of POMC mRNA in pituitary and ARC. The mRNA encodes a COOH-truncated POMC prohormone that is processed normally to ACTH and other melanocortin peptides (63). The mice have no detectable immunoreactive ßEND and have been used to infer the normal physiological function of this specific endogenous opioid peptide. Basal activity of the hypothalamic-pituitary-adrenal axis is normal and the βEND-deficient mice have normal corticosterone responses to a variety of stressors. βEND-deficient mice exhibit unaltered nociceptive thresholds and unaltered antinociceptive responses to i.p. morphine administration. However, βEND-deficient mice fail to exhibit endogenous opioid-mediated stress-induced antinociception and have increased levels of nonopioid (naloxone insensitive) stress-induced antinociception (63). Complimentary studies in enkephalin-deficient mice have demonstrated that they retain endogenous opioid stress-induced antinociception (64). Taken together, these data provide convincing genetic evidence that ßEND derived from the POMC prohormone, and not the more abundant enkephalin peptides, is the natural opioid mediating stress-induced antinociception. The most likely relevant neuroanatomical substrate is the projection of ARC POMC neurons to the central periaqueductal gray known to modulate descending antinociceptive pathways to the brainstem and spinal cord.

More recent studies have demonstrated that the ßEND-deficient mice have altered antinociceptive responses to morphine administered i.c.v. or i.t. The mutant mice are more sensitive to the antinociceptive action of morphine injected centrally into the lateral ventricle but less sensitive to morphine injected into the spinal compartment, compared to wild-type control mice (65). The μ-specific agonist DAMGO produced qualitatively identical results indicating the importance of the μ opioid receptor in the phenotype. Total binding sites for the μ, δ, and κ receptors, assessed by quantitative autoradiography, were normal in the ßEND-deficient mice suggesting that subtle alterations in μ receptor signaling may be the mechanism for the opposing alterations in central and spinal sensitivity to morphine. An electrophysiological study demonstrated a normal dose-response to DAMGO for the induction of an inwardly rectifying K^+ current (GIRK) in mediobasal hypothalamic neurons of the mutant mice (66) consistent with the suggestion that μ receptor number, affinity, and coupling to G proteins are not altered by the complete absence of ßEND.

Opioids, including βEND, have generally been shown to increase food intake in pharmacological studies (67). Paradoxically, β-endorphin deficient mice exhibit a sexually dimorphic obesity phenotype due to increased white fat mass in males (Appleyard et al, submitted). The male mice are slightly hyperphagic and have a normal basal metabolic rate. It is not yet clear if POMC derived melanocortins and ßEND are involved in a parallel, redundant anorexigenic pathway within the hypothalamus or if βEND additionally plays a role in an opioid reward pathway that is activated by feeding.

Null Mutation of the POMC Gene
Mice that are deficient in all POMC peptides (POMC Null) were generated by the Hochgeschwender laboratory by homologous recombination of a POMC allele containing a deletion of exon 3 (Figure 1, POMC Null). The mice develop hyperphagia and obesity similar to that seen in the lethal yellow (A^Y/a) mice and MC4-R deficient mice (68). In contrast to the other obesity models, POMC Null mice have no detectable levels of circulating corticosterone or epinephrine and their adrenal glands are one-third the size of wildtype littermates. The adrenal deficiencies in these mice strengthen the hypothesis that normal POMC expression is needed for complete adrenal maturation, although the role of NH_2-terminal peptides in addition to ACTH remains an open question that can now be approached experimentally *in vivo*. A partial perinatal lethality occurs in litters produced from heterozygous breeder pairs leading to significantly decreased numbers of homozygous POMC Null offspring compared to Mendelian predictions. Homozygous POMC Null adults are fertile and females carry their litters to term but all pups die in the immediate perinatal period (our unpublished data). The mechanism for this pattern of perinatal lethality from both heterozygous and homozygous POMC Null female mice has not yet been explained, but may result from the combined lack of glucocorticoids and epinephrine in the POMC Null pups.

In principle, the absence of all POMC peptides should recapitulate the sum of phenotypes observed in mice deficient in the various receptor subtypes for the peptides. Although not all the characteristics of the various receptor-deficient mice have been investigated, some of the predicted phenotypes are absent in the POMC Null mice. In particular, POMC Null mice do not exhibit the hair pigmentation that would be expected based on the phenotypes of A^Y/a yellow and null MC1-R mutants (see discussion in the following section). The original POMC Null mice bred on an agouti (A/A) 129 genetic background had patchy alterations in their coat color, but were clearly not a uniform yellow (68). We have crossed the POMC Null allele onto the C57BL/6 (a/a) genetic background and observed an unaltered black coat color in the mutant mice despite the absence of MSH, the only known natural agonist for the MC1-R (unpublished data). In sum, POMC deficient mice will be invaluable in deciphering the physiological roles of individual POMC peptides in the neuroendocrine and immune systems by transgenic strategies designed to rescue the expression of subsets of POMC products to either the pituitary and/or neural compartments.

MELANOCORTIN AND OPIOID RECEPTOR GENE MUTATIONS

Hypomorphic and Constitutively Active MC1-R Gene Alleles

Classical genetic studies indicated that two independent loci, extension (E) and agouti (A), primarily controlled the ratio of eumelanin and phaeomelanin in melanocytes, and hence the color of individual hairs. The extension locus is cell-autonomous and approximately a decade ago was demonstrated to encode the MCR-1 or MSH receptor (26). Numerous naturally occuring mutations at the extension locus have been characterized genetically because of the relative ease in tracing the mutant alleles during backcrosses and intercrosses by observation of coat color. Recessive mutations of the mouse extension locus (e) result in a yellow coat color due to low levels of tyrosinase activity in melanocytes and a high phaeomelanin/eumelanin pigment ratio. The e allele encodes a nonfunctional MCR-1 resulting from a frameshift mutation and premature translational stop codon between the fourth and fifth transmembrane domains. Dominant mutations of the mouse E locus result in black coat color and are epistatic to agouti. The sombre (E^{so}) allele encodes a MC1-R containing a single activating point mutation. This MC1-R is constitutively active in the absence of MSH and therefore not antagonized by agouti protein, accounting for the genetic epistasis. These original data and the analyses of other mutant E alleles have been the basis of structure-function models to explain receptor activation by agonist ligand binding. Interestingly, identical mutations introduced into the other highly homologous MCR subtypes have generally not resulted in receptor activation emphasizing the important functional differences among these related G-protein coupled receptors and the danger of over extrapolating data concerning ligand interactions from one receptor subtype to another.

Null Mutation of the MC4-R Gene

Targeted deletion of the MC4-R has shown it to play an important role in energy homeostasis. MC4-R is the predominant melanocortin receptor in the brain and genetically engineered mice homozygous for a null MC4-R allele are obese, hyperphagic, and exhibit increased longitudinal growth (69), all features shared with the A^Y/a yellow mouse. These data and the inability of the potent MSH analogue MTII to decrease food intake in MC4-R knockout mice (70) indicate that the MC4-R is essential for the anorectic effects of central melanocortins.

Null Mutation of the MC5-R Gene

Transgenic mice deficient in the MC5-R do not exhibit any phenotypes that are directly associated with weight regulation or feeding but they do exhibit thermoregulatory defects and fail to repel water from their coat. These phenotypes were shown to be a result of reduced production and secretion of lipids from sebaceous glands in MC5-R deficient mice (14). Additionally, the mice exhibited decreased production and attenuated stress-induced synthesis of porphyrins by the Harderian glands and decreased protein secretion from lacrimal glands. All of these exocrine glands, including the preputial gland, are normally sites of high MC5-R expression. The results suggest the existence of a hypothalamic-pituitary-exocrine axis that potentially regulates pheromone production from exocrine glands in coordination with the behavioral arousal state of the mouse.

Null Mutations of the MC3-R Gene

Null mutations of the POMC (68) and MC4-R (69) genes, and transgenic overexpression or misexpression of the MC4-R antagonists agouti and AGRP (28, 71) all produce phenotypes that can be attributed to diminshed MC4-R signaling. Although these genetic models have provided great insight into the role of POMC and MC4-R in energy homeostasis, little specific information can be extracted regarding a contributory role of MC3-R signaling. The MC3-R, like the MC4-R, is expressed in the CNS but in a more restricted distribution including ARC and VMN. Recently, two groups independently published reports of knockout mice lacking the MC3-R (72, 73). MC3-R deficient mice exhibit nearly normal growth curves and body weight, but carcass analyses reveal a significant increase in white adipose mass. Additionally, MC3-R deficient mice are more susceptible than controls to developing obesity on a high fat diet. In contrast to the obesity phenotype exhibited in MC4-R deficient mice, which can mostly be explained by alterations in feeding behavior, the MC3-R deficient mice apparently have a predominant metabolic defect involving energy partitioning. They have higher feed efficiency and decreased lean mass. Mice lacking both the MC3-R and MC4-R gain more weight than single MC4-R knockouts (73). Additional studies are needed to define the specific neural circuits responsible for the divergent actions of melanocortin peptides, MC3-R, and MC4-R on the regulation of food intake versus metabolic rate and energy expenditure.

Null Mutations of the Opioid Receptor Genes

Each of the three genes encoding the classical opioid receptor subtypes has been deleted by homologous recombination in mice (16). Surprisingly, there have been no published reports regarding the effects of these mutations on food intake and energy homeostasis. Based on our findings in the ßEND-deficient mice, we would predict that µ receptor knockout mice are the most likely to have a phenotype relevant to the physiologic regulation of body weight and fat mass.

SUMMARY AND FUTURE PROSPECTS

Complex interactions between the nervous system and the endocrine system are essential for the regulation of energy homeostasis. Transgenic and targeted gene mutational studies have demonstrated that neuropeptides derived from the POMC prohormone and their cognate G-protein coupled receptors are key elements of communication between and within these two important biological systems. Transgenic models have identified POMC neurons in the ARC as a prominent component of the hypothalamic neural circuitry that modulates energy homeostasis. Moreover, melanocortin and opioid peptides derived from POMC may have both additive and opposing actions in the balance of caloric intake and expenditure. Many of the mutations described in this chapter give rise to multisystem physiological disturbances and could also involve undetected developmental alterations, therefore one must be circumspect in the assignment of a specific peptide or receptor subtype as the major determinant in a disease. As a result, newer and more sophisticated molecular and genetic approaches to *in vivo* gene mutations, such as conditional gene deletion and regulation, are currently being explored. These newer techniques will permit a more detailed characterization of specific endocrine-disease pathophysiology, and in addition, help to reveal the function of each biologically active POMC peptide *in vivo*.

Acknowledgements

The authors are indebted to our colleagues and collaborators who have contributed to the ideas and experiments discussed in this chapter.

REFERENCES

1.   Eipper BA, Mains R.E. Structure and biosynthesis of pro-adrenocorticotropin/endorphin and related peptides. Endocr Rev 1980;1:1-27.
2.   Nakanishi S, Inoue A, Kita T, Nakamura M, Chang AC, Cohen SN, Numa S. Nucleotide sequence of cloned cDNA for bovine corticotropin-beta-lipotropin precursor. Nature 1979;278:423-427.

3. Castro MG, Morrison E. Post-translational processing of proopiomelanocortin in the pituitary and in the brain. Crit Rev Neurobiol 1997;11:35-57.
4. Smith AI, Funder JW. Proopiomelanocortin processing in the pituitary, central nervous system, and peripheral tissues. Endocr Rev 1998;9:159-179.
5. Finley JCW, Lindström P, Petrusz P. Immunocytochemical localization of β-endorphin-containing neurons in the rat brain. Neuroendocrinol 1981;33:28-42.
6. Elkabes S, Loh YP, Nieburgs A, Wray S. Prenatal ontogenesis of pro-opiomelanocortin in the mouse central nervous system and pituitary gland: an in situ hybridization and immunocytochemical study. Brain Res Dev Brain Res 1989;46:85-95.
7. Palkovits M, Mezey É, Eskay RL. Pro-opiomelanocortin-derived peptides (ACTH/β-endorphin/α-MSH) in brainstem baroreceptor areas of the rat. Brain Res 1987;436:323-328.
8. Seger MA, Bennett HP. Structure and bioactivity of the amino-terminal fragment of pro-opiomelanocortin. J Steroid Biochem 1986;25:703-710.
9. Denef C, Van Bael A. A new family of growth and differentiation factors derived from the N-terminal domain of proopiomelanocortin (N-POMC). Comp Biochem Physiol C Pharmacol Toxicol Endocrinol 1998;119:317-324.
10. Cone RD, Lu D, Koppula S, Vage DI, Klungland H, Boston B, Chen W, Orth DN, Pouton C, Kesterson RA. The melanocortin receptors: agonists, antagonists, and the hormonal control of pigmentation. Recent Prog Horm Res 1996;51:287-317.
11. Mountjoy KG, Robbins LS, Mortrud MT, Cone RD. The cloning of a family of genes that encode the melanocortin receptors. Science 1992;257:1248-1251.
12. Roselli-Rehfuss L, Mountjoy KG, Robbins LS, Mortrud M, Low MJ, Tatro J, Entwistle M, Simerly R, Cone RD. Identification of a receptor for γ-MSH and other proopiomelanocortin peptides in the hypothalamus and limbic system. Proc Natl Acad Sci USA 1993;90:8856-8850.
13. Mountjoy KG, Mortrud M, Low MJ, Simerly R, Cone RD. Characterization of a melanocortin receptor (MC4-R) localized in neuroendocrine and autonomic control circuits in the brain. Mol Endocrinol 1994;8:1298-1308.
14. Chen W, Kelly MA, Optiz-Araya X, Low MJ, Cone RD. Exocrine gland dysfunction in MC5-R deficient mice. Cell 1997;91:789-798.
15. Mansour A, Fox CA, Akil H, Watson SJ. Opioid-receptor mRNA expression in the rat CNS: anatomical and functional implications. Trends Neurosci 1995;18:22-29.
16. Hayward MD, Low MJ. Targeted mutagenesis of the murine opioid system. Results Probl Cell Differ 1999;26:169-191.
17. Pasternak GW. Pharmacological mechanisms of opioid analgesics. Clin Neuropharmacol 1993;16:1-18.
18. Elmquist JK, Elias CF, Saper CB. From lesions to leptin: hypothalamic control of food intake and body weight. Neuron 1999;22:221-232.
19. Tatro JB. Melanotropin receptors in the brain are differentially distributed and recognize both corticotropin and alpha-melanocyte stimulating hormone. Brain Res 1990;536:124-132.
20. Mountjoy KG, Wong J. Obesity, diabetes and functions for proopiomelanocortin-derived peptides. Mol Cell Endocrinol 1997;128:171-177.
21. Fan W, Boston BA, Kesterson RA, Hruby VJ, Cone RD. Role of melanocortinergic neurons in feeding and the agouti obesity syndrome. Nature 1997;385:165-168.
22. Reid LD. Endogenous opioid peptides and regulation of drinking and feeding. Am J Clin Nutr 1985;42:1099-1132.
23. Hope PJ, Chapman I, Morley JE, Horowitz M, Wittert GA. Food intake and food choice: the role of the endogenous opioid peptides in the marsupial Sminthopsis crassicaudata. Brain Res 1997;764:39-45.

24. Levine AS, Grace M, Portoghese PS, Billington CJ. The effect of selective opioid antagonists on butorphanol-induced feeding. Brain Res 1994;637:242-248.
25. Duhl DM, Stevens ME, Vrieling H, Saxon PJ, Miller MW, Epstein CJ, Barsh GS. Pleiotropic effects of the mouse lethal yellow (Ay) mutation explained by deletion of a maternally expressed gene and the simultaneous production of agouti fusion RNAs. Development 1994;120:1695-1708.
26. Robbins LS, Nadeau JH, Johnson KR, Kelly MA, Roselli-Rehfuss L, Baack E, Mountjoy KG, Cone RD. Pigmentation phenotypes of variant extension locus alleles result from point mutations that alter MSH receptor function. Cell 1993; 72:827-834.
27. Lu D, Willard D, Patel IR, Kadwell S, Overton L, Kost T, Luther M, Chen W, Woychik RP, Wilkison WO. Agouti protein is an antagonist of the melanocyte-stimulating-hormone receptor. Nature 1994;371:799-802.
28. Ollmann MM, Wilson BD, Yang YK, Kerns JA, Chen Y, Gantz I, Barsh GS. Antagonism of central melanocortin receptors in vitro and in vivo by agouti-related protein. Science 1997;278:135-138.
29. Comuzzie AG, Hixson JE, Almasy L, Mitchell BD, Mahaney MC, Dyer TD, Stern MP, MacCluer JW, Blangero J. A major quantitative trait locus determining serum leptin levels and fat mass is located on human chromosome 2. Nat Genet 1997;15:273-276.
30. Hixson JE, Almasy L, Cole S, Birnbaum S, Mitchell BD, Mahaney MC, Stern MP, MacCluer JW, Blangero J, Comuzzie AG. Normal variation in leptin levels in associated with polymorphisms in the proopiomelanocortin gene, POMC. J Clin Endocrinol Metab 1999;84:3187-3191.
31. Krude H, Biebermann H, Luck W, Horn R, Brabant G, Gruters A. Severe early-onset obesity, adrenal insufficiency and red hair pigmentation caused by POMC mutations in humans. Nat Genet 1998;19:155-157.
32. Farooqi IS, Yeo GS, Keogh JM, Aminian S, Jebb SA, Butler G, Cheetham T, O'Rahilly S. Dominant and recessive inheritance of morbid obesity associated with melanocortin 4 receptor deficiency. J Clin Invest 2000;106:271-279.
33. Yeo GS, Farooqi IS, Challis BG, Jackson RS, O'Rahilly S. The role of melanocortin signalling in the control of body weight: evidence from human and murine genetic models. QJM 2000;93:7-14.
34. Furth J, Gadsden EL, Upton AC. ACTH-secreting transplantable pituitary tumors. Proc Soc Exp Biol Med 1953;84:253-254.
35. Picon A, Bertagna X, de Keyzer Y. Analysis of proopiomelanocortin gene transcription mechanisms in bronchial tumour cells. Mol Cell Endocrinol 1999;147:93-102.
36. Tremblay Y, Tretjakoff I, Peterson A, Antakly T, Zhang CX, Drouin J. Pituitary-specific expression and glucocorticoid regulation of a proopiomelanocortin fusion gene in transgenic mice. Proc Natl Acad Sci USA 1988;85:8890-8894.
37. Hammer G, Fairchild-Huntress V, Low MJ. Pituitary-specific and hormonally regulated gene expression directed by the rat proopiomelanocortin promoter in transgenic mice. Mol Endo 1990;4:1689-1697.
38. Liu B, Hammer GD, Rubinstein M, Mortrud M, Low MJ. Identification of DNA elements cooperatively activating proopiomelanocortin gene expression in the pituitary gland of transgenic mice. Mol Cell Biol 1992;12:3978-3990.
39. Liu B, Mortrud M, Low MJ. DNA elements with AT-rich core sequences direct pituitary cell-specific expression of the proopiomelanocortin gene in transgenic mice. Biochemical J 1995;312:827-832.
40. Garrick D, Fiering S, Martin DI, Whitelaw E. Repeat-induced gene silencing in mammals. Nat Genet 1998;18:56-59.
41. Bousquet C, Ray DW, Melmed S. A common pro-opiomelanocortin-binding element mediates leukemia inhibitory factor and corticotropin-releasing hormone transcriptional synergy. J Biol Chem 1997;272:10551-10557.

42. Therrien M, Drouin J. Cell-specific helix-loop-helix factor required for pituitary expression of the pro-opiomelanocortin gene. Mol Cell Biol 1993;13:2342-2353.
43. Windle JJ, Weiner RI, Mellon PL. Cell lines of the pituitary gonadotrope lineage derived by targeted oncogenesis in transgenic mice. Mol Endocrinol 1990;4:597-603.
44. Efrat S, Linde S, Kofod H, Spector D, Delannoy M, Grant S, Hanahan D, Baekkeskov S. β-cell lines derived from transgenic mice expressing a hybrid insulin gene-oncogene. Proc Natl Acad Sci USA 1988;85:9037-9041.
45. Adams JM, Cory S. Transgenic models of tumor development. Science 1991;254:1161-1167.
46. Alarid E, Windle J, Whyte D, Mellon P. Immortalization of pituitary cells at descrete stages of development by directed oncogenesis in transgenic mice. Development 1996;122:3319-3329.
47. Low MJ, Liu B, Hammer GD, Rubinstein M, Allen RG. Post-translational processing of proopiomelanocortin (POMC) in mouse pituitary melanotroph tumors induced by a POMC-simian virus 40 large T antigen transgene. J Biol Chem 1993;268:24967-24975.
48. Low MJ, Kelly MA, Graham KE, Asa SL. Transgenic and induced-mutant mouse models of pituitary adenomas. In: The Management of Pituitary Tumors. (Ed S. Webb) pp 55-69. Bristol: BioScientifica Ltd, 1998.
49. Allen DL, Low MJ, Allen RG, Ben-Jonathan N. Identification of two classes of prolactin releasing factors in intermediate lobe tumors from transgenic mice. Endocrinology 1995;136:3093-3099.
50. Christofori G, Naik P, Hanahan D. Deregulation of both imprinted and expressed alleles of the insulin-like growth factor 2 gene during beta-cell tumorigenesis. Nat Genet 1995;10:196-201.
51. Grewal A, Bradshaw SL, Schuller AGP, Low MJ, Pintar JE. Expression of IGF system genes during T-antigen driven pituitary tumorigenesis. Horm Metab Res 1999;31:155-160.
52. Westphal CH, Muller L, Zhou A, Zhu X, Bonner-Weir S, Schambelan M, Steiner DF, Lindberg I, Leder P. The neuroendocrine protein 7B2 is required for peptide hormone processing in vivo and provides a novel mechanism for pituitary Cushing's disease. Cell 1999;96:689-700.
53. Furuta M, Yano H, Zhou A, Rouille Y, Holst JJ, Carroll R, Ravazzola M, Orci L, Furuta H, Steiner DF. Defective prohormone processing and altered pancreatic islet morphology in mice lacking active PC2. Proc Natl Acad Sci USA 1997;94:6646-6651.
54. Hnasko R, Khurana S, Shackleford N, Steinmetz R, Low MJ, Ben-Jonathan N. Two distinct pituitary cell lines from mouse intermediate lobe tumors: a cell that produces prolactin-regulating factor and a melanotroph. Endocrinology 1997;138:5589-5596.
55. Rubinstein M, Mortrud M, Liu B, Low MJ. Rat and mouse proopiomelanocortin gene sequences target tissue-specific expression to the pituitary gland but not to the hypothalamus of transgenic mice. Neuroendocrinology 1993;58:373-380.
56. Young JI, Otero V, Cerdán MG, Falzone TL, Chan EC, Low MJ, Rubinstein M. Authentic cell-specific and developmentally regulated expression of proopiomelanocortin genomic fragments in hypothalamic and hindbrain neurons of transgenic mice. J Neurosci 1998;18:6631-6640.
57. Elias CF, Lee C, Kelly J, Aschkenasi C, Ahima RS, Couceyro PR, Kuhar MJ, Saper CB, Elmquist JK. Leptin activates hypothalamic CART neurons projecting to the spinal cord. Neuron 1998;21:1375-1385.
58. Vrang N, Larsen PJ, Clausen JT, Kristensen P. Neurochemical characterization of hypothalamic cocaine-amphetamine-regulated transcript neurons. J Neurosci 1999;19:RC5.
59. Cheung CC, Clifton DK, Steiner RA. Proopiomelanocortin neurons are direct targets for leptin in the hypothalamus. Endocrinology 1997;138:4489-4492.

60. Elias CF, Aschkenasi C, Lee C, Kelly J, Ahima RS, Bjorbaek C, Flier JS, Saper CB, Elmquist JK. Leptin differentially regulates NPY and POMC neurons projecting to the lateral hypothalamic area. Neuron 1999;23:775-786.

61. Spergel DJ, Kruth U, Hanley DF, Sprengel R, Seeburg PH. GABA- and glutamate-activated channels in green fluorescent protein-tagged gonadotropin-releasing hormone neurons in transgenic mice. J Neurosci 1999;19:2037-2050.

62. Rubinstein M, Japon JA, Low MJ. Introduction of a point mutation into the mouse genome by homologous recomination in embryonic stem cells using a replacement type vector with a selectable marker. Nuc Acids Res 1993;21:2613-2617.

63. Rubinstein M, Mogil JS, Japon M, Chan EC, Allen RG, Low MJ. Absence of opioid stress-induced analgesia in mice lacking β-endorphin by site-directed mutagenesis. Proc Natl Acad Sci USA 1996;93:3995-4000.

64. König M, Zimmer AM, Steiner H, Holmes PV, Crawley JN, Brownstein MJ, Zimmer A. Pain responses, anxiety and aggression in mice deficient in pre-proenkephalin. Nature 1996;383:535-538.

65. Mogil JS, Grisel JE, Hayward MD, Bales JR, Rubinstein M, Belknap JK, Low MJ. Disparate spinal and supraspinal antinociceptive responses in ß-endorphin-deficient mutant mice. Neuroscience 2000;101:709-717.

66. Slugg RM, Hayward MD, Ronnekleiv OK, Low MJ, Kelly MJ. Effect of the μ-opioid agonist DAMGO on medial basal hypothalamic neurons in beta-endorphin knockout mice. Neuroendocrinology 2000;72:208-217.

67. Morley JE. Neuropeptide regulation of appetite and weight. Endocrine Rev 1987;8:256-287.

68. Yaswen L, Diehl N, Brennan MB, Hochgeschwender U. Obesity in the mouse model of pro-opiomelanocortin deficiency responds to peripheral melanocortin. Nat Med 1999;5:1066-1070.

69. Huszar D, Lynch CA, Fairchild-Huntress V, Dunmore JH, Fang Q, Berkemeier LR, Gu W, Kesterson RA, Noston BA, Cone RD. Targeted disruption of the melanocortin-4 receptor results in obesity in mice. Cell 1997;88:131-141.

70. Marsh DJ, Hollopeter G, Huszar D, Laufer R, Yagaloff KA, Fisher SL, Burn P, Palmiter RD. Response of melanocortin-4 receptor-deficient mice to anorectic and orexigenic peptides. Nat Genet 1999;21:119-122.

71. Klebig ML, Wilkinson JE, Geisler JG, Woychik RP. Ectopic expression of the agouti gene in transgenic mice causes obesity, features of type II diabetes, and yellow fur. Proc Natl Acad Sci USA 1995;92:4728-4732.

72. Butler AA, Kesterson RA, Khong K, Cullen MJ, Pelleymounter MA, Dekoning J, Baetscher M, Cone RD. A unique metabolic syndrome causes obesity in the melanocortin-3 receptor-deficient mouse. Endocrinology 2000;141:3518-3521.

73. Chen AS, Marsh DJ, Trumbauer ME, Frazier EG, Guan XM, Yu H, Rosenblum CI, Vongs A, Feng Y, Cao L. Inactivation of the mouse melanocortin-3 receptor results in increased fat mass and reduced lean body mass. Nat Genet 2000;26:97-102.

10

TRANSIENT TRANSGENESIS IN THE ENDOCRINE SYSTEM: VIRAL VECTORS FOR GENE DELIVERY

Anne David, Daniel Stone, Rachel L. Cowen, Maria G. Castro and Pedro R. Lowenstein

Molecular Medicine and Gene Therapy Unit, School of Medicine, University of Manchester, Manchester, M13 9PT, UK

INTRODUCTION

Transgenic animals and knockout technology have been of great importance to elucidate the molecular basis of many endocrine processes. In both these technologies, changes are introduced into germline cells, and animals develop with the addition or lack of individual genes. Germline alterations have two major shortcomings. Some will be embryonic lethal mutations, whilst others will induce compensatory changes during development, which will then contribute to the phenotype of mutant animals.

It has been more difficult to alter gene expression in an anatomically restricted and temporally regulated fashion specifically in adult endocrine tissues following germline manipulations. Viral vectors to transfer genes to endocrine tissues and into specific subtypes of endocrine cells *in vivo* would be of great use to achieve anatomically restricted transient changes in endocrine gene expression, selectively in adult animals in the absence of any developmental compensatory changes. Viral vector systems have been developed and used for gene transfer and gene therapy purposes, and in this chapter we will review the most relevant systems of interest to the experimental and clinical endocrinologist. Their most salient characteristics, as well as their advantages and shortcomings will also be highlighted (see table 1). We will also review applications of the different systems for generating transient transgenic endocrine models both *in vitro* and *in vivo*.

RETROVIRUS-DERIVED VECTORS

The retroviridae are classically subdivided into seven groups but despite this divergence a common virion structure is observed amongst all isolates. The outer structure of the virion consists of a host cell derived lipid envelope, which contains viral-specific envelope glycoproteins. The envelope glycoprotein is comprised of a transmembrane (TM) sub-unit and an external receptor-binding (SU) sub-unit. The viral envelope contains a protein capsid structure, composed of capsid (CA) proteins, that is attached to the envelope by a matrix protein (MA). The capsid houses the diploid single stranded RNA (ssRNA) genome along with the non-structural protease, reverse transcriptase, integrase and RNA associated nucleoprotein (NC).

The RNA genome is plus-stranded, between 7-10 kilobases, and is capped at the 5' end and polyadenylated at the 3' end by the cellular transcription machinery. The non-coding regulatory elements are located in the terminal regions which flank the four central coding regions termed *gag, pro, pol* and *env*. The *gag* gene produces a polyprotein, which is cleaved to produce the capsid proteins. The *pro* gene encodes the protease responsible for cleavage of *gag, pol* and *env* polyproteins. The *pol* gene encodes a polyprotein, which is cleaved to produce the reverse transcriptase and integrase proteins. The *env* gene encodes a polyprotein, which is cleaved to produce the two envelope proteins TM and SU (1).

The retrovirus life cycle begins when the SU domain of the envelope glycoprotein attaches to its cellular receptor (2). A conformational change in the envelope glycoprotein enables the viral envelope and cell membrane to fuse allowing release of the capsid core into the cytoplasm. Once inside the cytoplasm the viral reverse transcriptase converts the ssRNA genome to a double stranded proviral DNA genome within the capsid. A pre-integration complex containing the proviral genome and viral integrase is then transported to the cell nucleus where it can only enter following disruption of the nuclear membrane during mitosis (3). Following entry into the nucleus the proviral genome, consisting of *gag, pro, pol* and *env* genes flanked by two identical cis-acting regions termed long terminal repeats (LTRs), is randomly integrated into the host genome by the viral integrase (1). The host's transcriptional machinery gives rise later to all viral-specific RNAs (both mRNAs and viral genomes) using a strong promoter located within the LTR.

Retroviral vectors have been identified as attractive candidates for gene transfer and they are currently one of the vehicles of choice in ongoing gene therapy clinical trials. The ability of retroviruses to integrate their genome into that of the host cell, enabling subsequent gene transfer to daughter cells following mitosis, has led to the production of vector systems based on type C retroviruses that are non-pathogenic to humans, such as the Moloney murine leukaemia virus (MoMLV) or other oncoretroviruses. These vectors are replication defective, devoid of all viral genes, can carry up to 8 kilobases (kb) of foreign DNA and can be engineered to alter the natural viral tropism.

MoMLV based retroviral vectors have been developed so that all of the viral genes are deleted and replaced with marker and/or therapeutic genes (figure 1). The LTRs and packaging sequence (Ψ) are the only viral sequences that remain in the vector and in order to propagate them all the other viral genes are produced in *trans* by a packaging cell line. Initial packaging cell lines retained the viral genome within one construct that had either a single Ψ deletion (4, 5) or multiple deletions (6). Unfortunately the emergence of replication competent retroviruses (RCRs) has been detected in these cell lines (7, 8). Newer cell lines have been generated in which *gag* and *pol* sequences are integrated into separate chromosomal loci than *env* construct (9, 10). In this way, very few, if any, overlapping sequences are present, thus necessitating 2-4 recombination events to produce RCRs. Even so, RCRs have still been observed, but at a much lower incidence (11, 12).

A limitation to the use of retroviral vectors for gene transfer *in vivo* is the inactivation of MoMLV based vectors by human serum (13). Inactivation occurs through a non-lytic process, which relies upon complement activation (14) and the presence of natural antibodies recognizing α1-3 galactosyltransferase residues added to the envelope glycoprotein by murine packaging cell lines (15). MoMLV produced in mouse cells is readily inactivated by human serum whereas MoMLV produced in human or mink cells is not (16). Retroviral vectors generated in non-murine packaging cell lines are resistant to human serum (17) and the construction of vector systems based on these retroviruses would be advantageous for *in vivo* applications.

Another limitation of MoMLV derived retroviral vectors *in vivo* is the low titres currently attainable (10^5-10^7 colony forming units (cfu)/ml). By pseudotyping these vectors with the vesicular stomatitis virus envelope glycoprotein (VSV-G) it has been possible to generate titres of 10^9 cfu/ml using ultra-centrifugation (18). The inability of MoMLV vectors to transduce non-dividing cells is a further limitation of this system. The dependence on mitosis for transport of the pre-integration complex into the nucleus (3) limits the use of these vectors in applications where post-mitotic or slowly dividing cells need to be transduced.

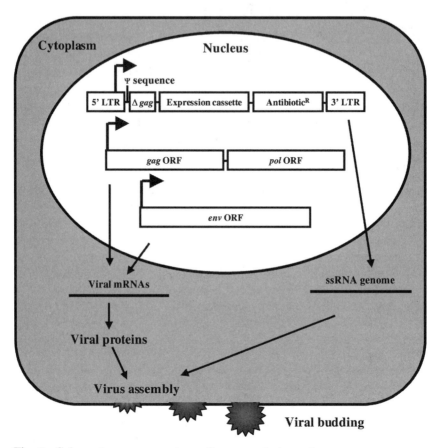

Fig 1. Schematic representation of a retroviral producer cell. The three constructs are stably transfected into a cell line such as NIH 3T3 mouse fibroblasts. The *gag/pol* ORFs are MLV derived from which the capsid proteins are produced by *gag*, and the reverse transcriptase and integrase by *pol*. The origin of the *env* ORF is dependent on the system being used and can be of varying viral origin (*e.g.* amphotropic MoMLV). The ssRNA genome of the viral vector is produced by the proviral construct. Assembly of retroviral vector particles occurs at the cell membrane, following insertion of envelope proteins into the membrane, and virus exits the cells by budding.

LENTIVIRUS

Lentiviruses are complex retroviruses, which have evolved the ability to actively infect cells in the absence of cell division (19). As this is a characteristic not present in type C retroviruses recent developments have led to the production of lentiviral vector systems with the hope of gaining long term expression in non-dividing or slowly dividing cell populations (20-22). The latest generations of lentiviral vectors are replication defective, devoid of most viral genes, carry up to 8 kilobases of foreign DNA, and can transduce non-dividing cell populations *in vitro* and *in vivo*.

The lentivirus with the most thoroughly characterised genetics, replication and pathogenesis is the human lentivirus HIV-1. Most work in the area of lentiviral vectors has been carried out using HIV-1, whose genome codes for 3 structural (*gag, pol* and *env*) and 6 regulatory (*Vif, Vpr, Vpu, Tat, Rev* and *Nef*) proteins.

First generation lentiviral vectors were based on a three-plasmid transfection system (23). The packaging plasmid, containing the whole genome with packaging signal and *env* ORF deleted, is under transcriptional control of the MIE-hCMV promoter and produces all the viral proteins needed *in trans*. A pseudotyped envelope protein, such as VSV-G was produced *in trans* by a second construct. The third transfer vector under transcriptional control of the HIV LTRs contains the packaging sequence, the minimal *gag* elements needed for optimal packaging efficiency, *rev*-responsive elements and an expression cassette. In order to reduce the risk of recombination producing replication competent lentivirus, second generation lentiviral vectors had the accessory proteins *Vif, Vpr, Vpu* and *Nef* deleted from the packaging construct without affecting the efficiency of gene transfer *in vitro* (24, 25) and in central nervous system (CNS) neurons *in vivo* (24). Third generation vectors can now be generated independently of the *Tat* transactivator (25, 26) or may have a self inactivating system by which transcriptional inactivation of the LTRs is achieved through deletion of the U_3 region of the LTR (27, 28).

The biosafety of an HIV-1 derived lentiviral vector system has been questioned. This has led to the development of alternative lentiviral vector systems derived from the less pathogenic HIV-2 strain (29, 30), or the primate simian immunodeficiency virus (SIV) (31). Although HIV-2 or SIV vectors may be theoretically safer they are still derived from primate lentiviruses. Lentiviral vectors based on the non-primate viruses, i.e., feline immunodeficiency virus (FIV), equine infectious anaemia virus (EIAV) and caprine arthritic encephalitis virus (CAEV) have also been developed. Both FIV (32, 33) and EIAV (34, 35) based vectors have shown reasonable transduction efficiency in human cells, but CAEV vectors have shown very poor levels of transduction (36). It is felt that the lack of knowledge regarding the molecular virology of these non-primate viruses may hinder further vector development in the short term.

ADENOVIRUS-DERIVED VECTORS

Adenoviruses (Ads) were discovered as infectious agents in 1953 (37) and more than 100 serotypes have been identified so far. In humans, Ad infection is mostly associated with respiratory diseases, conjunctivitis or gastro-enteritis (38). The adenovirus is a non-enveloped DNA virus of 60-90 nm diameter (figure 2) (39). Its capsid has a regular icosahedron shape mostly made up of hexon, penton and fiber proteins (39, 40). The core contains a linear double stranded DNA genome (of which the size and sequence vary between serotypes) associated with four proteins (40). Ad type 2 (Ad2) is the best characterised, and its complex genome (36 kb) encodes for a very large number of mRNAs through extensive splicing (40). The genome contains short inverted terminal repeat sequences (ITRs) necessary for replication, an encapsidation sequence (Ψ), and two major coding regions: early (E) and late (L), which are active, respectively, before and after the onset of viral DNA replication (figure 2). The coding region E1 is a very important region involved in the activation of all early viral promoters leading to the expression of proteins involved in viral replication. E2 mRNAs are produced from two different promoters and code for proteins involved in DNA replication such as the DNA binding protein (DBP) and DNA polymerase. E3 proteins are involved in the down-modulation of cellular immunity *in vivo* (41-43). The coding region E4 is required for viral DNA replication, activation of late viral gene transcription, host protein synthesis "shut off" and virion assembly (44-47). All late mRNAs are generated from the major late promoter (MLP) and code for structural proteins.

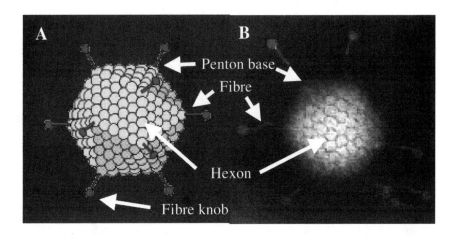

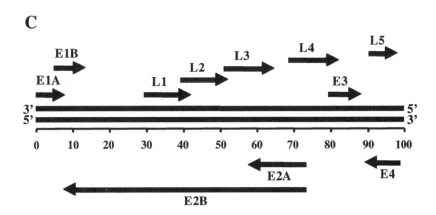

Fig 2. Schematic representation (A) and electron microscopy photograph (B) of human adenovirus type 5 (Ad5). Linear representation of the Ad5 genome (C) indicating relative map units plus early (red) and late (blue) transcription units.

Ads are capable of infecting both quiescent and dividing cells in a large variety of tissues from different species. The adenoviral life cycle is composed of several steps: (i) attachment of the fiber (knob domain) to the coxsackie and Ad receptor (CAR), a high affinity receptor expressed on the surface of target cells (48), (ii) interaction of the penton base with integrins ($\alpha_v\beta_3$ and $\alpha_v\beta_5$) leading to the internalisation of the Ad by endocytosis through clathrin coated pits (49), (iii) uptake into endosomes, (iv) acidification of the endosomes leading to (v) disassembly of the capsid, (vi) release of the virion into the cytoplasm, (vii) transport to the nuclear membrane, (viii) penetration of the genome into the nucleus through nuclear pores (50), (ix) stabilisation of the adenoviral genome as an episome within the nucleus and, finally, (x) initiation of virus replication.

Recombinant adenoviral vectors (rAds) are most commonly constructed from Ad serotype 5. First generation rAds contain deletions in the E1 (and sometimes E3) region(s) into which the transgene cassette (including promoter, transgene and polyadenylation sequences of up to ~8 kb) can be inserted (figure 3A). Deletions in E1 render rAds replication-deficient and deletions in E3 create more space for larger transgenes. However, E1A is required for rAd production and is therefore provided *in trans* by the host cell. RAds can be constructed by homologous recombination between a plasmid encoding the transgene cassette flanked by short Ad sequences and a plasmid containing an E1/E3 deficient Ad genome in 293 human embryonic kidney cells transfected with E1 (figure 3) (51, 52). RAd vectors have several advantages including ease of manipulation and amplification, to produce purified viruses of high titres (10^{12} plaque forming units (pfu)/ml) (53). The main drawback of these first generation viruses is that when they are used *in vivo*, they induce inflammatory and immune responses, characterised by inflammatory cell infiltration, cytotoxic T lymphocyte (CTL) activation and production of Ad-specific antibodies (54-56).

To decrease the immune response induced against the vector, different groups have produced second generation vectors by modifying the rAd genome backbone. A rAd with a thermosensitive mutation (ts 125) introduced in the E2 region (57-59) increased transgene expression, decreased inflammation, but did not abolish *in vivo* toxicity. Other groups have introduced total or partial deletions of E4 (60). However, to generate these vectors, a new cell line providing E4 *in trans* is required to allow rAd production. An *in vivo* reduction of vector-induced toxicity of such E1⁻/E4⁻ viruses has been shown (61, 62). It has been shown however that E4⁻ mutants down regulate some promoters driving transgenes and therefore do not achieve prolonged transgene expression (61, 63, 64). Recently, a rAd deleted for E1, E2a, E3 and E4 has been described (65). A third generation of rAds has been constructed by deletion of some of the late genes which have been replaced by a reporter gene (66).

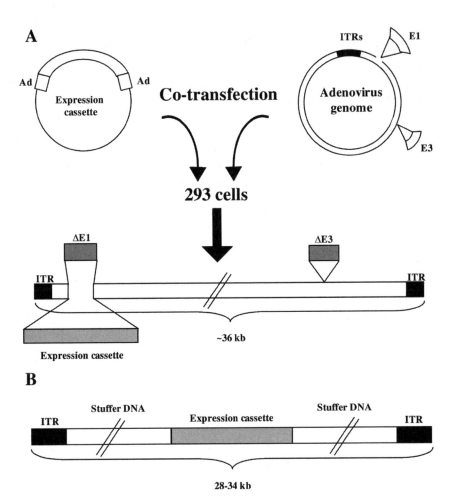

Fig 3. Schematic representations of the process used to construct first-generation rAds (A). Following co-transfection into 293 cells homologous recombination gives rise to a packagable genome which can be deleted in E1 and E3. Linear illustration of the helper-dependent adenovirus genome (B).

The latest adenoviral vectors are the "gutless", high capacity, or helper-dependent Ads (ΔrAds) (67-69). The main feature of these vectors is that all the adenoviral genes have been removed and replaced with stuffer DNA (figure 3B). Only the ITRs and packaging signals remain and a helper virus, which provides all necessary viral genes *in trans*, is required for virion production. The ΔrAd can be separated from helper virus by caesium chloride density gradient purification since their sizes are different. To increase the safety of ΔrAd, a helper virus has been constructed with *loxP* sites flanking the packaging signal sequence. Vectors are produced in 293 cell lines stably expressing Cre recombinase which excises the packaging signal (69). The ΔrAds are efficient for *in vivo* gene transfer (70, 71) and present the advantages of a large cloning capacity (~28 kb), reduced immunogenicity and prolonged transgene expression in rodents (72-76) and baboons (77).

In conclusion, rAds are very useful vectors which are able to infect quiescent and dividing cells. Furthermore, the generation of new non-immunogenic vectors is promising for numerous clinical gene therapy applications.

ADENO-ASSOCIATED VIRUS

Adeno-associated virus (AAV), a member of the parvoviridae family, has not been associated with any human disease. AAV can infect a large variety of cells including non-dividing cells. The non-enveloped virion is a small icosahedral particle (~20-24 nm). Its genome, a short linear single stranded DNA (~4.7 kb), is composed of two small ITRs (145 bp) and two open reading frames (ORFs), *rep* and *cap* coding for regulatory and structural proteins respectively (78-80).

The first 125 bp of the ITR, a palindromic sequence, form a hairpin structure enabling replication initiation. However, the replication process requires sequences provided *in trans* by a helper virus such as herpes simplex virus or adenovirus. The ITRs are also required for encapsidation and integration in a specific locus within human chromosome 19 (81-83). The entry into host cells occurs through binding with heparan sulphate proteoglycans and integrins (84, 85). AAV DNA is integrated into the host genome and remains in a latent phase since infection with a helper virus is needed to allow propagation and rescue of new virions. Unlike the ITRs, which are necessary for different steps of the life cycle, *rep* and *cap* are not required in *cis* and therefore can be deleted and replaced by sequences of interest in recombinant AAVs (rAAVs).

These rAAVs are constructed by co-transfection of two plasmids, one containing the transgene cassette flanked by the ITRs and the second providing *rep* and *cap*. The rAAV then needs the presence of a helper virus (classically an Ad) to be propagated (86). Infection in the absence of helper virus leads to latency. Infection with helper virus is needed for reactivation.

The bulk of a preparation obtained after a large-scale amplification is however a mixture of rAAV and helper virus. A two-step purification method consisting of heating to 56°C and caesium chloride gradient centrifugation eliminates the helper adenovirus and yields a preparation of up to 10^{10} infectious particles/ml. An alternative technique involves iodixanol gradients followed by ion exchange of heparin affinity chromatography (87).

Unlike adenoviral vectors, rAAVs do not induce a cellular immune response after *in vivo* administration, making them very promising vectors for therapeutic applications. Nevertheless, specific antibodies have been detected in the circulation (88), limiting potential re-administration. Apart from the complexity of the production procedures, another drawback is the relatively small capacity available for insertion of the transcription cassette (~4.7 kb).

HERPES SIMPLEX VIRUS

Herpes Simplex Virus type 1 (HSV-1) is a ubiquitous human pathogen possessing a large linear double-stranded DNA genome of approximately 150 kb, encoding for 70-80 genes. Like adenoviruses, HSV-1 can naturally infect a host of different cell types, including muscle (89), lung (90), liver (91), neurons (92, 93), pituitary cells (94), pancreatic islets (95) and many tumors. This broad host range is due to the ability of the HSV envelope glycoproteins gB and gC to bind to the extracellular heparan sulphate chains on cell surface proteoglycans (96). The interaction of a third viral glycoprotein gD is then required to precipitate the fusion of the viral envelope and cell membrane allowing viral entry (97). Glycoprotein gD binds to the receptor Nectin1, also named HveC (herpes virus entry mediator C) (98). The cellular function of Nectin 1 and its homologs is as an intercellular adhesion molecule and it has also been shown to mediate direct cell-to-cell transmission of wild-type HSV virus.

HSV-1 can infect both dividing and quiescent cells, with infection resulting in the degradation of host mRNA and shutoff in host protein synthesis (99). This promotes efficient viral replication and a viral lytic cycle. Infection of neurons with HSV1 can lead to the establishment of latency within an apparently healthy neurone. Long term transgene expression in the CNS of up to 18 months *in vivo* (100) has been witnessed, making HSV a potentially suitable vector for the treatment of neurological disorders such as Parkinson's disease.

Contained within the tegument of the HSV virion is the protein VP16 which is carried into the cell upon infection. VP16 induces the activation of the HSV immediate early (IE) genes, ICP0, 4, 6, 22 and 27. These genes then initiate expression of the early genes necessary for DNA replication, following which, late genes required for replication and encapsidation are expressed. The first recombinant HSV vectors were rendered replication defective by mutation of the IE gene ICP 4 and by creating an ICP4 expressing complementary cell line to allow vector propagation. Unfortunately these initial HSV vectors were toxic to many cell lines and therefore further vectors, deficient for most IE viral genes, have been developed (101). Paradoxically, the genetic manipulations required to eliminate toxicity and allow the genome to persist in cells for longer periods of time have been shown to dramatically lower the level of transgene expression from heterologous promoters (102).

Concurrently an alternative strategy has being developed utilising amplicons, to generate an HSV vector devoid of all HSV-1 genes. Amplicons are plasmid vectors containing HSV packaging and replication signals, which enable them to replicate and be packaged in the presence of a helper virus which provides all necessary HSV gene products *in trans*. As HSV DNA replication occurs by a rolling circle mechanism, the resulting amplicon vector is contained within a HSV-1 capsid (including the envelope, tegument and capsid) carrying multiple copies of the recombinant DNA plasmid, upto the upper packaging limit of 152 kb (103). First generation amplicon systems were limited by contaminating helper virus but newer methods of HSV production generate helper free preparations (104).

HSV vectors that are capable of a single round of infection (disabled infectious single-cycle [DISC]-HSV vectors) are currently being developed. These vectors have been used in cancer immunotherapy because in addition to carrying an immunostimulatory transgene such as GM-CSF they can also lyse the infected tumor cells (105). Other conditionally replicative HSV vectors have been used in an oncolytic approach. An HSV-1 variant deleted for the RL1 gene fails to produce the virulence factor ICP34.5 and replicates selectively within tumor cells. Using this variant the ability to restrict reporter gene expression to astrocytes, from which many intracerebral tumors develop, was demonstrated *in vivo* using the glial fibrillary acidic protein (GFAP) promoter (106). Another mutant HSV vector hrR3 preferentially replicates in dividing gliosarcoma tumor cells but not in postmitotic neural cells. This mutant also possesses an intact viral thymidine kinase (TK) gene, which makes infected cells chemosensitive to ganciclovir (107).

The size of the HSV-1 genome makes it possible to accommodate in theory as much as 50 kb of heterologous DNA in replication defective HSV-1 vectors without detrimentally affecting viral growth (108). This is extremely advantageous for gene therapy applications that require simultaneous expression of multiple gene products to achieve a therapeutic effect. In cancer therapy several transgenes have been co-expressed from an HSV vector in order to provide synergistic methods of tumor destruction and immune stimulation (109).

Despite evident advantages of HSV vectors including; large insert capacity, a wide host range, the ability to establish latency in neurons leading to long term transgene expression and improvements in vector construction, the problem of virus toxicity still remains, thus limiting the further use of HSV vectors *in vivo* (see table 1).

OTHER VIRUSES

We have highlighted the most widely used viral vectors but many other virus types have been investigated which may offer potential for endocrine gene transfer/therapy in the future. Conditional or restricted replicative virus vectors are being developed for cancer gene therapy. One example is the adenovirus system ONYX-15 (110), where an E1B-55 KDa gene deletion in the adenovirus vector restricts its replication to p53 deficient tumor cells. This vector is now being assessed in clinical trials (111).

Alternatively two viral vectors can be combined to try to harness the favourable attributes of each individual vector (112). Chimeric adenovirus/retrovirus vectors are being developed to produce a vector which can infect quiescent cells and then insert the transgene into the host cell genome. In this system simultaneous infection of a cell with two adenovirus vectors, which collectively encode for retrovirus vector and packaging signals, induces the target cell to function as a transient retroviral producer cell (113). Again using the concept of simultaneous infection of two complementing vectors AAV/adenovirus hybrids (114, 115) and HSV/AAV hybrids are being investigated (116). Viruses have also been combined with non-viral vectors. A well-characterised example of a viral/non viral hybrid is the Sendai virus (Hemagglutinating virus of Japan; HVJ)/ liposome complex. Here the F-protein from HVJ is fused with liposomes carrying the cDNA for the transgene. This HVJ/liposome complex has been further combined with the Epstein Barr Virus replicon apparatus (117). Other vector systems also exist based on Vaccinia virus (118), Sindbis and Semliki forest alphaviruses (119).

Table 1. Properties of the most commonly used viral vectors for gene transfer/therapy applications

Vector	Retrovirus	Lentivirus	Adenovirus	Gutless adenovirus	AAV	HSV
Insert capacity	8 Kb	8 Kb	8 Kb	28 Kb	4 Kb	50 Kb
Target specificity	Broad (dividing cells)	Broad	Broad	Broad	Broad	Broad
Nuclear localisation	Integration into host chromosome	Integration into host chromosome	Episomal	Episomal	Episomal or integration into host chromosome	Episomal
Toxicity	+/-	?	Yes	+/-	+/-	Yes
Immune responses	+/-	+/-	Yes	+/-	+/-	Yes
Stability of vector genome	Permanent	Permanent	Transient	Transient	Permanent	Transient
Clinical trials	Yes	No	Yes	No	Yes	Yes

TARGETED GENE DELIVERY

To localize the transgene to a specific target organ, several approaches have been developed. To efficiently deliver recombinant virus, the choice of route of administration and promoter is crucial. Furthermore, some modifications to viruses can render them capable of infecting non-permissive cell types or modify the natural tropism of the virus.

Administration Route

Some authors have shown that the choice of administration route is important for transduction efficiency and also for aiding in the selection of the target organ to be infected. Since retroviruses integrate only in dividing cells, they are mainly administered *ex vivo* to target cells, which are thereafter transplanted into the organ of interest. For other viruses, such as adenovirus vectors, intravenous injection preferentially leads to transgene expression in the liver (89). As systemic administration is not always efficient to target other tissues, local administration is commonly performed (120). Immune responses developed against viral vectors also vary depending on the route of vector administration (121-123).

Promoter

The choice of the promoter is crucial for each gene transfer/therapy strategy. Mainly, the promoter defines the strength and the distribution of transgene expression. Most common promoters are strong and ubiquitous (124, 125). Several promoters have been used for targeting different organs or cell populations (126-130). In this section, we will address promoters used for transgene expression in the endocrine system.

Somatostatin (131), amylase (132) and proglucagon (133) promoters can be used for targeting pancreatic cells. Cell type specific expression can be achieved in the neuroendocrine system with the chromogranin A promoter (134), in thyroid cells with the thyroglobulin promoter (135, 136) and in ovarian tumor cells with the L-plastin promoter (137). The GH α-glycoprotein promoter has been used to target pituitary tumor cells (138), whilst the prolactin promoter restricts expression to lactotrophic anterior pituitary cells *in vitro* and the anterior pituitary gland *in vivo* (52, 139). Another level of regulation can be added with an inducible system such as the tetracycline inducible system (140-144).

MODIFICATION OF VIRAL TROPISM

Two different problems are witnessed with the viruses currently available. The first one is the widespread non-specific infection elicited by most viruses and the second one is the resistance to infection of some cell types. Therefore, current viral vectors have been modified to render them more specific to selected cell types (145).

Retrovirus

Targeting strategies used for retroviral vectors differ depending on the tropism of the recombinant vectors used (146). Ecotropic MLV receptors are not expressed on the surface of human cells and therefore ecotropic viruses need to be targeted to allow productive infection. Amphotropic MLV receptors are present on the surface of all human cells justifying targeting strategies to restrict transgene expression to a specific cell subpopulation. Most retrovirus targeting strategies are based on modification of the envelope glycoproteins by genetic engineering, modification of viral particles or by combining strategies (figure 4A).

Direct cross-linking of vectors to target cells using antibody bridges has been used as a targeting strategy. Biotinylated antibodies against the viral envelope protein and against specific cell membrane markers were linked by streptavidin. Major histocompatibility complex (MHC) class I and II antigens (147) and receptors for transferrin (148), epidermal growth factor (EGF), insulin, high density lipoprotein, galactose or glycoconjugates (149) were used as cellular targets; however, the efficiency of productive infection was low. Chemical coupling of lactose to proteins can produce artificial asialoglycoproteins. This property has been used to target an ecotropic MLV to hepatocytes: retroviral particles modified with lactose can transduce hepatocyte cell lines expressing asialoglycoprotein receptors (150).

Insertions or substitutions in the receptor binding domain of the retroviral envelope have also been performed. With this method, different degrees of efficiency have been achieved. Modifications have been introduced to target integrin receptors. These mutants, however, were not very efficient since insertions of polypeptides impaired processing and incorporation of such proteins into the viral envelope (151). In contrast, insertion of erythropoietin or heregulin peptides in the receptor binding domain of ecotropic vectors allowed successful binding and infection through the erythropoeitin receptor and EGF receptor (EGFR) (152, 153). Some authors have described N-terminal fusion of single chain antibody fragments (ScFv) to the Mo-MLV envelope glycoprotein (154). The specificity of the ScFv allowed targeting into different populations expressing MHC class I (155) or cancer specific antigens such as the carcinoembryonic antigen (156) and the high-molecular-weight melanoma-associated antigen (157). Although these retroviruses were able to bind specifically to target cells no infectivity was observed. In order to maintain the endogenous conformation of the envelope glycoprotein, spacers of different lengths have been introduced between the retroviral proteins and the ligand or antibody used for targeting, but, despite these improvements, infectivity remained low, even if targeting was efficient (158, 159). Virions targeted to the cellular receptor for amphotropic murine retrovirus (Ram-1) were able to infect human cells whereas EGF hybrids showed binding to EGFR positive cells without completing infection (160). This suggested that infection was triggered through the interaction of the virion with its native receptor (160).

Finally, the most efficient attempts to achieve both targeting and infection of a specific population with amphotropic retrovirus have been obtained with a two-step targeting strategy. Chimeric viruses have been constructed by extension of the N terminal envelope glycoprotein with a specific ligand via a cleavable linker. The rational of this strategy is as follows: (i) the virus attaches specifically to the targeted cell through the high affinity interaction between ligand and receptor, (ii) the linker is cleaved by a specific activated protease (endogenous or exogenous), (iii) the endogenous viral binding sequence is released, (iv) the endogenous viral binding sequence binds to its receptor, (v) the virus enters into the host cell, and (vi) the infection cycle occurs. Using this method, several N-terminal extensions have been assessed using a variety of peptides such as EGF (161-163), leucine zipper (164), CD40 ligand (164) or insulin-like growth factor I (165). The cleavage of the linkers (factor Xa protease recognition signal, matrix metalloproteinase (MMP)-cleavage signal or plasmin sensitive linker) occurs after activation of their respective proteases such as factor Xa protease (161, 164, 165), MMP (163) or plasminogen (162), and efficient gene delivery is observed. An advantage of plasminogen is that the cleavage reaction can be reversed by addition of aprotinin, a serine protease inhibitor (162).

Very few studies have been performed with targeted retroviruses *in vivo*. Hall et al. reported efficient gene delivery to injured arteries with a collagen-targeted Mo-MLV vector (166).

Adenovirus
Attempts to target rAds have focused on the knob/CAR and penton base/integrin interactions of adenovirus type 5 (figure 4B). Through genetically altering the fiber or penton base of rAds it is thought that targeted gene transfer can be achieved. Alternatively, targeting may be achieved by crosslinking rAds and specific cell surface ligands through antibody conjugation.

Initial work showed that it is possible to genetically modify type 5 based rAds by replacing the knob domain with the knob domain of adenovirus serotype 3 (167). Subsequent work has shown that it is possible to modify the fiber directly by inserting a heterologous peptide into the HI loop of the fiber knob (168, 169). Alternatively peptide motifs can be attached to the C-terminus of the fiber knob (170-172). These fiber-modified vectors show an ability to enter cells through CAR independent pathways highlighting the rational behind fiber-modified vectors. However, none of these strategies abolishes natural viral tropism.

As an alternative to modifying the fiber knob, Wickham et al. have shown that it is possible to modify the penton base of rAds. By replacing the RGD motif responsible for the interaction with $\alpha_v\beta_3$ and $\alpha_v\beta_5$ integrins, with peptide motifs specific for other integrins, it is possible to alter viral tropism (173). Further work by Wickham et al. has shown that it is possible to target rAd vectors, via the penton base, in a fiber independent manner (174, 175). A FLAG peptide epitope was substituted for the RGD motif, within the penton base, and bispecific antibodies were then used to target the vector directly to α_v integrin or CD3 positive cells.

In order to abolish the native tropism of rAds it is necessary to block both the knob/CAR interactions and the penton base/integrin interactions. As mentioned previously this is possible by genetically altering the fiber knob and penton base directly. Several groups have used immunological cross-linking as an alternative strategy to achieve this. The use of bispecific antibodies, in which an anti-fiber Fab fragment is conjugated to a specific ligand for a cellular receptor, has shown promising results when trying to target specific cell populations. RAds conjugated with bispecific antibodies to the folate receptor, fibroblast growth factor receptor (FGFR) and epidermal growth factor receptor (EGFR) have been used to transduce cells positive for these receptors (176-178). It remains to be determined if antibody mediated re-targeting allows efficient targeted systemic gene delivery *in vivo*.

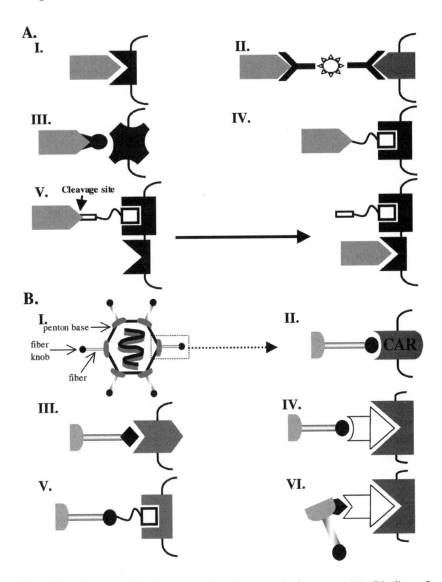

Fig 4. Virus/receptor binding strategies for retroviral vectors (A). Binding of the native envelope glycoprotein to its cellular receptor (I). Native envelope glycoprotein is targeted to an alternate receptor by antibody stretavidin/avidin cross-linking (II). The receptor-binding domain of the envelope glycoprotein is replaced with a motif that targets the retrovirus to an alternate receptor (III). An alternate binding domain is attached to N-terminus of the envelope

glycoprotein targeting it to an alternate receptor (IV). An alternate binding domain and a protease cleavable sequence are attached to the N-terminus of the envelope glycoprotein (V). The attached binding domain targets the vector to cells expressing a specific receptor and after protease cleavage internalisation is mediated by binding of the native envelope glycoprotein to its cellular receptor. Virus/receptor binding strategies for rAds (B). Schematic representation of an Ad (I). Wild type Ad 5 fiber knob binds to the CAR receptor (II). Ad 5 fiber pseudotyped with the Ad 3 knob is able to bind to the unidentified Ad 3 receptor, which is expressed on cell types not expressing CAR (III). A bispecific antibody, consisting of an anti-fiber Fab fragment conjugated to a specific ligand, is able to target the Ad fiber to the native receptor for the ligand (IV). A specific ligand is inserted within the HI loop or added to the C-terminus of the knob, enabling targeting to an alternate receptor (V). A peptide epitope replaces the RGD motif within the penton base and then bispecific antibodies to the peptide epitope target the rAd to an alternate receptor (VI).

Other viruses

Although the majority of work with regards to targeting viral vectors has concentrated on the more commonly used adenoviral and retroviral vectors, some work has investigated the potential of targeting other viruses. Some targeting experiments have been performed with AAV using a bispecific antibody to direct gene expression to megakaryocytes (179). The envelope of the Sindbis virus, a member of the alphavirus family, has also been modified to display the IgG-binding domain of protein A, thus, when coupled with a monoclonal antibody, the enveloped chimeric virus is able to target specific surface antigens and infect human cells with high efficiency (180). These results are important since the host cell-type specificity of these viruses is limited.

TRANSIENT TRANSGENESIS IN THE ENDOCRINE SYSTEM

We will now describe different endocrine models in which viral vectors have been used to achieve transient transgenesis. However, we will not cover the applications for the pituitary gland since these will be addressed in chapter 11.

RETROVIRUS VECTORS

Retroviruses can only infect dividing cells, therefore they are not ideal vectors for targeting the endocrine system. *In vitro*, retroviruses have been used for transducing pancreatic islets (181) or thyroid follicular cells (182-184), however transduction efficiency was low (1-30 %) and addition of growth factors was required (181). To circumvent this drawback, fibroblasts have been transduced *in vitro* with a retrovirus and then implanted *in vivo*. This technique proved successful in reversion of diabetes in mice, using a retrovirus encoding a genetically modified human proinsulin (185). Alternatively, retroviruses have been used to infect different endocrine tumor cells as they are actively dividing. Thyroid, pancreatic or gastrointestinal tumor cells have been transduced *in vitro* with different recombinant retroviruses containing prodrug, cytokine, proapoptic, enzymatic or suicide genes. *In vivo* models have also shown the efficiency of these methods (186, 187). The difficulty of this approach however is the need to target tumor cells *in vivo* in order to yield an effective treatment. Lentiviral vectors have been tested for transducing pancreatic islets and display promising results since they are able to transduce beta cells in contrast to retroviral vectors (181, 188).

ADENOVIRUS VECTORS

RAds have been the most widely used viral vectors in endocrine gene transfer/therapy, and have been used in a variety of research areas.

Diabetes

Pancreatic islets isolated from rat and mouse have been efficiently transduced *in vitro* with rAds encoding hexokinase I or β-galactosidase (189-191). *In vivo* experiments have also shown targeting of ductal epithelium, acinar cells and Langerhans islets after delivery via the pancreaticobiliary duct (192) although the expression was transient and associated with inflammation. Immune responses (inflammation and antibody production) against recombinant adenoviral vector have also been reported after direct rAd injection or after transplantation of *ex vivo* transduced pancreatic islets, but normal functionality was preserved, suggesting therapeutic transgenes can be tested (193). Administration of a rAd expressing leptin in ob/ob mice (a model for human non-insulin dependent diabetes mellitus) efficiently reversed diabetes symptoms such as food intake, body weight, serum insulin levels and glucose tolerance (194). Similarly, perfusion of the Zucker rat pancreas with a rAd encoding the leptin receptor corrected islet functions (195, 196). Furthermore, glycemia normalisation has been achieved in diabetic mice after intravenous injection with a rAd containing the proinsulin cDNA (197). Administration of a rAd encoding arginine vasopressin in the hypothalamus of the Brattleboro rat reversed diabetes insipidus with limited immune response (198).

Cancer

RAds have also been used to transduce a large variety of tumor cells. Treatment of a rat thyroid follicular cell line with a rAd encoding HSV-TK induced cell death in the presence of ganciclovir (136). Furthermore, in a medullary thyroid carcinoma model, administration of a rAd encoding IL-2 inhibited tumor growth in mice and rats without significant toxicity to other organs (199-201). A mouse model of congenital adrenal hyperplasia showed reversed phenotype (biochemistry, endocrinology and morphology) after injection of a rAd encoding the cytochrome P450 21-hydroxylase into the adrenal gland (202).

Cancer of the reproductive system has also been targeted with rAds. Ovarian or prostate cancer cell lines have been transduced with rAds containing reporter genes or therapeutic genes. *In vitro*, ovarian tumor cell lines treated with Ad p53 (203), Ad HSV-TK in association with ganciclovir (204), Ad p16 (205) or Ad PTEN (206) underwent cell death. Treatment of ovarian or cervical carcinoma cells has also been reported with rAds encoding anti-erbB2 ScFv (207) or B71 (208). *In vivo*, administration of Ad HSV-TK in association with ganciclovir to mice implanted with human ovarian tumor cells inhibited cell growth and prolonged animal survival (209-211). Interestingly, administration of Ad p53 did not show any beneficial effect (212). Both *in vitro* (induction of apotosis) and *in vivo* (reduction in tumor growth and prolongation of survival) effects have been shown in prostate cancer cells treated with Ad C-CAM (213), Ad p21 (214), Ad HSV-TK in association with ganciclovir (215), Ad PML (promyelocytic leukaemia gene encoding a growth transformation suppressor) (216), Ad IL-12 (217) or Ad FasL (218). However, *in vivo* treatment with Ad p53 did not prolong survival of animals bearing tumors despite cytotoxic effects *in vitro* (214, 219).

All these results are promising for treatment of cancers of the endocrine system with rAds. Furthermore, some phase I clinical studies have also been performed with different rAds in patients exhibiting ovarian (220-222) or prostate (223) cancers.

ADENO-ASSOCIATED VIRUS VECTORS

In vitro, human pancreatic carcinoma cells have been efficiently transduced with rAAV containing the rat preproinsulin II gene allowing insulin secretion for up to 6 months (224). In *ob/ob* mice, intramuscular injection of a rAAV encoding mouse leptin corrected endocrine and metabolic defects (body weight, food intake, leptin, insulin and glucose levels and glucose tolerance) for up to six months (225). Similarly, blood glucose levels were significantly decreased in diabetic mice after treatment (liver parenchyma injection) with a rAAV carrying the rat insulin gene (226).

HERPES SIMPLEX VIRUS TYPE 1 VECTORS

The application of HSV vectors to endocrine gene transfer/therapy is in its infancy. The use of HSV vectors for the treatment of endocrine tumors has been investigated, with *in vitro* studies demonstrating their ability to infect both normal and tumor anterior pituitary cell lines (94), prostate adenocarcinoma (227) and ovarian carcinoma (228). Walker et al. showed that a conditionally replicative HSV vector, which preferentially replicates in cancer cells, had oncolytic activity even at a low multiplicity of infection (MOI of 0.01). In athymic mice, intratumoral virus inoculation into established subcutaneous human prostate cancer tumors eradicated 22% of tumors and intravenous administration induced tumor regression in distant subcutaneous tumors. The therapeutic ability of an HSV vector encoding the HSV-TK gene was investigated in comparison to a comparable adenovirus vector. Wang et al. showed that HSV-1-mediated gene transfer is quantitatively higher than adenovirus-mediated in five human ovarian cancer cell lines at a 100-fold lower dose *in vitro*. Aside from their oncolytic use HSV vectors can infect pancreatic islets (229). For the treatment of diabetes pancreatic islet beta cells or islet grafts infected with virus expressing the anti-apoptotic gene bcl-2 were protected from immune mediated damage like that seen in type I diabetes (95).

CONCLUSIONS

Recent advances in molecular endocrinology have moved from biological correlation experiments, to attempt to prove causal relationships among the various factors, which determine endocrine functions. Disentangling such causal relationships requires experimental modifications to the molecular make up of the endocrine system. Transgenic animal technology allows such modifications. However, alterations are made to fertilised eggs, and such alterations are thus present throughout embryonic, foetal, and adult life, and often results in the induction of compensatory changes.

Gene transfer into adult endocrine system using viral vectors provides a powerful alternative to achieve transient genetic modifications of endocrine systems in adulthood. Viral vectors constitute versatile and powerful systems to achieve transient transgenesis to alter endocrine cell functions, or for the development of gene therapy strategies for endocrine diseases (139, 230).

Acknowledgements

The work in our laboratory is funded by the CRC (UK), The Wellcome Trust (UK), the BBSRC (UK), the MRC (UK), the British Heart Foundation (UK), EU grants (contract no's. B104-C798-0297, BMH4-CT98-3277, QLK3-CT-1999-00364), the Parkinson's Disease Society (UK), The Royal Society and REMEDI. AD is a post-doctoral fellow funded by a British Heart Foundation project grant. DS is recipient of a PhD studentship funded by the BBSRC (UK). RC is a BBSRC AstraZeneca CASE PhD student. PRL is a fellow of the Lister Institute of Preventive Medicine.

REFERENCES

1. Coffin, JM. Retroviridae: The viruses and their replication. In: Fundamental Virology (Eds: Fields BN, Knipe DM, Howley PM, et al.) Philadelphia: Lippincott-Raven Publishers, 1996.
2. Sommerfelt MA. Retrovirus receptors. J Gen Virol 1999;80:3049-64.
3. Roe TY, Reynolds TC, Yu G, Brown PO. Integration of murine leukaemia virus DNA depends on mitosis. EMBO J 1993;12:2099-2108.
4. Mann R, Mulligan RC, Baltimore D. Construction of a retrovirus packaging mutant and its use to produce helper-free defective retrovirus. Cell 1983;33:153-159.
5. Cone RD, Mulligan RC. High-efficiency gene transfer into mammalian cells: generation of helper free recombinant retrovirus with broad mammalian host range. Proc Natl Acad Sci USA 1984;81:6349-6353.
6. Miller AD, Buttimore C. Redesign of retrovirus packaging cell lines to avoid recombination leading to helper virus production. Mol Cell Biol 1986;6:2895-2902.
7. Muenchau DD, Freeman SM, Cornetta K, Zweibel JA, Anderson WF. Analysis of retroviral packaging lines for generation of replication-competent virus. Virology 1990;176:262-265.
8. Scarpa M, Cournoyer D, Muzny DM, Moore KA, Belmont JW, Caskey CT. Characterisation of recombinant helper retroviruses from moloney-based vectors in ecotropic and amphotropic packaging cell lines. Virology 1991;180:849-852.
9. Markowitz D, Goff S, Bank A. A safe packaging line for gene transfer: separating viral genes on two different plasmids. J Virol 1988;62:1120-1124.

10. Markowitz D, Goff S, Bank A. Construction and use of a safe and efficient packaging cell line. Virology 1988;167:400-406.
11. Chong H, Vile RG. Replication-competent retrovirus produced by a 'split-function' third generation amphotropic packaging cell line. Gene Ther 1996;3:624-629.
12. Chong H, Starkey W, Vile RG. A replication-competent retrovirus arising from a split-function packaging cell line was generated by recombination events between the vector, one of the packaging constructs, and endogenous retroviral sequences. J Virol 1998;72:2663-2670.
13. Welsh RM, Cooper NR, Jensen FC, Oldstone MBA. Human serum lyses RNA tumour viruses. Nature 1975;257:612-614.
14. Rother RP, Squinto SP, Mason JM, Rollins SA. Protection of retroviral vector particles in human blood through complement inhibition. Hum Gene Ther 1995;6:429-435.
15. Takeuchi Y, Porter CD, Strachan KM, Preece AF, Gustafsson K, Cosset F-L, Weiss RA, Collins MK. Sensitisation of cells and retroviruses to human serum by (α1-3) galactosyltransferase. Nature 1996;379:85-88.
16. Takeuchi Y, Cosset F-L, Lachmann PJ, Okada H, Weiss RA, Collins MK. Type C retrovirus inactivation by human complement is determined by both the viral genome and the producer cells. J Virol 1994;68:8001-8007.
17. Cosset F-L, Takeuchi Y, Battini JL, Weiss RA, Collins MK. High-titer packaging cells producing recombinant retroviruses resistant to human serum. J Virol 1995;69:7430-7436.
18. Burns JC, Friedmann T, Driever W, Burrascano M, Yee JK. Vesicular stomatitis virus G glycoprotein pseudotyped retroviral vectors: concentration to very high titer and efficient gene transfer into mammalian and nonmammalian cells. Proc Natl Acad Sci USA 1993;90:8033-8037.
19. Luciw PA. Human immunodeficiency viruses and their replication. In: Fundamental Virology (Eds: Fields BN, Knipe DM, Howley PM, et al.) Philadelphia: Lippincott-Raven Publishers, 1996.
20. Naldini L. Lentiviruses as gene transfer agents for delivery to non-dividing cells. Curr Opin Biotechnol 1998;9:457-463.
21. Federico M. Lentiviruses as gene delivery vectors. Curr Opin Biotechnol 1999;10:448-453.
22. Klimatcheva E, Rosenblatt JD, Planelles V. Lentiviral vectors and gene therapy. Front Biosci 1999;4:D481-496.
23. Naldini L, Blomer U, Gallay P, Ory D, Mulligan R, Gage FH, Verma IM, Trono D. In vivo gene delivery and stable transduction of nondividing cells by a lentiviral vector. Science 1996;272:263-267.
24. Zufferey R, Nagy D, Mandel RJ, Naldini L, Trono D Multiply attenuated lentiviral vector achieves efficient gene delivery in vivo. Nat Biotech 1997;15:871-875.
25. Kim VN, Mitrophanous K, Kingsmann SM, Kingsmann AJ. Minimal requirement for a lentivirus vector based on human immunodeficiency virus type 1. J Virol 1998;72:811-816.
26. Dull T, Zufferey R, Mandel RJ, Nguyen M, Trono D, Naldini L. A third-generation lentivirus vector with a conditional packaging system. J Virol 1998;72:8463-8471.
27. Miyoshi H, Blomer U, Takahashi M, Gage FH, Verma IM. Development of a self-inactivating lentivirus vector. J Virol 1998;72:8150-8157.

28. Zufferey R, Dull T, Mandel RJ, Bukovsky A, Quiroz D, Naldini L, Trono D. Self-inactivating lentivirus vector for safe and efficient in vivo gene delivery. J Virol 1998;72:9873-9880.

29. Poeschla E, Gilbert J, Li X, Huang S, Ho A, Wong-Stall F. Identification of a human immunodeficiency virus type 2 (HIV-2) encapsidation determinant and transduction of nondividing human cells by HIV-2 based lentivirus vectors. J Virol 1998;72:6527-6536.

30. Arya SK, Zamani M, Kundra P. Human immunodeficiency virus type 2 vectors for gene transfer: expression and potential for helper virus-free packaging. Hum Gene Ther 1998;9:1371-1380.

31. White SM, Renda M, Nam NY, Klimatcheva E, Zhu Y, Fisk J, Halterman M, Rimel BJ, Federoff H, Pandya S, Rosenblatt JD, Planelles V. Lentivirus vectors using human and simian immunodeficiency virus elements. J Virol 1999;73:2832-2840.

32. Poeschla E, Wong-Stall F, looney D. Efficient transduction of non-dividing human cells by feline immunodeficiency virus lentiviral vectors. Nat Med 1998;4:354-357.

33. Johnston JC, Gasmi M, Lim LE, Elder JH, Yee JK, Jolly DJ, Campbell KP, Davidson BL, Sauter SL. Minimum requirements for efficient transduction of dividing and nondividing cells by feline immunodeficiency virus vectors. J Virol 1999;73:4991-5000.

34. Olsen JC. Gene transfer vectors derived from equine infectious anemia virus. Gene Ther 1998;5:1481-1487.

35. Mitrophanous KA, Yoon S, Rohll JB, Patil D, Wilkes FJ, Kim VN, Kingsmann SM, Kingsmann AJ, Mazarakis ND. Stable gene transfer to the nervous system using a non-primate lentiviral vector. Gene Ther 1999;6:1808-1818.

36. Mselli-Lakhal L, Favier C, Da Silva Teixeira MF, Chettab K, Legras C, Ronfort C, Verdier G, Mornex JF, Chebloune Y. Defective RNA packaging is responsible for low transduction efficiency of CAEV-based vectors. Arch Virol 1998;143:681-695.

37. Rowe WP, Huebner RJ, Gilmore LK, Parrott RH, Ward TG. Isolation of a cytopathogenic agent from human adenoids undergoing spontaneous degeneration in tissue culture. Proc Soc Exp Biol Med 1953;84:570-573.

38. Enders JF, Bell JA, Dingle JH, Francis T, Hilleman MR, Huebner RJ, Payne AM-M. "Adenovirus" : group name proposed for new respiratory-tract viruses. Science 1956;124:119-120.

39. Horne RW, Brenner S, Waterson AP, Wildy P. The icosahedral form of an adenovirus. J Mol Biol 1959;1:84-86.

40. van Oostrum J, Burnett RM. Molecular composition of the adenovirus type 2 virion. J Virol 1985;56:439-448.

41. Signäs C, Katze MG, Persson H, Philipson L. An adenovirus glycoprotein binds chains of class I transplantation antigens from man and mouse. Nature 1982;299:175-178.

42. Gooding LR, Elmore LW, Tollefson AE, Brady HA, Wold WSM A. 14,700 MW protein from the E3 region of adenovirus inhibits cytolysis by tumor necrosis factor. Cell 1988;53:341-346.

43. Wold, WSM. Adenovirus genes that modulate the sensitivity of virus-infected cells to lysis by TNF. J Cell Biochem 1993;53:329-335.

44. Halbert DN, Cutt JR, Shenk T. Adenovirus early region 4 encodes functions required for efficient DNA replication, late gene expression, and host cell shutoff. J Virol 1985;56:250-257.

45. Weinberg DH, Ketner G. Adenoviral early region 4 is required for efficient viral DNA replication and for late gene expression. J Virol 1986;57:833-838.

46. Yoder SS, Berget SM. Role of adenovirus type 2 early region 4 in the early-to-late switch during productive infection. J Virol 1986;60:779-781.

47. Falgout B, Ketner G. Adenovirus early region 4 is required for efficient virus particle assembly. J Virol 1987;61:3759-3768.

48. Bergelson JM, Cunningham JA, Droguett G, Kurt-Jones EA, Krithivas A, Hong JS, Horwitz MS, Crowell RL, Finberg RW. Isolation of a common receptor for Coxsackie B viruses and Adenoviruses 2 and 5. Science 1997;275:1320-1323.

49. Wickham TJ, Mathias P, Cheresh DA, Nemerow GR. Integrins alpha v beta 3 and alpha v beta 5 promote adenovirus internalization but not virus attachment. Cell 1993;73:309-319.

50. Greber UF, Willets M, Webster P, Helenius A. Stepwise dismantling of adenovirus 2 during entry into cells. Cell 1993;75:477-486.

51. Graham FL, Smiley J, Russell WC, Nairn R. Characteristics of a human cell line transformed by DNA from human adenovirus type 5. J Gen Virol 1977;36:59-72.

52. Southgate T, Kingston PA, Castro MG. Gene transfer into neural cells *in vitro* using adenoviral vectors. Curr Prot Neurosci (In Press)

53. Graham FL, Prevec L. Manipulation of adenovirus vectors. Methods in Molecular Biology 1991;7:109-128.

54. Yang Y, Ertl HCJ, Wilson JM. MHC class I-restricted cytotoxic T lymphocytes to viral antigens destroy hepatocytes in mice infected with E1-deleted recombinant adenoviruses. Immunity 1994;1:433-442.

55. Yang Y, Nunes FA, Berensi K, Furth EE, Gönczöl E, Wilson JM. Cellular immunity to viral antigens limits E1-deleted adenoviruses for gene therapy. Proc Natl Acad Sci USA 1994;91:4407-4411.

56. Thomas CE, Birkett D, Castro MG, Lowenstein PR. Acute direct adenoviral vector cytotoxicity and chronic, but not acute inflammatory responses correlate with decreased vector-mediated transgene expression in the brain. Mol Ther (In Press)

57. Englehardt JF, Litzky L, Wilson JM. Prolonged transgene expression in cotton rat lung with recombinant adenoviruses defective for E2a. Hum Gene Ther 1994;5:1217-1229.

58. Englehardt JF, Ye X, Doranz B, Wilson JM. Ablation of E2a in recombinant adenoviruses improves transgene persistence and decreases inflammatory response in mouse liver. Proc Natl Acad Sci USA 1994;91:6196-6200.

59. Yang Y, Nunes FA, Berensi K, Furth EE, Gönczöl E, Engelhardt JF, Wilson JM. Inactivation of E2a in recombinant adenoviruses improves the prospect for gene therapy in cystic fibrosis. Nat Genet 1994;7:362-369.

60. Armentano D, Sookdeo CC, Hehir KM, Gregory RJ, George JA, Prince GA, Wadworth SC, Smith AE. Characterisation of an adenovirus gene transfer vector containing an E4 deletion. Hum Gene Ther 1995;6:1343-1353.

61. Gao G-P, Yang Y, Wilson JM. Biology of adenovirus vectors with E1 and E4 deletions for liver-directed gene therapy. J Virol 1996;70:8934-8943.

62. Wang Q, Greenburg G, Bunch D, Farson D, Finer MH. Persistent transgene expression in mouse liver following in vivo gene transfer with a ΔE1/ΔE4 adenovirus vector. Gene Ther 1997;4:393-400.

63. Armentano D, Zabner J, Sacks C, Sookdeo CC, Smith MP, George JA, Wadworth SC, Smith AE, Gregory RJ. Effect of the E4 region on the persistence of transgene expression from adenovirus vectors. J Virol 1997;71:2408-2416.

64. Dedieu J-F, Vigne E, Torrent C, Jullien C, Mahfouz I, Aubailly N, Orsini C, Guillaume J-M, Opolon P, Delaere P, Perricaudet M, Yeh P. Long-term gene delivery into the livers of immunocompetent mice with E1/E4-defective adenoviruses. J Virol 1997;71:4626-4637.

65. Gorziglia MI, Lapcevich C, Roy S, Kang Q, Kadan M, Wu V, Pechan P, Kaleko M. Generation of an adenovirus vector lacking E1, E2a, E3 and all of E4 except open reading frame 3. J Virol 1999;73:6048-6055.

66. Mitani K, Graham FL, Caskey CT, Kochanek S. Rescue, propagation, and partial purification of a helper virus-dependent adenovirus vector. Proc Natl Acad Sci USA 1995;92:3854-3858.

67. Fisher KJ, Choi H, Burda J, Chen S-J. Recombinant adenovirus deleted of all viral genes for gene therapy of cystic fibrosis. Virology 1996;217:11-22.

68. Kochanek S, Clemens PR, Mitani K, Chen H-H, Chan S, Caskey CT. A new adenoviral vector: Replacement of all viral coding sequences with 28 kb of DNA independently expressing both full-length dystrophin and beta-galactosidase. Proc Natl Acad Sci USA 1996;93:5731-5736.

69. Parks RJ, Chen L, Anton M, Sankar U, Rudnicki MA, Graham FL. A helper-dependant adenovirus vector system: removal of helper virus by Cre-medited excision of the viral packaging signal. Proc Natl Acad Sci USA 1996;93:13565-13570.

70. Clemens PR, Kochanek S, Sunada Y, Chan S, Chen H-H, Campbell KP, Caskey CT. In vivo muscle gene transfer of full-length dystrophin with an adenoviral vector that lacks all viral genes. Gene Ther 1996;3:965-972.

71. Haecker SE, Stedman HH, Balice-Gordon RJ, Smith DBJ, Greelish JP, Mitchell MA, Wells A, Sweeney HL, Wilson JM. *In vivo* expression of full-lenth human dystrophin from adenoviral vectors deleted of all viral gene. Hum Gene Ther 1996;7:1907-1914.

72. Chen H-H, Mack LM, Kelly R, Ontell M, Kochanek S, Clemens PR. Persistence in muscle of an adenoviral vector that lacks all viral genes. Proc Natl Acad Sci USA 1997;94:1645-1650.

73. Morral N, Parks RJ, Zhou H, Langston C, Schiedner G, Quinones J, Graham FL, Kochanek S, Beaudet AL. High doses of a helper-dependent adenoviral vector yield supraphysiological levels of α_1-antitrypsin with negligible toxicity. Hum Gene Ther 1998;9:2709-2716.

74. Morsy MA, Gu M, Motzel S, Zhao J, Lin J, Su Q, Allen H, Franklin L, Parks RJ, Graham FL, Kochanek S, Bett A, Caskey CT. An adenoviral vector deleted for all viral coding sequences results in enhanced safety and extended expression of a leptin transgene. Proc Natl Acad Sci USA 1998;95:7866-7871.

75. Schiedner G, Morral N, Parks RJ, Wu Y, Koopmans SC, Langston C, Graham FL, Beaudet AL, Kochanek S. Genomic DNA transfer with a high-capacity adenovirus vector results in improved in vivo gene expression and decreased toxicity. Nat Gen 1998;18:180-183.

76. Thomas CE, Schiedner G, Kochanek S, Castro MG, Lowenstein PR. Peripheral infection with adenovirus causes unexpected long-term brain inflammation in animals injected with first-generation, but not with high-capacity adenovirus vectors: towards realistic long-term neurological gene therapy for chronic diseases. Proc Natl Acad Sci USA 2000;97:7482-7487.

77. Morral N, O'Neal W, Rice K, Leland M, Kaplan J, Piedra PA, Zhou H, Parks RJ, Velji R, Aguilar-Cordova E, Wadworth S, Graham FL, Kochanek S, Carey KD, Beaudet AL. Administration of helper-dependent adenoviral vectors and sequential delivery of

different vector serotype for long-term liver-directed gene transfer in baboons. Proc Natl Acad Sci USA 1999;96:12816128-21.

78. Rose JA, Maizel JVJ, Inman JK, Shatkin AJ. Structural proteins of adenovirus-associated viruses. J Virol 1971;8:766-770.

79. Rose JA, Berns KI, Hoggan MD, Koczot FJ. Evidence for a single-stranded adenovirus-associated virus genome: formation of a DNA density hybrid on release of viral DNA. Proc Natl Acad Sci USA 1969;64:863-869.

80. Srivastava A, Lusby EW, Berns KI. Nucleotide sequence and organisation of the adeno-associated virus 2 genome. J Virol 1983;45:555-564.

81. Kotin RM, Berns KI. Organization of adeno-associated virus DNA in latently infected Detroit 6 cells. Virology 1989;170:460-467.

82. Kotin RM, Siniscalco M, Samulski RJ, Hunter L, Laughlin CA, McLaughlin S, Muzyczka N, Rocchi M, Berns KI. Site-specific integration by adeno-associated virus. Proc Natl Acad Sci USA 1990;87:2211-2215.

83. Linden RM, Winocour E, Berns KI. The recombination signals for adeno-associated virus site-specific integration. Proc Natl Acad Sci USA 1996;93:7966-7972.

84. Summerford C, Samulski, RJ. Membrane-associated heparan sulfate proteoglycan is a receptor for adeno-associated virus type 2 virions. J Virol 1998;72:1438-1445.

85. Summerford C, Bartlett JS, Samulski RJ. AlphaVbeta5 integrin: a co-receptor for adeno-associated virus type 2 infection. Nat Med 1998;5:78-82.

86. Samulski RJ, Chang LS, Shenk T. Helper-free stocks of recombinant adeno-associated viruses: normal integration does not require viral gene expression. J Virol 1989;63:3822-3828.

87. Zolotukhin S, Byrne BJ, Mason E, Zolotukhin I, Potter M, Chesnut K, Summerford C, Samulski RJ, Muzyczka N. Recombinant adeno-associated virus purification using novel methods improves infectious titer and yield. Gene Ther 1999;6:973-985.

88. Xiao X, Li J, Samulski RJ. Efficient long-term gene transfer into muscle tissue of immunocompetent mice by adeno-associated virus vector. J Virol 1996;70:8098-8108.

89. Huard J, Krisky D, Oligino T, Marconi P, Day CS, Watkins SC, Glorioso JC. Gene transfer to muscle using herpes simplex virus-based vectors. Neuromuscul Disord 1997;7:299-313.

90. D'Angelica M, Karpoff H, Halterman M, Ellis J, Klimstra D, Edelstein D, Brownlee M, Federoff H, Fong Y. In vivo interleukin-2 gene therapy of established tumors with herpes simplex amplicon vectors. Can Immunol Immunother 1999;47:265-271.

91. Miyanohara A, Johnson PA, Elam RL, Dai Y, Witztum JL, Verma IM, Friedmann T. Direct gene transfer to the liver with herpes simplex virus type 1 vectors: transient production of physiologically relevant levels of circulating factor IX. New Biol 1992;4:238-246.

92. Lachmann RH, Efstathiou S. Use of herpes simplex virus type 1 for transgene expression within the nervous system. Clin Sci 1999;96:533-541.

93. Lowenstein PR, Fournel S, Bain D, Tomasec P, Clissold PM, Castro MG, Epstein AL. Simultaneous detection of amplicon and HSV-1 helper encoded proteins reveals that neurons and astrocytoma cells do express amplicon-borne transgenes in the absence of synthesis of virus immediate early proteins. Brain Res Mol Brain Res 1995;30:169-175.

94. Goya RG, Rowe J, Sosa YE, Tomasec P, Lowenstein PR, Castro MG. Use of recombinant herpes simplex virus type 1 vectors for gene transfer into tumour and normal pituitary cells. Mol Cell Endocrinol 1998;139:199-207.

95. Rabinovitch A, Suarez-Pinzon W, Strynadka K, Ju Q, Edelstein D, Brownlee M, Korbutt GS, Rajotte RV. Transfection of human pancreatic islets with an anti-apoptotic gene (bcl-2) protects beta-cells from cytokine-induced destruction. Diabetes 1999;48:1223-1239.

96. Spear PG, Shieh MT, Herold BC, WuDunn D, Koshy TI. Heparan sulfate glycosaminoglycans as primary cell surface receptors for herpes simplex virus. Adv Exp Med Biol 1992;313:341-353.

97. Shukla D, Liu J, Blaiklock P, Shworak NW, Bai X, Esko JD, Cohen GH, Eisenberg RJ, Rosenberg RD, Spear PG. A novel role for 3-O-sulfated heparan sulfate in herpes simplex virus 1 entry. Cell 1999;99:13-22.

98. Cocchi F, Menotti L, Dubreuil P, Lopez M, Campadelli-Fiume G. Cell-to-cell spread of wild-type herpes simplex virus type 1, but not of syncytial strains, is mediated by the immunoglobulin-like receptors that mediate virion entry, nectin1 (PRR1/HveC/HIgR) and nectin2 (PRR2/HveB). J Virol 2000;74:3909-3917.

99. Kwong AD, Frenkel N. The herpes simplex virus virion host shutoff function. J Virol 1989;63:4834-4839.

100. Carpenter DE, Stevens JG. Long-term expression of a foreign gene from a unique position in the latent herpes simplex virus genome. Hum Gene Ther 1996;7:1447-1454.

101. Glorioso JC, DeLuca NA, Fink DJ. Development and application of herpes simplex virus vectors for human gene therapy. Annu Rev Microbiol 1995;49:675-710.

102. Samaniego LA, Neiderhiser L, DeLuca NA. Persistence and expression of the herpes simplex virus genome in the absence of immediate-early proteins. J Virol 1998;72:3307-3320.

103. Frenkel N, Singer O, Kwong AD. Minireview: the herpes simplex virus amplicon--a versatile defective virus vector. Gene Ther 1994;1:S40-46.

104. Geller AI, Yu L, Wang Y, Fraefel C. Helper virus-free herpes simplex virus-1 plasmid vectors for gene therapy of Parkinson's disease and other neurological disorders. Exp Neurol 1997;144:98-102.

105. Todryk S, McLean C, Ali S, Entwistle C, Boursnell M, Rees R, Vile R. Disabled infectious single-cycle herpes simplex virus as an oncolytic vector for immunotherapy of colorectal cancer. Hum Gene Ther 1999;10:2757-2768.

106. McKie EA, Graham DI, Brown SM. Selective astrocytic transgene expression in vitro and in vivo from the GFAP promoter in a HSV RL1 null mutant vector-potential glioblastoma targeting. Gene Ther 1998;5:440-450.

107. Boviatsis EJ, Park JS, Sena-Esteves M, Kramm CM, Chase M, Efird JT, Wei MX, Breakefield XO, Chiocca EA. Long-term survival of rats harboring brain neoplasms treated with ganciclovir and a herpes simplex virus vector that retains an intact thymidine kinase gene. Cancer Res. 1994;54: 5745-5751.

108. Wang X, Zhang G, Yang T, Zhang W, Geller AI. Fifty-one kilobase HSV-1 plasmid vector can be packaged using a helper virus-free system and supports expression in the rat brain. Biotechniques 2000;28:102-107.

109. Krisky D, Marconi P, Oligino T, Rouse RJ, Fink DJ, Cohen JB, Watkins SC, Glorioso JC. Development of herpes simplex virus replication-defective multigene vectors for combination gene therapy applications. Gene Ther 1998;5:1517-1530.

110. Bischoff JR, Kirn DH, Williams A, Heise C, Horn S, Muna M, Ng L, Nye JA, Sampson-Johannes A, Fattaey A, McCormick F. An adenovirus mutant that replicates selectively in p53-deficient human tumor cells. Science 1996;274:373-376.

111. Ganly I, Kirn D, Eckhardt SG, Rodriguez GI, Soutar DS, Otto R, Robertson AG, Park O, Gulley ML, Heise C, Von Hoff DD, Kaye SB. A phase I study of Onyx-015, an E1B attenuated adenovirus, administered intratumorally to patients with recurrent head and neck cancer. Clin Cancer Res 2000;6:798-806.

112. Reynolds PN, Feng M, Curiel DT. Chimeric viral vectors: the best of both worlds? Mol Med Today 1999;5:25-31.

113. Bilbao G, Feng M, Rancourt C, Jackson WHJ, Curiel DT. Adenoviral/retroviral vector chimeras: a novel strategy to achieve high-efficiency stable transduction in vivo. FASEB J 1997;11:624-634.

114. Fisher KJ, Kelley M, Burda JF, Wilson JM. Novel adenovirus-adeno associated virus hybrid vector that displays efficient rescue and delivery of the AAV genome. Hum Gene Ther 1996;7:2079-2087.

115. Recchia A, Parks RJ, Lamartina S, Toniatti C, Pieroni L, Palombo F, Ciliberto G, Graham FL, Cortese R, La Monica N, Colloca S. Site-specific integration mediated by a hybrid adenovirus/adeno-associated virus vector. Proc Natl Acad Sci USA 1999;96:2615-2620.

116. Johnston KM, Jacoby D, Pechan PA, Fraefel C, Borghesani P, Schuback D, Dunn RJ, Smith FI, Breakefield XO. HSV/AAV hybrid amplicon vectors extend transgene expression in human glioma cells. Hum Gene Ther 1997;8:359-370.

117. Kaneda Y, Saeki Y, Morishita R. Gene therapy using HVJ-liposomes: the best of both worlds? Mol Med Today 1999;5:298-303.

118. Peplinski GR, Tsung K, Norton JA. Vaccinia virus for human gene therapy. Surg Oncol Clin N Am 1998;7:575-588.

119. Polo JM, Belli BA, Frolov I, Sherrill S, Hariharan MJ, Townsend K, Perri S, Mento SJ, Jolly DJ, Chang SM, Schlesinger S, Dubensky TWJ. Stable alphavirus packaging cell lines for Sindbis virus and Semliki Forest virus-derived vectors. Proc Natl Acad Sci USA 1999;96:4598-4603.

120. Huard J, Lochmuller H, Ascadi G, Jani A, Massie B, Karpati G. The route of administration is a major determinant of the transduction efficiency of rat tissues by adenoviral recombinants. Gene Ther 1995;2:107-115.

121. Gahery-Segard H, Juillard V, Gaston J, Lengagne R, Pavirani A, Boulanger P, Guillet J-G. Humoral immune response to the capsid components of recombinant adenoviruses: routes of immunization modulate virus-induced Ig subclass shifts. Eur J Immunol 1997;27:653-659.

122. Harvey BG, Hackett NR, El-Sawy T, Rosengart TK, Hirschowitz EA, Lieberman MD, Lesser ML, Crystal RG. Variability of human systemic humoral immune responses to adenovirus gene transfer vectors administered to different organs. J Virol 1999;73:6729-6742.

123. Brockstedt DG, Podsakoff GM, Fong L, Kurtzman G, Mueller-Ruchholtz W, Engleman EG. Induction of immunity to antigens expressed by recombinant adeno-associated virus depends on the route of administration. Clin Immunol 1999;92:67-75.

124. Boshart M, Weber F, Jahn G, Dorsch-Hasler K, Fleckenstein B, Schaffne W. A very strong enhancer is located upstream of an immediate early gene of human cytomegalovirus. Cell 1985;41:521-530.

125. Gerdes CA, Castro MG, Lowenstein PR. Strong promoters are the key highly efficient, noninflammatory and noncytotoxic adenoviral-mediated transgene delivery into the brain in vivo. Mol Ther 2000;2:330-338.

126. Korfhagen TR, Glasser SW, Wert SE, Bruno MD, Daugherty CC, McNeish JD, Stock JL, Potter SS, Whitsett JA. Cis-acting sequences from a human surfactant protein gene confer pulmonary-specific gene expression in transgenic mice. Proc Natl Acad Sci USA 1990;87:6122-6126.

127. Potter JJ, Cheneval D, Dang CV, Resar LM, Mezey E, Yang VW. The upstream stimulatory factor binds to and activates the promoter of the rat class I alcohol dehydrogenase gene. J Biol Chem 1991;266:15457-15463.

128. Franz, WM, Brem, G, Katus, HA, Klingel, K, Hofschneider, PH, Kandolf, R Characterization of a cardiac-selective and developmentally upregulated promoter in transgenic mice. Cardioscience 1994;5:235-243.

129. Mayford M, Bach ME, Huang YY, Wang L, Hawkins RD, Kandel ER. Control of memory formation through regulated expression of a CaMKII transgene. Science 1996;274:1678-83.

130. Morelli AE, Larregina AT, Smith-Arica J, Dewey RA, Southgate TD, Ambar B, Fontana A, Castro MG, Lowenstein PR. Neuronal and glial cell type-specific promoters within adenovirus recombinants restrict the expression of the apoptosis-inducing molecule Fas ligand to predetermined brain cell types, and abolish peripheral liver toxicity. J Gen Virol 1999;3:571-578.

131. Leonard J, Peers B, Johnson T, Ferreri K, Lee S, Montminy MR. Characterization of somatostatin transactivating factor-1, a novel homeobox factor that stimulates somatostatin expression in pancreatic islet cells. Mol Endocrinol 1993;7:1275-1283.

132. DeMatteo RP, McClane SJ, Fisher K, Yeh H, Chu G, Burke C, Raper SE. Engineering tissue-specific expression of a recombinant adenovirus: selective transgene transcription in the pancreas using the amylase promoter. J Surg Res 1997;72:155-161.

133. Hussain MA, Lee J, Miller CP, Habener JF. POU domain transcription factor brain 4 confers pancreatic alpha-cell-specific expression of the proglucagon gene through interaction with a novel proximal promoter G1 element. Mol Cell Biol 1997;17:7186-7194.

134. Canaff L, Bevan S, Wheeler DG, Mouland AJ, Rehfuss RP, White JH, Hendy GN. Analysis of molecular mechanisms controlling neuroendocrine cell specific transcription of the chromogranin A gene. Endocrinology 1998;139:1184-1196.

135. Braiden V, Nagayama Y, Iitaka M, Namba H, Niwa M, Yamashita S. Retrovirus-mediated suicide gene/prodrug therapy targeting thyroid carcinoma using a thyroid-specific promoter. Endocrinology 1998;139:3996-3999.

136. Zeiger MA, Takiyama Y, Bishop JO, Ellison AR, Saji M, Levine MA. Adenoviral infection of thyroid cells: a rationale for gene therapy for metastatic thyroid carcinoma. Surgery 1996;120:921-925.

137. Chung I, Schwartz PE, Crystal RG, Pizzorno G, Leavitt J, Deisseroth AB. Use of L-plastin promoter to develop an adenoviral system that confers transgene expression in ovarian cancer cells but not in normal mesothelial cells. Cancer Gene Ther 1999;6:99-106.

138. Lee EJ, Anderson LM, Thimmapaya B, Jameson JL. Targeted expression of toxic genes directed by pituitary hormone promoters: a potential strategy for adenovirus-mediated gene therapy of pituitary tumors. J Clin Endocrinol Metabol 1999;84:786-794.

139. Castro MG, Windeatt S, Smith-Arica J, Lowenstein PR. Cell-type specific expression in the pituitary: physiology and gene therapy. Biochem Soc Transac 1999;27:858-863.

140. Gossen M, Bujard H. Tight control of gene expression in mammalian cells by tetracycline-responsive promoters. Proc Natl Acad Sci USA 1992;89:5547-5551.

141. Hwang JJ, Scuric Z, Anderson WF. Novel retroviral vector transferring a suicide gene and a selectable marker gene with enhanced gene expression by using a tetracycline-responsive expression system. J Virol 1996;70:8138-8141.

142. Fotaki ME, Pink JR, Mous J. Tetracycline-responsive gene expression in mouse brain after amplicon-mediated gene transfer. Gene Ther 1997;4:901-908.

143. Harding TC, Geddes BJ, Murphy D, Knight D, Uney JB. Switching transgene expression in the brain using an adenoviral tetracycline-regulatable system. Nat Biotechnol 1998;16:553-555.

144. Smith-Arica JR, Williams JC, Stone D, Smith J, Lowenstein PR, Castro MG. Switching on and off transgene expression within lactotrophic cells in the anterior pituitary gland *in vivo*. Endocrinology (Submitted)

145. Peng K-W, Russell SJ. Viral vector targeting. Curr Opin Biotechnol 1999;10:454-457.

146. Cosset F-L, Russell SJ. Targeting retrovirus entry. Gene Ther 1996;3:946-956.

147. Roux P, Jeanteur P, Piechaczyk M. A versatile approach to the targeting of specific cell types by retroviruses. Proc Natl Acad Sci USA 1989;86:9079-9083.

148. Goud B, Legrain P, Buttin G. Antibody-mediated binding of a murine ecotropic moloney retroviral vector to human cells allows internalisation but not establishment of the pro-viral state. Virology 1988;163:251-254.

149. Etienne-Julan M, Roux P, Carillo S, Jeanteur P, Piechaczyk M. The efficiency of cell targeting by recombinant retroviruses depends on the nature of the receptor and the composition of the artificial cell-virus linker. J Gen Virol 1992;73:3251-3255.

150. Neda H, Wu G. Chemical modification of an ecotropic murine leukaemia virus results in redirection of its target cell specificity. J Biol Chem 1991;266:14143-14149.

151. Valsesia-Wittmann S, Drynda A, Deleage G, Aumailley M, Heard JM, Danos O, Verdier G, Cosset F-L. Modifications of the binding domain of avian retrovirus envelope protein to redirect the host range of retroviral vectors. J Virol 1994;68:4609-4619.

152. Kasahara N, Dozy AM, Kan YW. Tissue-specific targeting of retroviral vectors through ligand-receptor interactions. Science 1994;25:1373-1376.

153. Han X, Kasahara N, Kan YW. Ligand-directed retroviral targeting of human breast cancer cells. Proc Natl Acad Sci USA 1995;92:9747-9751.

154. Russell SJ, Hawkins RE, Winter G. Retroviral vectors displaying functional antibody fragments. Nucl Acid Res 1993;21:1081-1085.

155. Marin M, Noel D, Valesia-Wittman S, Brockly F, Etienne-Julan M, Russell S, Cosset F-L, Piechaczyk M. Targeted infection of human cells via major histocompatibility complex class I molecules by Moloney murine leukaemia virus-derived viruses displaying single-chain antibody fragment-envelope fusion proteins. J Virol 1996;70:2957-2962.

156. Konishi H, Ochiya T, Chester KA, Begent RHJ, Muto T, Sugimura T, Terada M. Targeting strategy for gene delivery to carcinoembryonic antigen-producing cancer cells by retrovirus displaying a single-chain variable fragment antibody. Hum Gene Ther 1998;9:235-248.

157. Martin M, Kupsch J, Takeuchi Y, Russell S, Cosset F-L. Retroviral vector targeting to melanoma cells by single-chain antibody incorporation in envelope. Hum Gene Ther 1996;9:737-746.

158. Valsesia-Wittmann S, Morling FJ, Nilson BHK, Takeuchi Y, Russell SJ, Cosset F-L. Improvement of retroviral retargeting by using amino acid spacers between an additional binding domain and the N terminus of Moloney murine leukaemia virus SU. J Virol 1996;70:2059-2064.

159. Ager S, Nilson BHK, Morling FJ, Peng, KW, Cosset F-L, Russell SJ. Retroviral display of antibody fragments; interdomain spacing strongly influences vector infectivity. Hum Gene Ther 1996;7:2157-2164.

160. Cosset F-L, Morling FJ, Takeuchi Y, Weiss RA, Collins MK, Russell SJ. Retroviral retargeting by envelopes expressing an N-terminal binding domain. J Virol 1995;69:6314-6322.

161. Nilson BHK, Morling FJ, Cosset F-L, Russell SJ. Targeting of retroviral vectors through protease-substrate interactions. Gene Ther 1996;3:280-286.

162. Peng K-W, Morling FJ, Cosset F-L, Russell SJ. Retroviral gene delivery system activatable by plasmin. Tumour Targeting 1998;3:112-120.

163. Peng K-W, Morling FJ, Cosset F-L, Murphy G, Russell SJ. A gene delivery system activatable by disease-associated matrix metalloproteinases. Hum Gene Ther 1997;8:729-738.

164. Morling FJ, Peng K-W, Cosset F-L, Russell SJ. Masking of retroviral envelope functions by oligomerizing polypeptide adaptors. Virology 1997;234:51-61.

165. Chadwick MP, Morling FJ, Cosset F-L, Russell SJ. Modification of retroviral tropism by display of IGF-I. J Mol Biol 1999;285:485-494.

166. Hall FL, Gordon EM, Wu L, Zhu NL, Skotzko MJ, Starnes VA, Anderson WF. Targeting retroviral vectors to vascular lesions by genetic engineering of the MoMLV gp70 envelope protein. Hum Gene Ther 1997;8:2183-2192.

167. Krasnykh VN, Mikheeva GV, Douglas JT, Curiel DT. Generation of recombinant adenovirus vectors with modified fibers for altering viral tropism. J Virol 1996;70:6839-6346.

168. Dimitriev I, Krasnykh V, Miller CR, Wang M, Kashentseva E, Mikheeva G, Belousova N, Curiel DT. An adenovirus vector with genetically modified fibers demonstrates expanded tropism via utilisation of a coxsackievirus and adenovirus receptor-independent cell entry mechanism. J Virol 1998;72:9706-9713.

169. Krasnykh VN, Dmitriev I, Mikheeva GV, Miller CR, Belousova N, Curiel DT. Characterization of an adenovirus vector containing a heterologous peptide epitope in the HI loop of the fiber knob. J Virol 1998;72:1844-1852.

170. Michael SI, Hong JS, Curiel DT, Engler JA. Addition of a short peptide ligand to the adenovirus fiber protein. Gene Ther 1995;2:660-668.

171. Wickham TJ, Roelvink PW, Brough DE, Kovesdi I. Adenovirus targeted to heparan-containing receptors increases its gene delivery efficiency to multiple cell types. Nat Biotechnol 1996;14:1570-1573.

172. Wickham TJ, Tzeng E, Shears LLN, Roelvink PW, Li Y, Lee GM, Brough DE, Lizonova A, Kovesdi I. Increased in vitro and in vivo gene transfer by adenovirus vectors containing chimeric fiber proteins. J Virol 1997;71:8221-8229.

173. Wickham TJ, Carrion ME, Kovesdi I. Targeting of adenovirus penton base to new receptors through replacement of its RGD motif with other receptor-specific peptide motifs. Gene Ther 1995;2:750-756.

174. Wickham TJ, Segal DM, Roelvink PW, Carrion ME, Lizonova A, Lee GM, Kovesdi I. Targeted adenovirus gene transfer to endothelial and smooth muscle cells using bispecific antibodies. J Virol 1996;70:6831-6838.

175. Wickham TJ, Lee GM, Titus JA, Sconocchia G, Kovesdi I, Segal DM. Targeted adenovirus-mediated gene delivery to T cells via CD3 cells. J Virol 1997;71:7663-7669.

176. Douglas JT, Rogers BE, Rosenfeld ME, Michael SI, Feng M, Curiel DT. Targeted gene delivery by tropism-modified adenoviral vectors. Nat Biotech 1996;14:1574-1578.

177. Rogers BE, Douglas JT, Ahlem C, Buchsbaum DJ, Frincke J, Curiel DT. Use of a novel cross-linking method to modify adenovirus tropism. Gene Ther 1997;4:1387-1392.

178. Watkins SJ, Mesyanzhinov VV, Kurochkina LP, Hawkins RE. The 'adenobody' approach to viral targeting: specific and enhanced adenoviral gene delivery. Gene Ther 1997;4:1004-1012.

179. Bartlett JS, Kleinschmidt J, Boucher RC, Samulski RJ. Targeted adeno-associated virus vector transduction of nonpermissive cells mediated by a bispecific F(ab'gamma)2 antibody. Nat Biotechnol 1999;17:181-186.

180. Ohno K, Sawai K, Iijima Y, Levin B, Meruelo D. Cell-specific targeting of sindbis virus vectors displaying IgG-binding domains of protein A. Nat Biotechnol 1997;15:763-767.

181. Leibowitz G, Beattie GM, Kafri, T Cirulli V, Lopez AD, Hayek A, Levine F. Gene transfer to human pancreatic endocrine cells using viral vectors. Diabetes 1999;48:745-753.

182. O'Malley BWJ, Adams RM, Sikes ML, Sawada T, Ledley FD. Retrovirus-mediated gene transfer into canine thyroid using an ex vivo strategy. Hum Gene Ther 1993;4:171-178.

183. O'Malley BWJ, Ledley FD. DNA- and viral-mediated gene transfer in follicular cells:progress toward gene therapy of the thyroid. Laryngoscope 1993;103:1084-1092.

184. Ivan M, Ludgate M, Gire V, Bond JA, Wyndford-Thomas D. An amphotropic retroviral vector expressing a mutant gsp oncogene: effects on human thyroid cells in vitro. J Clin Endocrinol Metabol 1997;82:2702-2709.

185. Falqui L, Martinenghi S, Severini GM, Corbella P, Taglietti MV, Arcelloni C, Sarugeri E, Monti LD, Paroni R, Dozio N, Pozza G, Bordignon C. Reversal of diabetes in mice by implantation of human fibroblasts genetically engineered to release mature human insulin. Hum Gene Ther 1999;10:1753-1762.

186. Nishihara E, Nagayama Y, Mawatari F, Tanaka K, Namba H, Niwa M, Yamashita S. Retrovirus-mediated herpes simplex virus thymidine kinase gene transduction renders human thyroid carcinoma cell lines sensitive to ganciclovir and radiation in vitro and in vivo. Endocrinology 1997;138:4577-4583.

187. Kimura M, Yoshida Y, Narita M, Takenaga K, Takenouchi T, Yamaguchi T, Saisho H, Sakiyama S, Tagawa M. Acquired immunity in nude mice induced by expression of the IL-2 or IL-4 gene in human pancreatic carcinoma cells and anti-tumor effect generated by in vivo gene transfer using retrovirus. Int J Cancer 1999;82:549-555.

188. Ju Q, Edelstein D, Brendel MD, Brandhorst D, Brandhorst H, Bretzel RG, Brownlee M. Transduction of non-dividing adult human pancreatic beta cells by an integrating lentiviral vector. Diabetologia 1998;41:736-739.

189. Becker TC, BeltrandelRio H, Noel RJ, Johnson JH, Newgard CB. Overexpression of hexokinase I in isolated islets of Langerhans via recombinant adenovirus. Enhancement of glucose metabolism and insulin secretion at basal but not stimulatory glucose levels. J Biol Chem 1994;269:21234-21238.

190. Csete ME, Benhamou PY, Drazan KE, Wu L, McIntee DF, Afra R, Mullen Y, Busuttil RW, Shaked A. Efficient gene transfer to pancreatic islets mediated by adenoviral vectors. Transplantation 1995;59:263-268.

191. Csete ME, Afra R, Mullen Y, Drazan KE, Benhamou PY, Shaked A. Adenoviral-mediated gene transfer to pancreatic islets does not alter islet function. Transplant Proc 1994;26:756-757.

192. Raper SE, DeMatteo RP. Adenovirus-mediated in vivo gene transfer and expression in normal rat pancreas. Pancreas 1996;12:401-410.

193. Sigalla J, David A, Anegon I, Fiche M, Huvelin JM, Boeffard F, Cassard A, Soulillou JP, Le Mauff B. Adenovirus-mediated gene transfer into isolated mouse adult pancreatic islets: normal beta-cell function despite induction of an anti-adenovirus immune response. Hum Gene Ther 1997;8:1625-1634.

194. Muzzin P, Eisensmith RC, Copeland KC, Woo SLC. Correction of obesity and diabetes in genetically obese mice by leptin gene therapy. Proc Natl Acad Sci USA 1996;93:14804-14808.

195. Wang M-Y, Koyama K, Shimabukuro M, Newgard CB, Unger RH. OB-Rb gene transfer to leptin-resistant islets reverses diabetogenic phenotype. Proc Natl Acad Sci USA 1998;95:714-718.

196. Wang M-Y, Koyama K, Shimabukuro M, Mangelsdorf D, Newgard CB, Unger RH. Overexpression of leptin receptors in pancreatic islets of Zucker diabetic fatty rats restores GLUT-2, glucokinase, and glucose-stimulated insulin secretion. Proc Natl Acad Sci USA 1998;95:11921-11926.

197. Short DK, Okada S, Yamauchi K, Pessin JE. Adenovirus-mediated transfer of a modified human proinsulin gene reverses hyperglycemia in diabetic mice. Am J Physiol 1998;275:E748-E756.

198. Geddes BJ, Harding TC, Lightman SL, Uney JB. Long-term gene therapy in the CNS: reversal of hypothalamic diabetes insipidus in the Brattleboro rat by using an adenovirus expressing arginine vasopressin. Nat Med 1997;3:1402-1404.

199. Zhang R, Minemura K, De Groot LJ. Immunotherapy for medullary thyroid carcinoma by a replication-defective adenovirus transducing murine interleukin-2. Endocrinology 1998;139:601-608.

200. Zhang R, Baunoch D, De Groot LJ. Genetic immunotherapy for medullary thyroid carcinoma: destruction of tumors in mice by in vivo delivery of adenoviral vector transducing the murine interleukin-2 gene. Thyroid 1998;8:1137-1146.

201. Zhang R, Straus FH, De Groot LJ. Effective genetic therapy of established medullary thyroid carcinomas with murine interleukin-2: dissemination and cytotoxicity studies in a rat tumor model. Endocrinology 1999;140:2152-2158.

202. Tajima T, Okada T, Ma XM, Ramsey W, Bornstein S, Aguilera G. Restoration of adrenal steroidogenesis by adenovirus-mediated transfer of human cytochrome P450 21-hydroxylase into the adrenal gland of 21-hydroxylase-deficient mice. Gene Ther 1999;6:18981-903.

203. Santoso JT, Tang DC, Lane SB, Hung J, Reed DJ, Muller CY, Carbone DP, Lucci JA, Miller DS, Mathis JM. Adenovirus-based p53 gene therapy in ovarian cancer. Gynecol Oncol 1995;59:171-178.

204. Rosenfeld ME, Feng M, Micheal SI, Siegal GP, Alvarez RD, Curiel DT. Adenoviral-mediated delivery of the herpes simplex virus thymidine kinase gene selectively sensitizes human ovarian carcinoma cells to ganciclovir. Clin Cancer Res 1995;1:1571-1580.

205. Wolf JK, Kim TE, Fightmaster D, Bodurka D, Gerhenson D, Mills G, Wharton JT. Growth suppression of human ovarian cancer cell lines by the introduction of a p16 gene via a recombinant adenovirus. Gynecol Oncol 1999;73:27-34.

206. Minaguchi T, Mori T, Kanamori Y, Matsushima M, Yoshikawa H, Taketani Y, Nakamura Y Growth suppression of human ovarian cancer cells by adenovirus-mediated transfer of the PTEN gene. Cancer Res. 1999;59:6063-6067.

207. Deshane J, Siegal GP, Wang M, Wright M, Bucy RP, Alvarez RD, Curiel DT. Transductional efficacy and safety of an intraperitoneally delivered adenovirus encoding an erbB-2 intracellular single-chain antibody for ovarian cancer gene therapy. Gynecol Oncol 1997;64:378-385.

208. Gilligan MG, Knox P, Weedon S, Barton R, Kerr DJ, Searle P, Young LS. Adenoviral delivery of B7-1 (CD80) increases the immunogenicity of human ovarian and cervical carcinoma cells. Gene Ther 1998;5:965-974.

209. Behbakht K, Benjamin I, Chiu HC, Eck SL, Van Deerlin PG, Rubin SC, Boyd J. Adenovirus-mediated gene therapy of ovarian cancer in a mouse model. Amer J Obstet Gynec 1996;175:1260-1265.

210. Rosenfeld ME, Wang M, Siegal GP, Alvarez RD, Mikheeva G, Krasykh V, Curiel DT. Adenoviral-mediated delivery of herpes simplex virus thymidine kinase results in tumor reduction and prolonged survival in a SCID mouse model of human ovarian carcinoma. J Mol Med 1996;74:455-462.

211. Tong XW, Block A, Chen SH, Contant CF, Agoulnik I, Blankenburg K, Kaufman RH, Woo SL, Kieback DG. In vivo gene therapy of ovarian cancer by adenovirus-mediated thymidine kinase gene transduction and ganciclovir administration. Gynecol Oncol 1996;61:175-179.

212. Von Gruenigen VE, Santoso JT, Coleman RL, Muller CY, Miller DS, Mathis JM. In vivo studies of adenovirus-based p53 gene therapy for ovarian cancer. Gynecol Oncol 1998;69:197-204.

213. Kleinerman DI, Zhang WW, Lin SH, Nguyen TV, von Eschenbach AC, Hsieh JT. Application of a tumor suppressor (C-CAM1)-expressing recombinant adenovirus in androgen-independent human prostate cancer therapy: a preclinical study. Cancer Res 1995;55:2831-2836.

214. Eastham JA, Hall SJ, Sehgal I, Wang J, Timme TL, Yang G, Connell-Crowley L, Elledge SJ, Zhang WW, Harper JW. In vivo gene therapy with p53 and p21 adenovirus for prostate cancer. Cancer Res. 1995;55:5151-5155.

215. Hall SJ, Mutchnik SE, Chen SH, Woo SL, Thompson TC. Adenovirus-mediated herpes simplex virus thymidine kinase gene and ganciclovir therapy leads to systemic activity against spontaneous and induced metastasis in an orthotopic mouse model of prostate cancer. Int J Cancer 1997;70:183-187.

216. He D, Mum ZM, Le X, Hsieh JT, Pong RC, Chung LW, Chang KS. Adenovirus-mediated expression of PML suppresses growth and tumorigenicity of prostate cancer cells. Cancer Res 1997;57:1868-1872.

217. Nasu Y, Bangma CH, Hull GW, Lee HM, Hu J, Wang J, McCurdy MA, Shimura S, Yang G, Timme TL, Thompson TC. Adenovirus-mediated interleukin-12 gene therapy for prostate cancer: suppression of orthotopic tumor growth and pre-established lung metastases in an orthotopic model. Gene Ther 1999;6:338-349.

218. Hedlund TE, Meech DJ, Srikanth S, Kraft AS, Miller GJ, Schaack JB, Duke RC. Adenovirus-mediated expression of Fas ligand induces apoptosis of human prostate cancer cells. Cell Death Differ 1999;6:175-182.

219. Yang C, Cirielli C, Capogrossi MC, Passaniti A. Adenovirus-mediated wild-type p53 expression induces apoptosis and suppresses tumorigenesis of prostatic tumor cells. Cancer Res 1995;55:4210-4213.

220. Alvarez RD, Curiel DT. A phase I study of recombinant adenovirus vector-mediated delivery of an anti-erbB-2 single-chain (sFv) antibody gene for previously treated ovarian and extraovarian cancer patients. Hum Gene Ther 1997;8:229-242.

221. Alvarez RD, Curiel DT. A phase I study of recombinant adenovirus vector-mediated intraperitoneal delivery of herpes simplex virus thymidine kinase (HSV-TK) gene and intravenous ganciclovir for previously treated ovarian and extraovarian cancer patients. Hum Gene Ther 1997;8:597-613.

222. Hortobagyi GN, Hung MC, Lopez-Berestein G. A phase I multicenter study of E1A gene therapy for patients with metastatic breast cancer and epithelial ovarian cancer that overexpresses HER-2/neu or epithelial ovarian cancer. Hum Gene Ther 1998;9:1775-1798.

223. Herman JR, Adler HL, Aguilar-Cordova E, Rojas-Martinez A, Woo S, Timme TL, Wheeler M, Thompson TC, Scardino PT. In situ gene therapy for adenocarcinoma of the prostate: a phase I clinical trial. Hum Gene Ther 1999;10: 1239-1249.

224. Peng L, Sidner RA, Bochan MR, Burton MM, Cooper ST, Jindal RM. Construction of recombinant adeno-associated virus vector containing the rat preproinsulin II gene. J Surg Res 1997;69:193-198.

225. Murphy JE, Zhou S, Giese K, Williams LT, Escobedo JA, Dwarki VJ. Long-term correction of obesity and diabetes in genetically obese mice by a single injection of recombinant adeno-associated virus encoding mouse leptin. Proc Natl Acad Sci USA 1997;94:13921-13926.

226. Sugiyama A, Hattori S, Tanaka S, Isoda F, Kleopoulos S, Rosenfeld M, Kaplitt M, Sekihara H, Mobbs C. Defective adeno-associated viral-mediated transfection of insulin gene by direct injection into liver parenchyma decreases blood glucose of diabetic mouse. Horm Metabol Res 1997;28:599-603.

227. Walker JR, McGeagh KG, Sundaresan P, Jorgensen TJ, Rabkin SD, Martuza RL. Local and systemic therapy of human prostate adenocarcinoma with the conditionally replicating herpes simplex virus vector G207. Hum Gene Ther 1999;10:2237-2243.

228. Wang M, Rancourt C, Navarro JG, Krisky D, Marconi P, Oligino T, Alvarez RD, Siegal GP, Glorioso JC, Curiel DT. High-efficacy thymidine kinase gene transfer to ovarian cancer cell lines mediated by herpes simplex virus type 1 vector. Gynecol Oncol 1998;71:278-287.

229. Liu Y, Rabinovitch A, Suarez-Pinzon W, Muhkerjee B, Brownlee M, Edelstein D, Federoff HJ. Expression of the bcl-2 gene from a defective HSV-1 amplicon vector protects pancreatic beta-cells from apoptosis. Hum Gene Ther 1996;7:1719-1726.

230. Windeatt S, Southgate TD, Dewey RA, Bolognani F, Perone MJ, Larregina AT, Maleniak TC, Morris PD, Goya RG, Klatzman D, Lowenstein PR, Castro MG. Adenovirus-mediated herpes simplex virus type-1 thymidine kinase gene therapy suppresses oestrogen-induced pituitary prolactinomas. J Clin Endocrinol Metabol 2000;85:1296-1305.

11

CELL TYPE SPECIFIC AND INDUCIBLE TRANSGENESIS IN THE ANTERIOR PITUITARY GLAND

Maria G. Castro, Judith C. Williams, Tom D. Southgate,
Joseph Smith-Arica, Daniel Stone, Andres Hurtado-Lorenzo,
Pablo Umana and Pedro R. Lowenstein
*Molecular Medicine and Gene Therapy Unit, School of Medicine,
University of Manchester, Manchester, M13 9PT, UK*

INTRODUCTION

With the recent emergence of the sequence for the human genome, and the inevitable wealth of DNA sequences that will be gathered from it, the use of genetic manipulation as a tool in uncovering the molecular basis of many physiological processes within the AP gland will become more apparent. The use of knockout and transgenic technologies has already been extremely useful in elucidating roles for many gene products. It is now well established that these approaches have their drawbacks, i.e. although the introduction or deletion of a particular gene may be useful for uncovering a particular role *in vivo*, the compensatory changes that are likely to occur during development may alter the animals phenotype. This would in turn affect the physiological process, which might occur in wild type animals. Viral vectors offer the possibility of creating transgenics in specific organs, tissues or even regions within a larger organ (i.e. brain nuclei) during adulthood. This offers a very powerful tool to not only express genes of interest, but also to ablate specific cell populations by expressing toxins or depleting specific mRNAs by expressing ribozymes. The possibilities of this technology are enormous both to uncover physiological effects, also for the development of experimental therapeutic strategies, for the treatment of human disease. In this chapter we will explore the possibility of expressing transgenes within specific cell populations in the anterior pituitary gland and the switching "on" and "off" of these exogenous genes.

ANTERIOR PITUITARY (AP) GLAND: HORMONES AND RECEPTORS

The anterior lobe of the pituitary gland contains different types of hormone-producing cell populations and each cell type is characterized by the particular hormone it produces. Specific trophic factors are responsible for the precise temporal and spatial development of hormone producing cells within the AP gland. They include lactotrophs (prolactin, PRL), somatotrophs (growth hormone, GH), thyrotrophs (thyroid-stimulating hormone, TSH), gonadotrophs (follicle-stimulating hormone, FSH, and luteinizing hormone, LH), and corticotrophs (pro-opiomelanocortin, POMC). Production of these hormones is induced and regulated, respectively, by specific hypothalamic releasing hormones and inhibitors, all of which act on distinct receptors present on the surface of the different pituitary cells. In addition to the hormone secreting endocrine cells, the AP gland also contains fibroblasts, endothelial cells and follicular stellate cells which express the S-100 protein. The function of these later cells within the AP is not clearly understood.

Receptors for hormone releasing factors, within their corresponding cell types, include: estrogen and thyrotropin-releasing hormone (TRH) receptors on lactotrophs, growth hormone-releasing hormone receptors (GHRH-R) on somatotrophs, TRH receptors on thyrotrophs, gonadotropin-releasing hormone receptors (GnRH-R) and estrogen receptors on gonadotrophs, and corticotrophin-releasing hormone receptors (CRH-R) and vasopressin receptors on corticotrophs. The receptors that bind factors that inhibit hormone release include those for; dopamine on lactotrophs and thyrotrophs, somatostatin on gonadotrophs and thyrotrophs, and glucocorticoids on corticotrophs and thyrothrophs. Some of these receptors are over expressed in disease, sometimes altering the cell-type expression pattern found in the normal pituitary gland. For example, vasopressin and CRH receptors are over expressed in corticotroph tumors (1), GnRH-R is preferentially expressed in functioning rather than non-functioning gonadotroph adenomas (2), estrogen receptor expression in macroadenomas mimics that found in normal pituitary (3), and GHRH-R is more widely expressed across different cell types in pituitary tumors e.g., prolactinomas, non-functional adenomas, and in ACTH-secreting adenomas, than in normal pituitary gland (4).

IN VIVO AND *IN VITRO* GENE DELIVERY TO THE ADULT ANTERIOR PITUITARY GLAND: METHODOLOGICAL CONSIDERATIONS

The AP gland can be readily infected using recombinant adenovirus vectors both *in vitro* (5-9) and *in vivo* (7-9). For *in vitro* studies, primary AP cultures can be infected using a range of viral doses, i.e., mutiliplicity of infection (MOI; infectious viral particles per cell) of 5-100 to obtain a gradual increase of transgene expression with increasing viral input. Transgene expression from RAds can be sustained *in vitro* within endocrine AP cells for a long time, i.e. in long-term cultures for up to three weeks. After this time, the viability of the cells in culture decreases dramatically and therefore experiments beyond this time point are not feasible under standard conditions (5). For *in vivo* expression of transgenes within the AP gland, we have developed a stereotaxic technique to give maximum transgene expression with minimum damage to the AP gland or central nervous system structures. This is crucial both for pre-clinical gene therapy developments and also for studies involving physiological manipulations of the AP gland. We used adult rats for gene delivery *in vivo* into the AP *in situ*. The detailed protocol for performing the AP transcranial surgery has been described in detail previously by our group (figure 1) (8) and others (9). Using this technique we are able to transduce a large area of the anterior pituitary gland with no serious adverse side effects. Survival after surgery is normally 90-100%. The utility of this model to achieve transcriptional targeting of transgene expression to pre-determined cell-types, reversion of a pathological phenotype and duration of transgene expression will be discussed in the following sections.

CELL-TYPE SPECIFIC EXPRESSION OF TRANSGENES USING VIRAL VECTORS

The use of promoters to transcriptionally target transgene expression from viral vectors has been very successful. Our increasing understanding of transcriptional control suggests that we will eventually be able to design assemblies of hybrid promoter/enhancer units and thus direct transcriptional expression to whichever cell-type we choose. However, several problems must be overcome before transcriptional targeting is optimized in terms of restricting cell-type specificity and maximizing expression efficiency. For instance, the relevant locus control regions/enhancer/silencer/promoter sequences that control expression can be distributed over several kilobases and within chromatin domains that are difficult to reproduce within the context of vector systems. Also careful consideration and possible genetic manipulation of the structure of viral vectors, and the competing promoter/enhancer elements, must be made to optimize the generation of vectors in which promoter/enhancer specificity is retained without compromising levels of transgene expression. In addition, most vectors are strictly limited for cloning space, whereas the cis-acting control elements that provide transcriptional specificity can amount to several kilobases.

Recombinant adenoviral vectors are particularly amenable for this purpose. Initially it was thought that the adenoviral E1 enhancer, which flanks the site of insertion in E1, would result in cell-specific promoters expressing in a non cell-specific fashion. However, numerous studies using adenoviral vectors and cell-specific promoters have shown that cell-type specific transcriptional targeting can be maintained from these vectors. Numerous cell-specific promoters have been used to express transgenes from adenoviral vectors. The neuronal specific enolase (NSE) promoter, which has previously been shown to direct transgene expression to neurons (10), has been shown to maintain this cell-specific expression when used in adenoviral vectors *in vitro* and *in vivo* (11,12), in recombinant adeno-associated viral vectors (13), but not herpes simplex type 1 (HSV1) viral vectors (14,15). The lack of specificity in HSV-1 may be in part due to interference from strong transcriptional transactivators present within the viral genome. Also, the glial fibrillary acidic protein (GFAP) promoter, which restricts gene expression to astrocytes (16), produced highly restricted expression when expressed from adenoviral and retroviral vectors both *in vitro* and *in vivo* (14,17). Other promoters that have met with great success include the myosin light chain II (MLCII) promoter, directing transgene expression in cardiomyocytes (18) from adenovirus.

From herpes viral vectors the tyrosine hydroxylase promoter has been shown to direct expression *in vitro* in rat peripheral neurons (19). Hepatoma-specific replication has been achieved using the albumin enhancer/promoter to drive expression in a replication-competent herpes simplex virus system both *in vitro* and *in vivo* (20). In the case of retroviral vectors, transcriptional targeting has been successfully achieved by replacing viral regulatory sequences within the long terminal repeat (LTR) with the desired enhancer/promoter elements. Liver specific promoters, such as phosphoenolpyruvate carboxy kinase (PEPCK) promoter, the human α-antitrypsin (hAAT) promoter and the human alpha-fetoprotein have been shown to drive hepatocyte specific expression from retroviral vectors *in vivo* (21-24). Endothelial cells have also been transcriptionally targeted with retroviruses using the human pre-proendothelin-1 promoter (25) and the E-selectin and KDR promoters (26). Muscle specific expression has been achieved using the muscle creatine kinase promoter in adenoviral vectors (27) and retroviral vectors (28). Recently, neuronal restricting elements have been used to restrict the expression of a pancellular promoter to neurons with reasonable success (29). Adeno-associated recombinant viral vectors have also been exploited for efficient, transcriptionally targeted gene expression *in vitro* and *in vivo*. For instance, oligodendrocytes have been transduced by means of a myelin-forming cell-type specific promoter (30) and primary rat brain neuronal cultures have been transduced using the NSE promoter (31). Enhancer elements have been incorporated into AAV vectors to increase the expression from cell-type specific promoters. The addition of a woodchuck hepatitis virus post-transcriptional regulatory element (WPRE) into a vector containing the neuron-specific platelet-derived growth factor-beta chain promoter resulted in efficient expression in dopaminergic neurons and their projections within the striatum for at least 41 weeks after virus injection (32). Liver-specific expression was achieved using a thyroid hormone-binding globulin promoter combined with microglobulin/bikunin enhancer sequences (33). The parvovirus B19p6 promoter has been used to restrict expression to erythroid-lineage cells (34).

TRANSCRIPTIONAL TARGETING TO THE PITUITARY GLAND

Transgene expression may be restricted to particular anterior pituitary cell-types either by use of a cell-type specific promoter controlling transcription of the transgene, or by targeting the vector used for transgene delivery to a cell-type specific receptor. Candidates for the first strategy include promoters of the respective hormone genes, e.g., PRL, GH or POMC promoters, and promoters of the genes encoding receptors for the hormone releasing or inhibitory factors, *e.g.*, gonadotropin-releasing hormone receptor (GnRH-R) or V3 vasopressin receptor (35,36). For the second strategy, vectors may be targeted to bind to particular hormone releasing and/or inhibitory factor receptors. Cell-surface receptors that are relatively over expressed in a particular cell-type in a disease state include the epidermal growth factor receptor (EGF-R) (37) or the polysialylated neural cell adhesion molecule (PSA-NCAM) in aggressive macroprolactinomas (38), and these may also be exploited for vector targeting.

For gene transfer/therapy applications within the anterior pituitary gland, an important goal is not only to restrict transgene expression to a desired cell-type, but also to achieve adequate levels of transgene expression. Using cell-type specific hormone promoters to drive the expression of a desired transgene, all cell populations within the gland will be infected, but expression should be restricted to the desired cell-type (figure 2). Recently, Lee et al. (9,39) have shown cell-type specific transgene expression of β-galactosidase (β-gal) and herpes simplex virus thymidine kinase (HSV1-TK) driven by the growth hormone (GH) and the human glycoprotein hormone α-subunit (αT3) promoters. Adenoviral vectors expressing the marker gene β-galactosidase driven by the human growth hormone promoter or the α-glycoprotein subunit were injected directly into the pituitary with a stereotactic device and this lead to cell-type specific expression of the transgene within somatotrophic and gonadotrophic or thyrotrophic anterior pituitary cells respectively (9). These promoters were also used to restrict transgene expression within AP cells *in vitro*, and *in vivo* in a nude mouse transplantable tumor model (39). Using GH3 cells subcutaneously implanted in nude mice, a significant reduction in tumor size was observed after delivering HSV1-TK under control of the GH promoter combined with the administration of ganciclovir (39).

Southgate et al. (8) using first generation adenoviruses harboring the human prolactin promoter driving the expression of transgenes showed cell-type specific expression in lactotrophic and mamo-somatotrophic cells (8,40). Recombinant adenoviruses encoded the marker gene, β-galactosidase and the therapeutic gene HSV1-TK under the control of the prolactin promoter (hPrl) (8). *In vitro* testing, in GH3 (lactotroph) tumor cells showed expression of both transgenes. Apoptosis was induced in GH3 cells infected with the recombinant adenovirus expressing HSV1-TK under the control of the human prolactin promoter (RAd-hPrl-HSV1-TK) after administration of the pro-drug, ganciclovir (8). No transgene expression or cell death was observed in the corticotrophic tumor cell line AtT20 which expresses pro-opiomelanocortin (POMC). RAd-hPrl-HSV1-TK virus has also been used to infect AP primary cultures, where the prolactin promoter restricted expression of HSV1-TK almost exclusively to prolactin and a subpopulation of growth hormone expressing AP cells (figure 3). These cells showed no signs of apoptosis in the absence or presence of GCV. The *in situ* testing of these promoters within the AP gland has been limited, with the human cytomegalovirus promoter being tested in the AP gland *in vivo* in a model of lactroph induced hyperplasia using estrogen/sulpiride implants (8). Using this approach a 30% reduction in the pituitary mass and a four-fold reduction in circulating prolactin levels was achieved (7). HSV1-TK driven by the human prolactin promoter was expressed almost exclusively in lactrophic cells *in situ* and in a sub population of growth hormone expressing cells, probably somatomamotrophs (8) (figure 3). However, the level of therapeutic transgene expression achieved with this system was not sufficient to induce a measurable beneficial therapeutic outcome when combined with the administration of GCV. This is in contrast to the results obtained when the expression of HSV1-TK was driven by the strong ubiquitous hCMV promoter in the same model (figure 4) (7).

In addition to the promoters mentioned above, other pituitary hormone promoters could be used to achieve cell-type specific expression of transgenes. These include the pro-opimelanocortin promoter (41), which expresses predominantly in the corticotrophic cells of the anterior pituitary gland. Using deletion analysis Kraus et al. (41) have determined negative regulatory elements between nucleotides -676 and -414 and positive regulatory elements between nucleotides -414 and -93 of this promoter. When placed in front of the heterologous thymidine kinase promoter, nucleotides -414/-223 enhanced transcription in AtT20 cells and in primary cultures of human pituitary tumor cells, but not in various non-pituitary cell lines (41). Thyrotropin (TSH) is a glycoprotein hormone that is produced only by thyrotrope cells of the AP gland (42). Haugen et al. (43) have shown that within the thyrotropin β-promoter there is an element in the 5' flanking region that is required for promoter expression in thyrotropes as it interacts with the pituitary specific transcription factor Pit-1. Another promoter, which could be used to restrict transgene expression within viral vectors, is the gonadotropin-releasing hormone receptor promoter (GnRH-R). The 1.2Kb region of the 5' flanking region of the mouse GnRH receptor can target expression specifically to gonadotrophic cells within the AP *in vitro* and *in vivo* in transgenic mice (35). The mGnRH-R gene promoter directs transgene expression, which appears to be cell-type specific for both FSHβ- and LHβ-subunit expressing cells within the AP gland (35).

REGULATED TRANSGENE EXPRESSION

Both for basic and putative clinical applications it would be very desirable to control levels of transgene expression and also to switch transgene expression "on" and "off" by using small molecules that could be delivered systemically. Early attempts to regulate the expression of gene products were based on inducible endogenous promoters, which were responsive to heat-shock (44), heavy metals (45), interferon or dsRNA (46), or steroids (47,48). The inherent advantage of these approaches is that they rely on endogenous transcription factors, obviating the need for additional gene regulation. The disadvantages, *e.g.*, toxicity and pleiotropic effects of the inducers, make these systems unsuitable for most *in vivo* applications. Another disadvantage is the heterogeneity of basal expression levels and induction ratios for these promoters due to tissue-specific differences in endogenous 'inducing factors' (49).

Engineered expression systems based on heterologous prokaryotic or microbial regulatory elements have proved more successful than the endogenous, inducible promoters. The *lac* system, based on the bacterial *lac* operator-repressor system, has been shown to regulate gene expression in mammalian cells *in vitro*, providing low basal expression and high induction with the addition of the inducer IPTG. However this system has not yet been shown to function *in vivo*. Another regulatory system, the RU486 (mifepristone) system has been shown to generate tightly regulated gene expression both *in vitro* and *in vivo* (50-53). This progesterone regulatory system relies on the inducer RU486 to modulate transcription. The inductor binds to the ligand-binding domain of a mutant human progesterone receptor fused to the VP16 transactivator domain and the yeast GAL4 DNA-binding domain resulting in transcriptional activation of the target gene. Limitations of this system are the slow re-induction *in vivo* (54) as consequence of the long half-life of RU486 and poor diffusion within tissues (55). Other systems include the ecdysone system, based upon a *Drosophila melanogaster* receptor hormone (56) and the rapamycin system (57) based upon the human rapamycin-binding domain, FKBP, of the human rapamycin binding protein, FRAP. This Interaction has been shown to be highly functional, generating tightly regulated transgene expression *in vitro* and *in vivo* (56-60). However, its application in the field of gene therapy can be hampered by the immunosuppressive properties of rapamycin (57).

In this section, we will concentrate on the tetracycline gene regulatory system (tet system) (61), which to date is the most widely used and characterized system for the regulation of transgene expression. The tetracycline regulatable system has several advantages, i.e., this system requires only two components (in contrast with ecdysone or rapamycin systems which require three) that can be delivered from a single vector without affecting its efficiency. Also the system can be switched "off" and then "on" within 72 hours by the addition or removal of either tetracycline or its analog doxycycline, and thereby can be used to suppress or induce transgene expression. The pharmacokinetics, pharmacodynamics and side effects of tetracycline and derivates are well known and because of the high affinity of the tetracycline regulatable system for the tetracycline repressor the antibiotic can be used in low doses, reducing the possibility of adverse side effects on mammalian cells and transgenic animals. This system is also very well-situated for work involving transgenic animals as some tetracycline analogs, like doxycycline, can be secreted into the milk and cross the placenta barrier (62). We will therefore expand on the tet regulatory system, since this is the system which we have used both *in vitro* and *in vivo* to achieve regulatable and cell-type specific transgene expression within the AP gland.

The tet system is based on the previously characterized tetracycline resistance gene (63-65). This system possesses two fundamental elements, a transactivator (tTA) and a tetracycline responsive element (TRE) (figure 5). The tTA is a chimeric protein possessing two domains, a DNA binding domain, containing unique helix-turn-helix motifs that binds to the TRE, and an activator domain (VP16) which is a strong eukaryotic transactivator from herpes simplex virus type 1. The TRE possesses seven repeats of the tetracycline operator sequence, upstream of a TATA box. The heptads of operator sequences, and TATA box lie upstream of the transgene initiation codon. In the presence of tetracycline (or its analogue), the antibiotic binds to the dimerised tTA and prevents tTA/TRE interaction, blocking transgene expression. However, in the absence of tetracycline, the dimerised tTA binds to its operator sequences within the TRE and initiates transgene expression (figure 5). This system was termed the "tet off" system. Thus, the simple addition of a regulatory molecule to a biological system is sufficient to control transgene expression. Further, four point mutations in the tTA producing the rtTA, lead to the development of the "tet on" system (66). In this system tetracycline activates transgene expression. More recently, a dual regulatory system for switching the expression of two genes has been proposed. It is based on the utilization of the "tet off" tTA for regulation of one gene and "tet on" rtTA for regulation of the second. The former represses gene expression in the presence of doxycycline and the latter induces expression in presence of doxycycline, resulting in a mutually exclusive control of regulation. This system could be use in gene therapy e.g., for the control of a therapeutic gene with one regulatory circuit and termination of the regime using the other (67). In comparison with tTA the rtTA has several limitations such as: background activity in the un-induced state due to the low but residual affinity toward TRE, instability *in vivo* and relative low sensitivity toward doxycycline. To circumvent these limitations, Urlinger and colleagues reported 5 new mutants of the rtTA from which rtTA2s-M2, has a reduced basal activity, increased doxycycline sensitivity, require less concentration of doxycycline to work, and is more stable in eukaryotic cells (68).

The development of these tetracycline inducible systems by Gossen and Bujard (61) has led to their rapid integration into the field of gene transfer and therapy. Numerous groups have used the tet system to regulate transgene expression from viral vectors, transiently, stably, and also in transgenic animals. Such a diversity of functional paradigms, which have stringently tested the tet system, underlines its robust nature and potential uses. Several studies using the tet system in viral vectors have been reported, providing strong support for the utilization of this system. Initial studies using recombinant adenoviral vectors showed the "tet off" system to be functional from such vectors *in vitro* (69). Tight regulation and high induction levels were achieved when the tet system was expressed from two separate adenoviral vectors and used to infect Hela cells (69). Moreover, this study points to an important requirement of the tet system when expressing from episomal vectors such as adenoviruses. In the initial studies by Gossen and

Bujard (70), they suggested that the use of a nuclear localisation sequence on the transactivator was not a requirement, and could result in increased basal expression of the transgenes. These studies however were performed on stable cell lines and not from episomal vectors. Studies using adenoviral vectors showed that the nuclear localisation sequence might be beneficial when using the "tet off" system with adenoviral vectors as its omission resulted in increased basal expression (69). Harding et al., however, used adenoviruses with a nuclear localised "tet on" system to infect hippocampal cultures *in vitro* and showed that the nuclear localisation sequence reduced induction (71). Moreover, the tet "off" system provided higher levels of induction than the tet "on" system. Alternative studies *in vivo*, again suggest that when expressed from adenoviral vectors, the tet system provides an efficient method for regulating gene expression with high induction levels and very low basal levels of expression. Using a dual adenoviral system tight regulation and high induction levels were observed, when injected into the hippocampus (72). Alternative studies expressing tyrosine hydroxylase from the tet system in a single adenoviral vector showed similar results in neural progenitors (73,74).

The initial functional tests using the tet system have led many groups to express potential therapeutic transgenes from this system. Studies expressing toxic genes from the tet system in adenoviruses underline the tight regulation observed when using this system. Expression of tumor necrosis factor α (75), herpes simplex virus type 2-ribonucleotide reductase (76), and more recently the apoptosis inducer, Fas ligand (FasL) (77), gave tight regulation of transgene expression with little or no toxicity. In the case of FasL, tight control of expression was achieved by creating a novel double recombinant adenovirus vector, in which the tet-responsive element and the transactivator element are built into opposite ends of the same vector to avoid enhancer interference (77). The tet systems have been shown to be functional from other viral vectors, including retroviral vectors. Initial *in vitro* studies suggested that high induction levels of transgene expression could be achieved when expressing the tet system from a single retrovirus (78,79). Other viral vectors that have been used to express the tet system include herpes simplex virus type 1 (80), adeno-associated viral vectors (81) and human cytomegalovirus (82). In addition the tet system has shown functionality from plasmid DNA (83-86) and has also been stringently tested in transgenic animals (87-90).

COMBINING REGULATED AND CELL TYPE SPECIFIC TRANSGENE EXPRESSION

Recent efforts in combining cell-specificity with regulation in viral vectors have met with success both *in vitro* and *in vivo*. Initial proof of principle has come from transgenic studies where cell-type specific promoters were used to direct the expression of the tetracycline system. Two such promoters include the liver specific LAP promoter (89) and the neuronal specific CAMK II promoter (91). Recent studies have shown that, by combining the tetracycline system with the cell-type specific GFAP promoter in adenoviruses, cell-type specific and regulated expression was achieved for the first time within cell lines of glial origin (92). Both the GFAP and the NSE promoter have recently been used to drive expression of transgenes within glial and neuronal cells both *in vitro* in primary cortical cultures and also *in vivo* when injected into the hippocampus (93). In a similar study using an adenovirus expressing the tTA under the control of the neuronal-specific synapsin I promoter and GFAP it was possible to induce specific and regulated transgene expression in neuronal and glial cell populations respectively *in vitro* and *in vivo* (94). Further regulated liver specific expression has been achieved *in vivo* using helper-dependent (HD) adenoviral vectors, which have improved safety and long-term expression *in vivo* due to the absence of adenoviral genes (50). Transgenic technologies have also been beneficiary of regulated cell type specific expression. Utomo et al. (1999) reported the development of a universal system for temporal, spatial and cell type specific control of gene expression in transgenic mice which combines the tet system and the Cre-recombinase-loxP system. They describe tetracycline controlled Cre-mediated recombination controlled by the hCMV promoter. The authors suggest that this technology could be useful with a cell-type specific promoter to control gene expression in a specific sub-population of cell types (95).

Hormone gene expression is tightly regulated within the anterior pituitary gland. This in combination with inducible expression systems could be harnessed for developing cell-type specific and regulated transgene expression both *in vitro* and *in vivo*. Recently, by driving the expression of the tetracycline system using the prolactin promoter we have been able to achieve lactotrophic cell-type specific and inducible expression of transgenes from adenoviral vectors both *in vitro* in primary anterior pituitary cultures, in anterior pituitary tumor cell lines and within the anterior pituitary gland *in vivo* (96). The pancellular Chicken β-actin/hCMV promoter (CAG) driving expression of β-gal under control of the tetracycline inducible expression system (RAd-CAG-tTA(nls)/RAd-TRE-β-gal) in primary anterior pituitary cultures did not restrict expression to any particular cell type, but did produce highly regulatable transgene expression. After the addition of doxycycline there was no detectable transgene expression either *in vitro* or *in vivo* (figure 6) (96). The human prolactin promoter, driving the expression of the tetracycline system (RAd-hPrl-tTA(nls)/RAd-TRE-β-gal), *in vitro* produced highly regulatable β-galactosidase expression which was restricted to the lactotrophic tumor cell line, GH3 and lactotrophic cells in primary anterior pituitary culture (96). Although regulated transgene expression was observed in primary anterior pituitary cell cultures, such expression was not exclusive to lactotrophic cells, 13% of β-gal positive cells in the absence of doxycycline were not lactotrophic in origin. *In vivo* delivery into the AP gland with RAd-CAG-tTA(nls) and RAd-TRE-β-gal resulted in high level of transgene expression, which was not restricted to any particular population of cells within the pituitary (96). *In vivo* delivery of RAd-hPrl-tTA(nls) and RAd-TRE-β-gal resulted in moderate expression within the anterior pituitary gland with expression restricted mainly to lactotrophic cells and a subpopulation of cells expressing GH that are likely to be mammosomatotrophs which also express prolactin. This will now offer the possibility of not only expressing transgenes specifically within lactrophic cells, but it will also allow very tight regulation of the levels of transgene expression, both *in vitro* and *in vivo*.

LONG TERM TRANSGENE EXPRESSION WITHIN THE ANTERIOR PITUITARY GLAND *IN SITU*

As mentioned previously successful virus mediated gene delivery to the anterior pituitary gland can be achieved through several methods although most studies involved short-term gene transfer. The need for extended transgene expression either for therapeutic or basic physiological studies has led to investigation into the longevity of transgene expression obtainable within the anterior pituitary gland *in situ* and the consequences of this invasive form of gene delivery.

Recent investigations in our laboratory have explored the duration of transgene expression seen within the anterior pituitary up to 3 months after adenovirus delivery when transgene expression is driven by constitutive (hCMV) or cell type specific (hPrl) promoters. Expression of HSV1-TK transgene is seen in multiple hormone producing cell types with the hCMV promoter but remains restricted to prolactin producing cells with the hPrl promoter (Figure 7). Also transgene expression from both the hCMV and hPrl promoters can be seen at 3 months post injection although a significant decrease in transgene expression is seen over time (98).

The exact cause of the decline in transgene expression seen in the anterior pituitary over time is unclear although activation of virus induced immune responses or decreased levels of adenoviral genome may play a part. PCR analysis of pituitary sections at 3 months revealed the presence of HSV1-TK transgene (98) but whether a reduced level of genome is present, thus resulting in reduced expression, remains to be determined. Whether the decline in expression was a consequence of a virus induced immune response is also unclear although increased levels of ED1 (activated macrophages), CD8β (T cells) and CD161 (NK cells) positive cells are seen following adenovirus delivery (98). Although within the cranium, the pituitary gland lies outside of the blood brain barrier and upon infection it has been shown that, unlike infection of the brain parenchyma (99), an adenovirus specific humoral immune response can be mounted (98). The induction of adenovirus specific cellular immune responses is also known to play a role in the reduction of transgene expression following adenovirus-mediated delivery to other organs (100) but unlike in the pituitary, expression is eliminated within 7 days. Whether an effective cellular immune response is mounted following adenovirus delivery to the anterior pituitary is unknown and the exact role of the cellular immune response in the reduction of transgene expression in the pituitary remains undetermined.

The long-term effects on circulating hormone levels following direct injection into the anterior pituitary gland have also been monitored (98). No statistically significant difference in circulating LH, FSH, GH, TSH or prolactin levels was seen at any time point (3 days, 14 days, 1, 2 and 3 months) although ACTH levels were elevated at 3 days before returning to normal at 14 days. These results are encouraging as adenovirus-mediated transgene delivery does not appear to affect regulatory hormone production with the exception of ACTH, and an increase in ACTH levels would be expected at early time points due to its role as a stress hormone.

CONCLUSIONS

The expression of transgenes within the AP gland *in vitro* and *in vivo* has become a reality with the use of viral vectors. Also, the possibility of turning transgene expression "on" and "off" using small molecules such as the tetracycline analogue doxycycline, has enabled an array of experimental paradigms, which could be harnessed to uncover the function of newly discovered genes, to allow the temporal regulation of transgenes expressed during adulthood, to specifically delete mRNA molecules in a cell-type specific and temporal fashion. This technology offers new avenues of experimental research both for basic and potentially clinical applications.

Acknowledgements

Supported by grants from the BBSRC (UK), The Wellcome Trust (UK), The Royal Society, and European Union-Biomed grants, contract no. BMH4-CT98-3277, BMH4-CT98-0297 and QLK3-CT-1999-00364 to PRL and MGC. PRL is a fellow of The Lister Institute of Preventive Medicine. TS is a fellow supported by Action research (UK). DS is funded by a BBSRC (UK) PhD studentship. JSA is funded by a BBSRC (UK) PhD industrial case studentship We are grateful to Mrs. R Poulton and Ms T Maleniak for expert secretarial and technical assistance respectively. We also wish to thank Profs AM Heagerty, R Green, F Creed and D. Gordon for their continuous support and encouragement.

REFERENCES

1. de Keyzer Y, Rene P, Beldjord C, Lenne F, Bertagna X. Overexpression of vasopressin (V3) and corticotrophin-releasing hormone receptor genes in corticotroph tumours. Clin Endocrinol 1998;49:475-82.

2. Kottler ML, Seret-Begue D, Lahlou N, Assayag M, Carre MC, Lagarde JP, Ajzenberg C, Christin-Maitre S, Bouchard P, Mikol J, Counis R, Warnet A. The GnRH receptor gene is preferentially expressed in functioning gonadotroph adenomas and displays a Mae III polymorphism site. Clin Endocrinol 1998;49:115-23.

3. Friend KE, Chiou YK, Lopes MB, Laws ER Jr., Hughes KM, Shupnik MA. Estrogen receptor expression in human pituitary: correlation with immunohistochemistry in normal tissue, and immunohistochemistry and morphology in macroadenomas. J Clin Endocrinol Metab 1994;78:1497-504.

4. de Keyzer Y, Lenne F, Bertagna X. Widespread transcription of the growth hormone-releasing peptide receptor gene in neuroendocrine human tumors. Eur J Endocrinol 1997;137:715-8.

5. Castro MG, Goya RG, Sosa YE, Rowe J, Larregina A, Morelli A, Lowenstein PR. Expression of transgenes in normal and neoplastic anterior pituitary cells using recombinant adenoviruses: long term expression, cell cycle dependency, and effects on hormone secretion. Endocrinology 1997;138:2184-94.

6. Neill JD, Musgrove LC, Duck LW, Sellers JC. High efficiency method for gene transfer in normal pituitary gonadotropes: adenoviral-mediated expression of G protein-coupled receptor kinase 2 suppresses luteinizing hormone secretion. Endocrinology 1999;140:2562-9.

7. Windeatt S, Southgate TD, Dewey RA, Bolognani F, Perone MJ, Larregina AT, Maleniak TC, Morris ID, Goya RG, Klatzmann D, Lowenstein PR, Castro MG. Adenovirus-mediated herpes simplex virus type-1 thymidine kinase gene therapy suppresses oestrogen-induced pituitary prolactinomas. J Clin Endocrinol Metab 2000;85:1296-305.

8. Southgate TD, Windeatt S, Smith-Arica J, Gerdes CA, Perone MJ, Morris I, Davis JR, Klatzmann D, Lowenstein PR, Castro MG. Transcriptional targeting to anterior pituitary lactotrophic cells using recombinant adenovirus vectors in vitro and in vivo in normal and estrogen/sulpiride induced hyperplastic anterior pituitaries. Endocrinology 2000;141:3493-505.

9. Lee EJ, Thimmapaya B, Jameson JL. Stereotactic injection of adenoviral vectors that target gene expression to specific pituitary cell types: implications for gene therapy. Neurosurgery 2000;46:1461-9.

10. Forss-Petter S, Danielson PE, Catsicas S, Battenberg E, Price J, Nerenberg M, Sutcliffe JG. Transgenic mice expressing beta-galactosidase in mature neurons under neuron-specific enolase promoter control. Neuron 1990;5:187-97.

11. Morelli AE, Larregina AT, Smith-Arica J, Dewey RA, Southgate TD, Ambar B, Fontana A, Castro MG, Lowenstein PR. Neuronal and glial cell type-specific promoters within adenovirus recombinants restrict the expression of the apoptosis-inducing molecule Fas ligand to predetermined brain cell types, and abolish peripheral liver toxicity. J Gen Virol 1999;80:571-83.

12. Navarro V, Millecamps S, Geoffroy MC, Robert JJ, Valin A, Mallet J, Gal La Salle G. Efficient gene transfer and long-term expression in neurons using a recombinant adenovirus with a neuron-specific promoter. Gene Ther 1999;6:1884-92.

13. Peel AL, Zolotukhin S, Schriomsher GW, Muzyczka N, Reier PJ. Efficient transduction of green fluorescent protein in spinal cord neurons using adeno-associated virus vectors containing cell type-specific promoters. Gene Ther 1997;4:16-24.

14. Andersen JK, Frim DM, Isacson O, Breakfield XO. Herpesvirus-mediated gene delivery into the rat brain: specificity and efficiency of the neuron-specific enolase promoter. Cell Mol Neurobiol 1993;13:503-15.

15. Roemer K, Johnson PA, Friedmann T. Transduction of foreign regulatory sequences by a replication-defective herpes simplex virus type 1: the rat neuron-specific enolase promoter. Virus Res 1995;35:81-9.

16. Brenner M, Kisseberth WC, Su Y, Besnard F, Messing A. GFAP promoter directs astrocyte-specific expression in transgenic mice. J Neurosci 1994;14:1030-7.

17. Miyao Y, Shimizu K, Moriuchi S, Yamada M, Nakahira K, Nakajima K, Nakao J, Kuriyama S, Tsujii T, Mikoshiba K, Hayakawa T, Ikenaka K. Selective expression of foreign genes in glioma cells: Use of mouse myelin basic protein gene promoter to direct toxic gene expression. J Neurosci Res 1993;36:472-9.

18. Griscelli F, Gilardi-Hebenstreit P, Hanania N, Franz WM, Opolon P, Perricaudet M, Ragot T. Heart-specific targeting of beta-galactosidase by the ventricle-specific cardiac myosin light chain 2 promoter using adenovirus vectors. Hum Gene Ther 1998;9:1919-28.

19. Oh YJ, Moffat M, Wong S, Ullrey D, Geller AI, O'Malley KL. A herpes simplex virus-1 vector containing the rat tyrosine hydroxylase promoter directs cell type-specific expression of beta-galactosidase in cultured rat peripheral neurons. Brain Res Mol Brain Res 1996;35:227-36.

20. Miyatake SI, Tani S, Feigenbaum F, Sundaresan P, Toda H, Narumi O, Kikuchi H, Hashimoto N, Hangai M, Martuza RL, Rabkin, SD. Hepatoma-specific antitumor activity of an albumin enhancer/promoter regulated herpes simplex virus in vivo. Gene Ther 1999;6:564-72.

21. Hatzoglou M, Lamers W, Bosch F, Wynshaw-Boris A, Clapp DW, Hanson RW. Hepatic gene transfer in animals using retroviruses containing the promoter from the gene for phosphoenolpyruvate carboxykinase. J Biol Chem 1990;265:17285-93.

22. Hatzoglou M, Park E, Wynshaw-Boris A, Kaung HL, Hanson RW. Hormonal regulation of chimeric genes containing the phosphoenol pyruvate carboxykinase promoter regulatory region in hepatoma cells infected by murine retroviruses. J Biol Chem 1988;263:17798-808.

23. Hafenrichter DG, Ponder KP, Rettinger SD, Kennedy SC, Wu X, Saylors RS, Flye MW. Liver-directed gene therapy: evaluation of liver specific promoter elements. J Surg Res 1994;56:510-7.

24. Uto H, Ido A, Hori T, Hirono S, Hayashi K, Tamaoki T, Tsubouchi H. Hepatoma-specific gene therapy through retrovirus-mediated and targeted gene transfer using an adenovirus carrying the ecotropic receptor gene. Biochem Biophys Res Commun 1999;265:550-5.

25. Jager U, Zhao Y, Porter CD. Endothelial cell-specific transcriptional targeting from a hybrid long terminal repeat retrovirus vector containing human prepro-endothelin-1 promoter sequences. J Virol 1999;73:9702-9.

26. Modlich U, Pugh CW, Bicknell R. Increasing endothelial cell specific expression by the use of heterologous hypoxic and cytokine-inducible enhancers. Gene Ther 2000;7:896-902.

27. Karpati G. Possible in vivo gene therapy of skeletal muscle in Duchenne muscular dystrophy. Gene Ther 1995;2:583-4.

28. Ferrari G, Salvatori G, Rossi C, Cossu G, Mavilio F. A retroviral vector containing a muscle-specific enhancer drives gene expression only in differentiated muscle fibers. Hum Gene Ther 1995;6:733-42.

29. Millecamps S, Kiefer H, Navarro V, Geoffroy MC, Robert JJ, Finiels F, Mallet J, Barkats M. Neuron-restrictive silencer elements mediate neuron specificity of adenoviral gene expression. Nat Biotech 1999;17:865-9.

30. Chen H, McCarty DM, Bruce AT, Suzuki K. Oligodendrocyte-specific gene expression in mouse brain: use of a myelin-forming cell type-specific promoter in an adeno-associated virus. J Neurosci Res 1999;55:504-13.

31. Wang S, Bui V, Hughes JA, King MA, Meyer EM. Adeno-associated virus mediated gene transfer into primary rat brain neuronal and glial cultures: enhancement with the pH-sensitive surfactant dodecyl 2-(1'-imidazolyl) propionate. Neurochem Int 2000;37:1-6.

32. Paterna JC, Moccetti T, Mura A, Feldon J, Bueler H. Influence of promoter and WHV post-transcriptional regulatory element on AAV-mediated transgene expression in the rat brain. Gene Ther 2000;7:1304-11.

33. Wang L, Takabe K, Bidlingmaier SM, Ill CR, Verma IM. Sustained correction of bleeding disorder in hemophilia B mice by gene therapy. Proc Natl Acad Sci USA 1999;96:3906-10.

34. Kurpad C, Mukherjee P, Wang XS, Ponnazhagan S, Li L, Yoder MC, Srivastava A. Adeno-associated virus 2-mediated transduction and erythroid lineage-restricted expression from parvovirus B19p6 promoter in primary human hematopoietic progenitor cells. J Hematother Stem Cell Res 1999;8:585-92.

35. Albarracin CT, Frosch MP, Chin WW. The gonadotropin-releasing hormone receptor gene promoter directs pituitary-specific oncogene expression in transgenic mice. Endocrinology 1999;140:2415-21.

36. Rene P, Lenne F, Ventura M, Bertagna X, de Keyzer Y. Nucleotide sequence and structural organization of the human vasopressin pituitary receptor (V3) gene. Gene 2000;241:57-64.

37. Jaffrain-Rea ML, Petrangeli E, Lubrano C, Minniti G, Di Stefano D, Sciarra F, Frati L, Tamburrano G, Cantore G, Gulino A. Epidermal growth factor binding sites in human pituitary macroadenomas. J Endocrinology 1998;158:425-33.

38. Daniel L, Trouillas J, Renaud W, Chevallier P, Gouvernet J, Rougon G, Figarella-Branger D. Polysialylated-neural cell adhesion molecule expression in rat pituitary transplantable tumors (spontaneous mammotropic transplantable tumor in Wistar-Furth rats) is related to growth rate and malignancy. Cancer Res 2000;60:80-5.

39. Lee EJ, Anderson LM, Thimmapaya B, Jameson, JL. Targeted expression of toxic genes directed by pituitary hormone promoters: a potential strategy for adenovirus-mediated gene therapy of anterior pituitary tumors. J Clin Endocrinol Metab 1999;84:786-94.

40. Castro MG, Windeatt S, Smith-Arica J, Lowenstein PR. Cell-type specific expression in the pituitary: physiology and gene therapy. Biochem Soc Trans 1999;27:858-63.

41. Kraus J, Buchfelder M, Hollt V. Regulatory elements of the human proopiomelanocortin gene promoter. DNA Cell Biol 1993;12:527-36.

42. Pierce JG, Parsons TF. Glycoprotein hormones: structure and function. Annu Rev Biochem 1981;50:465-95.

43. Haugen BR, McDermott MT, Gordon DF, Rupp CL, Wood WM, Ridgeway EC. Determinants of thyrotrope-specific thyrotropin beta promoter activation: Cooperation of Pit-1 with another factor. J Biol Chem 1996;271:385-9.

44. Wurm FM, Gwinn KA, Kingston RE. Inducible overproduction of the mouse c-myc protein in mammalian cells. Proc Natl Acad Sci USA, 1986;83:5414-8.

45. Mayo KE, Warren R, Palmiter RD. The mouse metallothionein-I gene is transcriptionally regulated by cadmium following transfection into human or mouse cells. Cell 1982;29:99-108.

46. Ryals J, Dierks P, Ragg H, Weissmann C. A 46-nucleotide promoter segment from an IFN-alpha gene renders an unrelated promoter inducible by virus. Cell 1985;41:497-507.

47. Lee F, Mulligan R, Berg P, Ringold G. Glucocorticoids regulate expression of dihydrofolate reductase cDNA in mouse mammary tumour virus chimaeric plasmids. Nature 1981;294:228-32.

48. Hynes NE, Groner B. Mammary tumor formation and hormonal control of mouse mammary tumor virus expression. Curr Top Microbiol Immunol 1982;101:51-74.

49. Palmiter RD, Chen HY, Brinster RL. Differential regulation of metallothionein-thymidine kinase fusion genes in transgenic mice and their offspring. Cell , 1982;29:701-10.

50. Wang Y, O'Malley BW Jr, Tsai SY, O'Malley BW. A regulatory system for use in gene transfer. Proc Nat Acad Sci USA 1994;91:8180-4.

51. Molin M, Shoshan MC, Ohman-Forslund K, Linder S, Akusjarvi G. Two novel adenovirus vector systems permitting regulated protein expression in gene transfer experiments. J Virol 1998;72:8358-61.

52. Abruzzese RV, Godin D, Burcin M, Mehta V, French M, Li Y, O'Malley BW, Nordstrom JL. Ligand-dependent regulation of plasmid-based transgene expression in vivo. Hum Gene Ther 1999;10:1499-1507.

53. Burcin MM, Schiedner G, Kochanek S, Tsai SY, O'Malley BW. Adenovirus-mediated regulable target gene expression in vivo. Proc Natl Acad Sci USA 1999;96 355-60.

54. Wan Y, DeMayo FJ, Tsai SY, O'Malley BW. Ligand-inducible and liver specific target gene expression in transgenic mice. Nat Biotechnol 1997;15:239-43.

55. Serguera C, Bohl D, Rolland E, Prevost R, Herard TM. Control of erythropoietin secretion by doxycycline or mifepristone in mice bearing polymer-encapsulated engineered cells. Hum Gene Ther 1999;10:375-83.

56. No D, Yao TP, Evans RM. Ecdysone-inducible gene expression in mammalian cells and transgenic mice. Proc Natl Acad Sci USA 1996;93:3346-51.

57. Rivera VM, Clackson T, Natesan S, Pollock R, Amara JF, Keenan T, Magari SR, Phillips T, Courage NL, Cerasoli F Jr, Holt DA, Gilman M. A humanized system for pharmacologic control of gene expression. Nat Med 1996;2:1028-32.

58. Rivera VM, Ye X, Courage NL, Sachar J, Cerasoli F Jr, Wilson JM, Gilman M. Long-term regulated expression of growth hormone in mice after intramuscular gene transfer. Proc Natl Acad Sci USA 1999;96:8657-62.

59. Dunlop J, Lou Z, McIlvain HB. Steroid Hormone-Inducible Expression of the GLT-1 Subtype of High-Affinity l-Glutamate Transporter in Human Embryonic Kidney Cells. Biochem Biophys Res Commun 1999;265:101-5.

60. Johns DC, Marx R, Mains RE, O'Rourke B, Marban E. Inducible genetic suppression of neuronal excitability. J Neurosci 1999;19:1691-7.

61. Gossen M, Bujard H. Tight control of gene expression in mammalian cells by tetracycline-responsive promoters. Proc Natl Acad Sci USA 1992;89:5547-51.

62. Agha-Mohammadi S, Lotze MT. Regulatable systems: applications in gene therapy and replicating viruses. J Clin Invest 2000;105:1177- 83.

63. Beck CF, Mutzel R, Barbé J, Müller W. A multifunctional gene (tetR) controls Tn10-encoded tetracycline resistance. J Bacteriol 1982;150:633-42.

64. Takahashi M, Altschmied L, Hillen W. Kinetic and equilibrium characterisation of the Tet repressor-tetracycline complex by fluorescence measurements. Evidence for divalent metal ion requirement and energy transfer. J Mol Biol 1986;187:341-8.

65. Hinrichs W, Kisker C, Duvel M, Muller A, Tovar K, Hillen W, Saenger W. Structure of the Tet repressor-tetracycline complex and regulation of antibiotic resistance. Science 1994;264:418-420.

66. Gossen M, Freundlieb S, Bender G, Muller G, Hillen W, Bujard H. Transcriptional activation by tetracyclines in mammalian cells. Science 1995;268:1766-9.

67. Baron U, Schnappinger D, Helbl V, Gossen M, Hillen W, Bujard H. Generation of conditional mutants in higher eukaryotes by switching between the expression of two genes. Proc Natl Acad Sci USA 1999;96:1013-8.

68. Urlinger S, Baron U, Thellmann M, Hasan MT, Bujard H, Hillen W. Exploring the sequence space for tetracycline-dependent transcriptional activators: novel mutations yield expanded range and sensitivity. Proc Natl Acad Sci USA 2000;97:7963-8.

69. Yoshida Y, Hamada H. Adenovirus-mediated inducible gene expression through tetracycline-controllable transactivator with nuclear localization signal. Biochem Biophys Res Commun 1997;230:426-30.

70. Gossen M, Bujard H. Efficacy of tetracycline-controlled gene expression is influenced by cell type: commentary. Biotechniques 1995;19:213-6.

71. Harding TC, Geddes JB, Noel JD, Murphy D, Uney JB. Tetracycline-regulated transgene expression in hippocampal neurones following transfection with adenoviral vectors. J Neurochem 1997;69:2620-3.

72. Harding TC, Geddes BJ, Murphy D, Knight D, Uney JB. Switching transgene expression in the brain using an adenoviral tetracycline-regulatable system. Nat Biotech 1998;16:553-5.

73. Corti O, Horellou P, Colin P, Cattaneo E, Mallet J. Intracerebral tetracycline-dependent regulation of gene expression in grafts of neural precursors. NeuroReport 1996;7:1655-9.

74. Corti O, Sabate O, Horellou P, Colin P, Dumas S, Buchet D, Buc-Caron MH, Mallet J. A single adenovirus vector mediates doxycycline-controlled expression of tyrosine hydroxylase in brain grafts of human neural progenitors. Nat Biotech 1999;17:349-54.

75. Hu SX, Ji W, Zhou Y, Logothetis C, Xu HJ. Development of an adenoviral vector with tetracycline-regulatable human tumour necrosis factor alpha gene expression. Cancer Res 1997;57:3339-43.

76. Massie B, Couture F, Lamoureux L, Mosser DD, Guilbault C, Jolicoeur P, Belanger F, Langelier Y. Inducible overexpression of a toxic protein by an adenovirus vector with a tetracycline-regulatable expression cassette. J Virol 1998;72:2289-96.

77. Rubinchik S, Ding R, Qiu AJ, Zhang F, Dong J. Adenoviral vector which delivers FasL-GFP fusion protein regulated by the tet-inducible expression system. Gene Ther 2000;7:875- 85.

78. Hofmann A, Nolan GP, Blau HM. Rapid retroviral delivery of tetracycline-inducible genes in a single autoregulatory cassette. Proc Natl Acad Sci USA 1996;93:5185-90.

79. Paulus W, Baur I, Boyce FM, Breakefield XO, Reeves SA. Self-contained, tetracycline-regulated retroviral vector system for gene delivery to mammalian cells. J Virol 1996;70:62-7.

80. Ho DY, McLaughlin JR, Sapolsky RM. Inducible gene expression from defective herpes simplex virus vectors using the tetracycline-responsive promoter system. Brain Res Mol Brain Res 1996;41:200-9.

81. Rendahl KG, Leff SE, Otten GR, Spratt SK, Bohl D, Van Roey M, Donahue BA, Cohen LK, Mandel RJ, Danos O, Snyder RO. Regulation of gene expression in vivo following transduction by two separate rAAV vectors. Nat Biotech 1998;16:757-61.

82. Kim HJ, Gatz C, Hillen W, Jones TR. Tetracycline repressor-regulated gene repression in recombinant human cytomegalovirus. J Virol 1995;69:2565-73.

83. Paulus W, Baur I, Oberer DM, Breakefield XO, Reeves SA. Regulated expression of the diphtheria toxin A gene in human glioma cells using prokaryotic transcriptional control elements. J Neurosurg 1997;87:89-95.

84. Agha-Mohammadi S, Alvarez-Vallina L, Ashworth LJ, Hawkins RE. Delay in resumption of the activity of tetracycline-regulatable promoter following removal of tetracycline analogues. Gene Ther 1997;4:993-7.

85. Lowenstein PR, Southgate TD, Smith-Arica JR, Smith J, Castro MG. Gene therapy for inherited neurological disorders: towards therapeutic intervention in the Lesch-Nyhan syndrome. Prog Brain Res 1998;117:485-501.

86. Huang CJ, Spinella F, Nazarian R, Lee MM, Dopp JM, de Vellis J. Expression of green fluorescent protein in oligodendrocytes in a time and level-controllable fashion with a tetracycline-regulated system. Mol Med 1999;5:129-37.

87. Efrat S, Fusco-DeMane D, Lemberg H, al Emran O, Wang X. Conditional transformation of a pancreatic beta-cell line derived from transgenic mice expressing a tetracycline-regulated oncogene. Proc Natl Acad Sci USA 1995;92:3576-80.

88. Yu JS, Sena-Esteves M, Paulus W, Breakefield XO, Reeves SA. Retroviral delivery and tetracycline-dependent expression of IL-1beta converting enzyme (ICE) in a rat glioma model provides controlled induction of apoptotic death in tumor cells. Cancer Res 1996;56:5423-7.

89. Kistner A, Gossen M, Zimmermann F, Jerecic J, Ullmer C, Lubbert H, Bujard H. Doxycyline-mediated quantitative and tissue-specific control of genome expression in transgenic mice. Proc Natl Acad Sci USA 1996;93:10933-8.

90. Mansuy IM, Winder DG, Moallem TM, Osman M, Mayford M, Hawkins RD, Kandel ER. Inducible and reversible gene expression with the rtTA system for the study of memory. Neuron 1998;21:257-65.

91. Mansuy IM, Mayford M, Jacob B, Kandel ER, Bach ME. Restricted and regulated overexpression reveals calcineurin as a key component in the transition from short-term to long-term memory. Cell 1998;92:39-49.

92. Chen J, Bezdek T, Chang J, Kherzai AW, Willingham T, Azzara M, Nisen PD. A glial-specific, repressible, adenovirus vector for brain tumor gene therapy. Cancer Res 1998;58:3504-7.

93. Smith-Arica JR, Morelli AE, Larregina AT, Smith J, Lowenstein PR, Castro MG. Cell-type specific and regulatable transgenesis in the adult brain: adenovirus-encoded combined transcriptional targeting and inducible transgene expression. Mol Ther 2000;2:579-587

94. Ralph GS, Bienemann A, Harding TC, Hopton M, Henley J, Uney, JB. Targeting of tetracycline-regulatable transgene expression specifically to neuronal and glial cell populations using adenoviral vectors. Neuroreport 2000;11:2051-5.

95. Utomo AR, Nikitin AY, Lee, WH. Temporal, spatial and cell type-specific control of Cre-mediated DNA recombination in transgenic mice. Nat Biotech 1999;17:1091-6.

96. Smith-Arica JR, Williams JC, Stone D, Smith J, Lowenstein PR, Castro MG. Switching on and off transgene expression within lactotrophic cells in the anterior pituitary gland in vivo. Endocrinology 2000 (Submitted).

97. Ghersa P, Gobert RP, Sattonnet-Roche P, Richards CA, Merlo Pich E, Hooft van Huijsduijnen R. Highly controlled gene expression using combinations of a tissue-specific promoter, recombinant adenovirus and a tetracycline-regulatable transcription factor. Gene Ther 1998;5:1213-20.

98. Southgate TD, Stone D, Williams JC, Lowenstein PR, Castro MG. Long term transgene expression within the anterior pituitary gland in situ: impact on circulating hormone levels, cellular and antibody mediated immune reponses. Endocrinology 2001;142:464-476

99. Kajiwara K, Byrnes AP, Ohmoto Y, Charlton HM, Wood MJ, Wood KJ. Humoral immune responses to adenovirus vectors in the brain. J Neuroimmunol 2000;103:8-15.

100. Yang Y, Nunes FA, Berencsi K, Furth EE, Gonczol E, Wilson JM. Cellular immunity to viral antigens limits E1-deleted adenoviruses for gene therapy. Proc Natl Acad Sci USA 1994;91:4407-11.

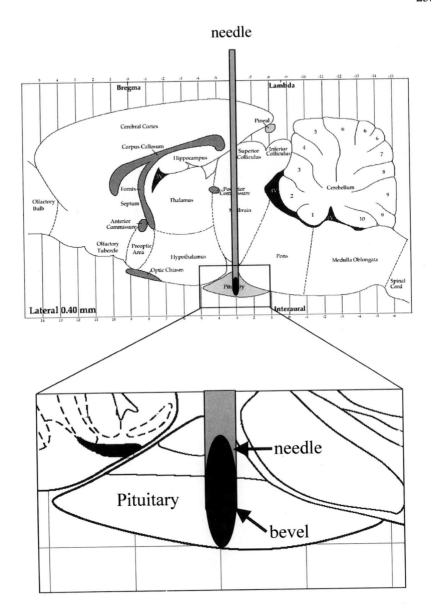

Figure 1

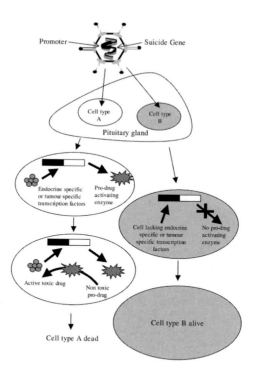

Figure 2

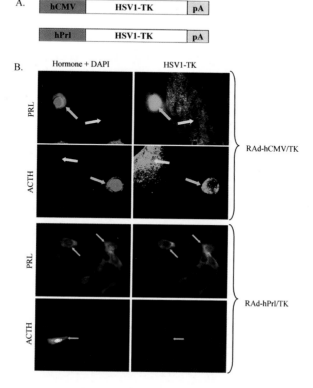

Figure 3

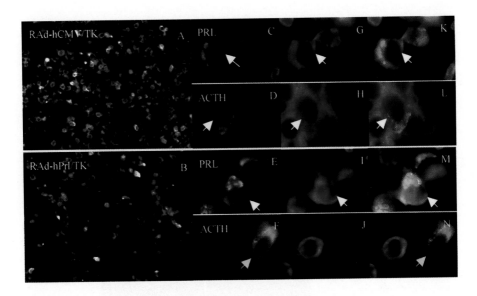

Figure 4

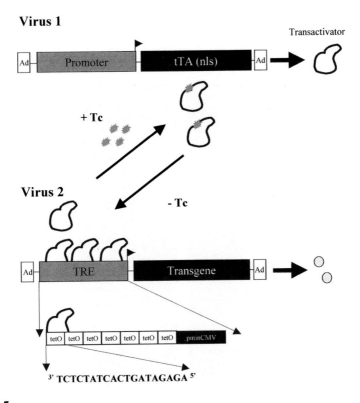

Figure 5

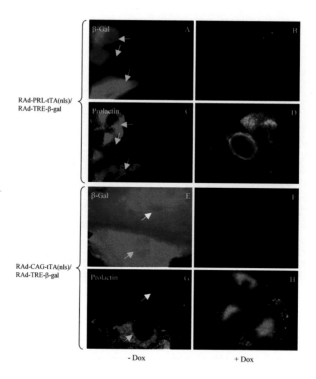

RAd-PRL-tTA(nls)/
RAd-TRE-β-gal

RAd-CAG-tTA(nls)/
RAd-TRE-β-gal

- Dox   + Dox

Figure 6

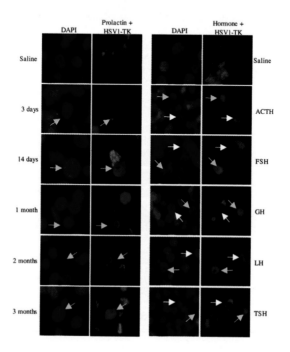

Figure 7

Index